College Mathematics

College Mathematics

Eugene D. Nichols
Florida State University

RINEHART PRESS / HOLT, RINEHART AND WINSTON
SAN FRANCISCO

Library of Congress Cataloging in Publication Data

Nichols, Eugene Douglas, 1923–
    College mathematics: (everything you always wanted to know about mathematics but were afraid you couldn't understand)

    Includes index.
    1. Mathematics—1961–   I. Title.
QA39.2.N52   510   74–23584
ISBN 0–03–089247–3

©1975 by Rinehart Press
      5643 Paradise Drive
      Corte Madera, California 94925

A division of Holt, Rinehart and Winston, Inc.

All rights reserved
PRINTED IN THE UNITED STATES OF AMERICA

5 6 7 8   038   9 8 7 6 5 4 3 2 1

Contents

Preface		ix
Chapter 1	**MATHEMATICS FOR FUN AND NUMERATION SYSTEMS**	1
1-1	Why Study Mathematics?	1
1-2	Mathematics for Fun	3
1-3	Our Decimal Numeration System	6
1-4	Base-Five Numeration	8
1-5	Adding in Base-Five	11
1-6	Subtracting in Base-Five	14
1-7	Binary Numeration System	16
1-8	Adding and Subtracting in Base-Two	19
Chapter 2	**SETS**	25
2-1	Some Familiar Sets	25
2-2	Equivalent Sets	31
2-3	Subsets and Equality	34
2-4	Operations on Sets	38
2-5	Sets of Points	41
2-6	Venn Diagrams	44

Chapter 3   SOME REMARKABLE FEATURES OF NUMBERS   48

3-1	Even, Odd, Prime, Composite Numbers	48
3-2	Primes—How Many and How to Find Them	54
3-3	The Amazing Primes	57
3-4	Some Fantastic Numbers	62
3-5	Greatest Common Divisor	64
3-6	Least Common Multiple	68
3-7	Dividing the Product of Two Numbers by a Prime Number	70
3-8	Fundamental Theorem of Arithmetic	73
3-9	Divisibility Rules	76
3-10	Casting out Nines	81

Chapter 4   WHOLE NUMBERS, INTEGERS, and RATIONALS   87

4-1	The System of Whole Numbers	87
4-2	The System of Integers, Addition	90
4-3	Subtraction of Integers	95
4-4	Multiplication of Integers	98
4-5	Rational Numbers	103
4-6	Ordering Rational Numbers	109
4-7	Decimals	112
4-8	Density of Rational Numbers	116

Chapter 5   NUMBERS AS EXPONENTS   121

5-1	Whole Number Exponents	121
5-2	The Power of a Power and the Power of a Product	127
5-3	The Power of a Quotient and the Quotient of Powers	130
5-4	Integers as Exponents	134
5-5	Square Root	139

Chapter 6   REAL NUMBERS, EQUATIONS, and INEQUALITIES   146

6-1	Numbers That Aren't Rational	146
6-2	Proving a Number Irrational	151
6-3	Irrational Numbers and Decimals	153
6-4	Two Important Relations: Is Equal to ($=$), is Less than ($<$)	157
6-5	The System of Real Numbers	160
6-6	Inequalities	163
6-7	Solving Equations and Inequalities	168

Chapter 7 GEOMETRY 177

7-1	A Variety of Geometries	177
7-2	Nonmetric Geometry: Figures in a Plane	178
7-3	Sets of Points in Space	183
7-4	Metric Geometry and the Metric System	186
7-5	Areas and Volumes	191
7-6	Geometry and Deductive Thinking	196
7-7	Motion Geometry	199

Chapter 8 FINITE SYSTEMS and MATRICES 203

8-1	Addition in a Three-Number System	203
8-2	Other Operations and Their Properties in S_3	207
8-3	Another System—S_5	211
8-4	The System S_4	215
8-5	Solving Equations in S_5	220
8-6	Two-By-Two Matrices	223
8-7	Addition and Subtraction of 2-By-2 Matrices	227
8-8	Multiplication	232
8-9	Applications of Matrices	236

Chapter 9 GRAPHS, FUNCTIONS and LINEAR PROGRAMMING 240

9-1	Graphing on a Number Line	240
9-2	Graphing in a Plane	246
9-3	Lines	252
9-4	Slope of a Line	256
9-5	Linear Programming	261
9-6	Relations and Functions	268

Chapter 10 LOGIC 276

10-1	Statements and Their Truth-Values	276
10-2	Quantifiers	279
10-3	Negating Statements	281
10-4	Conjunctions and Disjunctions	285
10-5	Implications	289
10-6	Converses and Inverses	293
10-7	Contrapositives and Tautologies	296
10-8	Validity and Arguments	299

Chapter 11 PROBABILITY 306

11-1	Introduction to Probability	306
11-2	Multiple Tosses of a Coin	309
11-3	Sample Spaces for Various Experiments	311
11-4	Occurrence of More Than One Event	316
11-5	Some Counting Problems	323
11-6	Odds and Mathematical Expectation	328

Chapter 12 STATISTICS 333

12-1	Introduction to Statistics	333
12-2	Mean, Median, Mode	334
12-3	Standard Deviation	336
12-4	Percentile Rank	339
12-5	Representing Data by Graphs	342

Answers to Selected Problems 348

Index 379

Preface

This is a substantially revised version of *College Mathematics for General Education* published in 1970. It is offered in response to the demand for a course in the elements of mathematics for those students who do not intend to major in mathematics or natural sciences. The topics included in the text are those that the users of the original version of the book found most relevant for a course of this type. These topics were selected because it is believed that they contribute to the students' appreciation of mathematics in that it provides them with insights into the structural properties of mathematical systems and gives them experience with some fun parts of mathematics. While mathematical skills are given some attention, they do not constitute the major emphasis.

The first chapter provides students with the lighter side of mathematics and gives them a chance to experience some joys of discovering relationships. Chapter 2 gives the foundation in the use of the language of sets, which is so helpful in expressing many mathematical relationships.

Since the mathematical background of students enrolled in courses of this nature is quite varied, the text is organized so that all students can profit by the topics of wide applicability and the instructor can have some options as to which parts to cover and which to omit. Some chapters stand by themselves and can be omitted without a loss of continuity. Starred Exercises indicate particularly challenging problems.

Chapter 5, dealing with exponents and powers, for example, can be omitted for an abbreviated course. Chapter 8, dealing with finite systems and matrices, can be omitted as well, although this chapter may be of great

Preface

interest to many students. Each of the last four chapters is quite independent of the rest of the book and can be omitted or covered according to the choice of the instructors to fit the interests and background of their students or the demands of the course.

The style of the text is informal. Although the students are introduced to mathematical proofs, they are selected for their simplicity. Furthermore, the proofs can be omitted without jeopardizing the student's ability to continue in the course.

An instructor's manual to accompany the text is available to substantially assist in the teaching of the course. It contains answers to selected exercises, several quizzes for each chapter, and an additional test for each chapter.

I wish to express my sincere thanks to Elva M. Brandt for her aid in the preparation of the *Instructor's Manual*. I am also grateful for her insightful observations, which undoubtedly will enhance the readability and teachability of the text.

I also wish to acknowledge and thank many reviewers, who have, on the basis of their experiences with the original version of the text, suggested many changes which resulted in this new edition. In response to their suggestions, the chapter on matrices was deleted; but a section on structural properties of two-by-two matrices with illustrations of some of the applications of these matrices is now included in Chapter 8 which deals with systems. Also, due to their suggestions, a section on linear programming was added, and the chapter on trigonometry was deleted.

Of course, I am grateful to the many students who have used the previous edition of the book and whose reactions to the course were reflected in the reviewers' comments.

E. D. николs

College Mathematics

1

Mathematics for Fun and Numeration Systems

1.1 WHY STUDY MATHEMATICS?

REVIEW

None

OBJECTIVES

- Describe your own need for mathematics
- Determine what mathematical concepts and skills are needed by all people
- Decide how mathematics is used in a particular industry
- Speculate how knowing more mathematics could help an individual
- Determine what concepts are necessary for further study of mathematics

Mathematics is different things to different people. Mathematics is a way of making a living, an unloved but required college course, a game to be played and enjoyed, a helpful tool in conducting one's life. However you view mathematics, everyone in today's highly technological and computer-oriented world must have some understanding of basic mathematical concepts and must have facility with some basic computational skills. Getting along without these is very difficult, if not impossible.

The task of deciding what specific mathematical knowledge an individual should have is extremely difficult. However, it is easy to conclude that citizens collectively need a vast amount of mathematical knowledge for their very survival. We depend on electricity, on the car, on the airplane, on the computer, and so on. And all of these depend on mathematics. Someone must know this sophisticated mathematics in order to keep us going.

We are also honest enough to admit that some people can manage with very little, if any, knowledge of mathematics. But, there is always the nagging question of whether they could have done more had they known more mathematics. Were some doors closed to them because they lacked some mathematical insight and were not in command of some skill? The question remains unanswered in most cases, but the individual query: "Would I have been better off?" persists.

In this book, we have selected some fundamental concepts and skills that occupy a central role in mathematics. They can serve as a basis for further study of mathematics, and they can be used in dealing with ordinary, daily mathematical problems.

Exercises 1.1

1. Try to describe as accurately as you can your own need for mathematics.
2. List some mathematical concepts and skills that you believe every citizen of today's world must know in order to function in an enlightened way.
3. Name as many ways as you can think of in which mathematics enters into the making of a car.
4. Can you think of anyone you know who might have achieved more had he known more mathematics?
5. Name some mathematical concepts that you consider essential for further study of mathematics.
6. What particular mathematical skills would you find it extremely difficult to get along without?
7. A farmer with 50 cows had to order their yearly supply of vitamin supplement. He was told that one sack of supplement would last each cow one year. Unlike you, the farmer had never taken any math and did not know how many sacks to order. As a result, he ordered 100 sacks because he thought it was better to have too much rather than too little. Unfortunately, all the leftover sacks had to be thrown away at the end of the year because the vitamins had lost their potency. If you were the farmer (who had now obtained his knowledge of math the

1.2 Mathematics for Fun

hard way), how many sacks of vitamin supplement would you order for the next year?

1.2 MATHEMATICS FOR FUN

REVIEW

- 4×6 means 4 sixes or $6 + 6 + 6 + 6$
- The area of a rectangle with length ℓ and width w is the product $\ell \times w$
- $1 + 2 + 3 + \cdots + 98 + 99 + 100$ means the sum of all whole numbers from 1 through 100
- 2^3 is read "two cubed" or "two to the third power"; it is equal to $2 \times 2 \times 2$ or 8
- A quadrilateral is a polygon with four sides
- A pentagon is a polygon with five sides
- A hexagon is a polygon with six sides
- A diagonal of a polygon is a line segment connecting two nonconsecutive vertices. *Example:* $\overline{AC}$ is a diagonal in the pentagon $ABCDE$

OBJECTIVES

- Add whole numbers from 1 through 100 using a shortcut
- Discover a pattern for adding n consecutive whole numbers
- Discover a pattern for adding consecutive odd numbers
- Discover a pattern for adding consecutive even numbers
- Discover a pattern for adding the cubes of consecutive whole numbers
- Discover a pattern in adding numbers made up of nines
- Discover a pattern that relates the number of diagonals to the number of sides in a polygon

A ten-year-old boy was given the problem of adding the whole numbers from 1 through 100. To the surprise of his teacher, he came up with an answer in a very short time. He took a very clever approach to finding this sum. Here is how it went.

$$
\begin{array}{r}
1 + 2 + 3 + \cdots + 98 + 99 + 100 \\
100 + 99 + 98 + \cdots + 3 + 2 + 1 \\
\hline
101 + 101 + 101 + \cdots + 101 + 101 + 101
\end{array}
$$

1 Mathematics for Fun and Numeration Systems

He added the numbers from 1 through 100 to the numbers from 100 through 1. It is then a simple matter to find this sum and divide it by 2.

We see that adding the numbers from 1 through 100 twice is the same as taking 101 one hundred times; that is,

$$100 \times 101$$

which is 10,100. Dividing this by 2 gives 5050 for the sum of all whole numbers from 1 through 100.

Another way to look at this problem brings in some geometry. Think of a rectangle, say 5 units by 6 units. Its area is 5×6. But, as can be seen from Figure 1.1, its area is also

$$1 + 2 + 3 + 4 + 5$$

taken twice. So, we conclude that

$$1 + 2 + 3 + 4 + 5 = \frac{5 \times 6}{2}$$

Let's consider another case.

In Figure 1.2 we have a 6 by 7 rectangle. As before, we read from the rectangle that

$$1 + 2 + 3 + 4 + 5 + 6 = \frac{6 \times 7}{2}$$

which is 21.

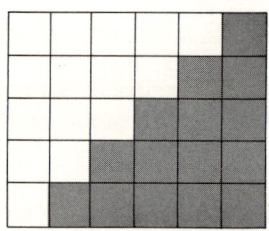
FIGURE 1.1

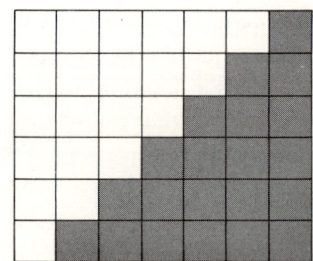
FIGURE 1.2

Now we can continue to add 1 through 7, 1 through 8, and so on. At some point we become aware of a pattern:

$$1 + 2 + 3 + \cdots + (n-2) + (n-1) + n = \frac{n(n+1)}{2}$$

1.2 Mathematics for Fun

Let's use this pattern and replace n by 100 in it:

$$1 + 2 + 3 + \cdots + 98 + 99 + 100 = \frac{100 \times 101}{2}$$
$$= 5050$$

This is the answer we found before.

Let's take a look at another problem. What is the sum of 100 consecutive odd numbers starting with 1? A good procedure is to first work out some simple cases. So, let's examine a few beginning cases:

$1 + 3 = 4$
$1 + 3 + 5 = 9$
$1 + 3 + 5 + 7 = 16$

We see the following:

adding the first two odd numbers—the sum is 4 ($2 \times 2 = 4$)
adding the first three odd numbers—the sum is 9 ($3 \times 3 = 9$)
adding the first four odd numbers—the sum is 16 ($4 \times 4 = 16$)

The pattern appears to be: multiply the number of odd numbers being added by itself. So, the sum of 100 consecutive odd numbers starting with 1 is 100×100 or 10,000.

Exercises 1.2

1. Add the even numbers from 2 through 100 using the method illustrated in this section; that is, add the even numbers from 2 through 100 to the even numbers from 100 through 2, find the sum of these numbers, and divide by 2.
2. By adding 2 and 4, then 2 and 4 and 6, and so on, discover the pattern for adding consecutive even numbers. Write the formula for the sum of the first n consecutive even numbers.
3. Compute the following sums:

    $1^3 + 2^3$
    $1^3 + 2^3 + 3^3$
    $1^3 + 2^3 + 3^3 + 4^3$
    and so on

    Discover the pattern for adding the cubes of the consecutive numbers. Write the formula for the sum of the cubes of n consecutive numbers.
4. In his last will and testament, an old European king ordered his two sons to have two sons each by their twentieth birthdays. He also de-

clared that his grandsons and their sons should do the same, continuing the procedure for 80 years. The king wanted to be sure there would always be a ruler for his empire. How many potential rulers would there be during the 80 years if all direct male descendants of the old king were able to follow his order? Can you develop a general formula explaining the way you determined the number?

5. Compute the following products:

    9×9
    99×9
    999×9
    and so on

    Write the product of 9 and a number consisting of 20 nines. Describe the product of 9 and a number consisting of n nines.

6. Draw a quadrilateral and all of its diagonals. How many diagonals are there? Now do the same for a pentagon, hexagon, and so on. Write the formula for the number of diagonals in an n-sided polygon in terms of n.

7. Congratulations! You've just won the most recent contest sponsored by Meaty Dog Food. You have a choice of two prizes. You can either have $100 now, or you can have $2 today, $4 tomorrow, $6 the next day, and so on for 20 days. Which prize would you take? Can you develop a general formula to guide future winners?

1.3 OUR DECIMAL NUMERATION SYSTEM

REVIEW

- 10^4 is the fourth power of 10
- In 10^4, 10 is the *base* and 4 is the *exponent*

OBJECTIVES

- Know what the phrase *numeration system* means
- Know what the word *numeral* means
- Know the ten digits used in the decimal numeration system
- Show a number as a sum of products of powers of 10
- Show why any nonzero number raised to the power 0 is equal to 1

1.3 Our Decimal Numeration System

When speaking of a *numeration system*, we are concerned with ways of writing names for numbers. A symbol that names a number is called a *numeral*. The most commonly used numeration system is the *decimal system*. Its name comes from the use of ten digits: 0, 1, 2, 3, 4, 5, 6, 7, 8, and 9.

To see how the powers of the number ten enter into the decimal system, examine the following:

$$69{,}253 = 6 \times 10{,}000 + 9 \times 1000 + 2 \times 100 + 5 \times 10 + 3 \times 1$$
$$= 6 \times 10^4 + 9 \times 10^3 + 2 \times 10^2 + 5 \times 10^1 + 3 \times 10^0$$

Notice that $10^1 = 10$ and $10^0 = 1$. The zero power of ten deserves a little more attention. Why it is logical to consider 10^0 equal to 1 is apparent from the pattern displayed below.

$10^4 = 10{,}000$
$10^3 = 1000$
$10^2 = 100$
$10^1 = 10$
$10^0 = ?$

In the last line at the left, we are asking what 10^0 should equal. To decide, we observe the pattern: each time the exponent is decreased by 1, we divide the power by 10. So, to know what 10^0 should be in order to fit this pattern, we must divide 10 by 10, which is 1. Thus, we conclude that $10^0 = 1$.

By the way, you may have decided that the same argument would hold for any nonzero number raised to the power 0. And you are right! Look at the pattern for 5 raised to the various powers.

$5^4 = 625$
$5^3 = 125$
$5^2 = 25$
$5^1 = 5$
$5^0 = ?$

Here, each time we decrease the exponent by 1, we divide the power by 5. So, 5^0 must be 5 divided by 5 or 1. Thus, $5^0 = 1$. In all of this we assume the pattern continues. That's often the way decisions are made in mathematics.

Exercises 1.3

1. What is the meaning of a numeration system?
2. What is a numeral?
3. What digits are used in the decimal numeration system?
4. Show 72,517 as a sum of products of powers of ten.
5. Unfortunately, some teachers suffer from the incurable disease "long-numberitis." This affliction makes it impossible to say any number

larger than three digits. Can you help these teachers by making the following numbers into products of two factors so that no factor is longer than three digits?
   a. 36,000,000   b. 22,100   c. 6290   d. 398,000,000,000
6. Build a sequence of powers of 8 to show that $8^0 = 1$.
7. You are taking your aunt on a plane ride in your newly acquired supercharged twin-engine banger special when suddenly the door falls off and your aunt falls out. Within two seconds, you dive your plane to catch her before she hits the ground. If you are 100 ft above the ground when the fateful incident occurs and the two rates of descent are as given below, quickly calculate whether or not you will save your falling aunt.

   Your aunt falls 10 ft every 2 seconds.

   Your plane dives 20 ft every 2 seconds.

★ 8. Give an argument why 0^0 should be left undefined.
9. Imagine that you have been appointed head of the World Math Department of the United Nations. Your job is to invent a new numeration system or improve upon the old one. Would you use the same symbols? The same system? Briefly describe what you would do.

1.4 BASE-FIVE NUMERATION SYSTEM

REVIEW

- Any nonzero number raised to the zero power is equal to 1

OBJECTIVES

- Show the place values in the base-ten numeration system in terms of powers of ten
- Display the pattern for the consecutive powers of ten
- Show the place values in the quinary numeration system in terms of powers of five
- Write expanded forms of numbers in base-ten and base-five numeration systems
- Find base-ten numerals for numbers given in base five
- Write the numbers from 1 to 625 in base five
- Write base-five and base-ten numerals for numbers 5^5 and 5^6

1.4 Base-Five Numeration System

Just as the number ten is the key number in the base-ten numeration system, the number five is the key number in the base-five numeration system. It is also called the *quinary system*. In Figure 1.3 the values of the places in the base-ten numeration system in terms of powers of ten are displayed. Note that this display can continue on and on to the left and to the right, indicated by three dots.

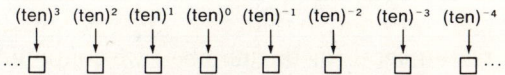

FIGURE 1.3

Using the powers of ten as shown in Figure 1.3, we can write 7534.2189 in the *expanded* form as follows:

$$7534.2189 = 7 \times 10^3 + 5 \times 10^2 + 3 \times 10^1 + 4 \times 10^0 + 2 \times 10^{-1} \\ + 1 \times 10^{-2} + 8 \times 10^{-3} + 9 \times 10^{-4}$$

That the powers of ten do fall into a consistent pattern can be seen from the following:

.
.
.

$10^3 = 1000$
$10^2 = 100$
$10^1 = 10$
$10^0 = 1$
$10^{-1} = \frac{1}{10}$
$10^{-2} = \frac{1}{100}$
$10^{-3} = \frac{1}{1000}$

.
.
.

Numbers can be expanded in a similar way in the quinary numeration system. We use only five digits: 0, 1, 2, 3, 4, and the place values are powers of five now, as shown in Figure 1.4.

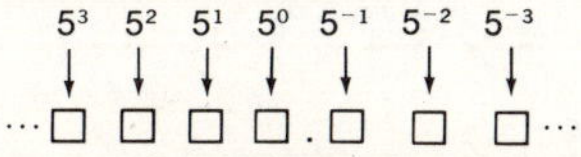

FIGURE 1.4

We use the subscript $_{\text{five}}$ when writing numerals in the quinary system, and we read each digit individually: 3410.214_{five} is read "three-four-one-zero-point-two-one-four in base five." The numeral 3410.214_{five} would be expanded as follows:

$$3410.214_{\text{five}} = 3 \times 5^3 + 4 \times 5^2 + 1 \times 5^1 + 0 \times 5^0 + 2 \times 5^{-1}$$
$$+ 1 \times 5^{-2} + 4 \times 5^{-3}$$

To find the base-ten numeral for this number, we continue:

$$= 3 \times 125 + 4 \times 25 + 1 \times 5 + 2 \times \tfrac{1}{5} + \tfrac{1}{25} + 4 \times \tfrac{1}{125}$$
$$= 375 + 100 + 5 + \tfrac{2}{5} + \tfrac{1}{25} + \tfrac{4}{125}$$
$$= 480 + \tfrac{4}{10} + \tfrac{4}{100} + \tfrac{32}{1000}$$
$$= 480 + \tfrac{4}{10} + \tfrac{7}{100} + \tfrac{2}{1000}$$
$$= 480.472$$

Therefore, $3410.214_{\text{five}} = 480.472_{\text{ten}}$.

A partial record of counting in base five would look like this.

1	21	41	111
2	22	42	.
3	23	43	.
4	24	44 (twenty-four)	.
10 (five)	30	100 (twenty-five)	444 (one hundred twenty-four)
11	31	101	1000 (one hundred twenty-five)
12	32	102	.
13	33	103	.
14	34	104	.
20 (ten)	40	110 (thirty)	4444 (six hundred twenty-four)
↑			10000 (six hundred twenty-five)

Read: two-zero in base five

We will now find the base-ten numeral for 432_{five}. (Remember to read 432_{five} as "four-three-two in base five," since it is *not* four hundred thirty-two.)

$$432_{\text{five}} = 4 \times 5^2 + 3 \times 5^1 + 2 \times 5^0$$
$$= 4 \times 25 + 3 \times 5 + 2 \times 1$$
$$= 100 + 15 + 2$$
$$= 117_{\text{ten}}.$$

1.5 Adding in Base Five

Exercises 1.4

Expand each of the following by powers of ten:

1. 37
2. 492
3. 5137
4. 58,032
5. 601,984
6. 4,902,864
7. 4835.396
8. 57,001.3974
9. In the last section, you met the terrible disease "longnumberitis." Now you will meet an even more terrible disease, "shortnumberitis," which also afflicts many teachers. This disease is so terrible that any teacher who has it cannot say any number of less than three digits. Arrange the numbers below so they can be said by someone who has the disease.
   a. 3.6×10^3 b. 2.93×10^6 c. 1.21×10^{-1} d. 9.36×10^2

Expand each of the following by powers of five:

10. 34_{five}
11. 312_{five}
12. 4013_{five}
13. 40.3_{five}
14. 23014.113_{five}
15. 11400.0013_{five}
16. Find the base-ten names for each of the numbers in exercises 10 through 15.
17. The fourth power of five is equal to six hundred twenty-five, that is, $5^4 = 625$. And $10000_{five} = 625_{ten}$. Write the base-five numeral for 5^5 and 5^6. What are the base-ten numerals for these two numbers?
18. How would you read 110324_{five}? What is the base-ten numeral for this number?
19. You bring back plans for a gold synthesizer from the base-five planet of Pentavion. Unfortunately, all the dimensions are in base five. Convert the dimensions to base ten so we can build this remarkable machine on Earth.
    a. length = 20_{five} in.   b. height = 34_{five} in.   c. width = 43_{five} in.
★20. Make a display like the one shown in Figure 1.4 for the place values in the base-eight numeration system.
★21. Make a display to show the place values in the base-four numeration system.

1.5 ADDING IN BASE FIVE

REVIEW

- Adding in base ten:
  ```
    11
   397
  +468
  ────
   865
  ```

OBJECTIVE

- Add in base five

1 Mathematics for Fun and Numeration Systems

To illustrate the point behind the procedure for adding in base five, think of adding in base ten by grouping to get tens. For example, think of adding 8 and 7 following these steps:

$$\begin{aligned} 8 + 7 &= 8 + (2 + 5) \quad \textit{Think: } 7 = 2 + 5 \\ &= (8 + 2) + 5 \\ &= 10 + 5 \\ &= 15 \end{aligned}$$

We can follow the same principle when adding in base five, except that we will group to get fives instead of tens. The examples below illustrate the procedure.

Example 1 $\quad 12_{five}$
$ +24_{five}$
$ \overline{41_{five}}$

$$\begin{aligned} \textit{Think: } 2 + 4 &= 2 + (3 + 1) \\ &= (2 + 3) + 1 \\ & \!\!\!\uparrow\!\text{---one group of five} \\ &= 1 \text{ five} + 1 \\ &= 10_{five} + 1 \\ &= 11_{five} \end{aligned}$$

Now we have 1 five + 2 fives + 1 five = 4 fives. Thus, $12_{five} + 24_{five} = 41_{five}$.

Example 2 $\quad 33_{five}$
$ + \; 24_{five}$
$ \overline{112_{five}}$

$$\begin{aligned} \textit{Think: } 3 + 4 &= 3 + (2 + 2) \\ &= (3 + 2) + 2 \\ & \!\!\!\uparrow\!\text{---one group of five} \\ &= 1 \text{ five} + 2 \\ &= 10_{five} + 2 \\ &= 12_{five} \end{aligned}$$

Next we have 3 fives + 2 fives + 1 five = (3 + 2) fives + 1 five

$$\begin{aligned} &\phantom{= 3 \text{ fives} + 2 \text{ fives} + 1 \text{ five} = } \!\!\!\uparrow\!\text{---one group of twenty-fives} \\ &= 1 \text{ twenty-five} + 1 \text{ five} \\ &= 100_{five} + 10_{five} \\ &= 110_{five} \end{aligned}$$

Thus, $33_{five} + 24_{five} = 112_{five}$.

Example 3 $\quad 144_{five}$
$ +144_{five}$
$ \overline{343_{five}}$

$$\begin{aligned} \textit{Think: } 4 + 4 &= 4 + (1 + 3) \\ &= (4 + 1) + 3 \\ & \!\!\!\uparrow\!\text{---one group of five} \\ &= 1 \text{ five} + 3 \\ &= 10_{five} + 3 \\ &= 13_{five} \end{aligned}$$

1.5 Adding in Base Five

Next we have 4 fives + 4 fives + 1 five = 4 fives + (1 five + 3 fives) + 1 five
= (4 fives + 1 five) + 3 fives + 1 five
⎣—one group of twenty-fives
= 1 twenty-five + 4 fives
= 140_{five}

Next we have
1 twenty-five + 1 twenty-five + 1 twenty-five = 3 twenty-fives
= 300_{five}

Thus, $144_{\text{five}} + 144_{\text{five}} = 343_{\text{five}}$.

Exercises 1.5

Show the groupings to get tens:

1. 6 + 9
2. 7 + 5
3. 4 + 8

Add in base five:

4. 23_{five}
   $+11_{\text{five}}$

5. 31_{five}
   $+23_{\text{five}}$

6. 103_{five}
   $+\ 34_{\text{five}}$

7. 3230_{five}
   $+\ 342_{\text{five}}$

8. 4002_{five}
   $+\ 444_{\text{five}}$

9. 444_{five}
   $+444_{\text{five}}$

10. As the Sherlock Holmes of the math department, you are faced with the task of solving the terrible "Base Murders." At the scene of each murder (each victim was a math teacher), a completed sum has been found. As your assistant, Watson, it is my belief that we will find the murderer if we discover what base the problem is in. The suspects are Mr. Black, who teaches base 5; Mr. Grown, who teaches base 8; and Mr. Toad, who teaches base 6. Here is exhibit *A*, found at the scene of the first murder:

    ```
       453
    + 405
    ─────
      1302
    ```

    "Who dunnit?"

11. Your great-great-granduncle Scrooge has just died, leaving a fortune in his personal safe. He was a firm believer in math, and he left the combination to the safe in code. He also left a message: "Just discover the

1 Mathematics for Fun and Numeration Systems 14

base that each sum is in; then open the door and have the goodies within." Determine the bases of the problems below to solve the code and obtain the combination.

a.   6283   b.   1121   c.   2433
    + 188   + 1222   +1342
     6482   10120   4330

★12. Develop a multiplication table in base five.

★13. Choose an example to illustrate a two-digit by two-digit multiplication in base five.

14. It has been argued that we probably developed the base-ten system because our hands have ten fingers. Recent star probes indicate that people on a planet orbiting the star Tau Ceti have just one arm and one finger. Discuss what their numeration system might be like, if they have one.

1.6 SUBTRACTING IN BASE FIVE

REVIEW

- Subtracting in base ten:

$$\begin{array}{r} \overset{12}{6\cancel{2}15} \\ \cancel{7}\cancel{3}5 \\ -248 \\ \hline 487 \end{array}$$

OBJECTIVE

- Subtract in base five

When subtracting in base ten, regrouping by tens is used, as illustrated below:

$$\begin{array}{r} 63 \\ -25 \\ \hline 38 \end{array}$$

Think: Since 5 cannot be subtracted from 3, regroup 63 as 5 tens + 13 ones.

$$\begin{array}{r} 5 \text{ tens} + 13 \text{ ones} \\ -2 \text{ tens} + 5 \text{ ones} \\ \hline 3 \text{ tens} + 8 \text{ ones} = 38_{\text{ten}} \end{array}$$

The same pattern is followed when subtracting in base five, except regrouping is done by fives. Study the examples on the next page.

1.6 Subtracting in Base Five

Example 1

$$\begin{array}{r} 1 \\ \cancel{2}3_{\text{five}} \\ -14_{\text{five}} \\ \hline 4 \end{array}$$

FIRST STEP:
Regroup 2 fives as 1 five and five ones. Five ones and 3 ones give eight ones. And 4 subtracted from eight is 4.
SECOND STEP:
1 five − 1 five = 0.
Thus, $23_{\text{five}} - 14_{\text{five}} = 4$.
It is not necessary to indicate the base when writing a single numeral.

Example 2

$$\begin{array}{r} 32 \\ \cancel{4}\cancel{3}1_{\text{five}} \\ -143_{\text{five}} \\ \hline 233_{\text{five}} \end{array}$$

FIRST STEP:
Regroup 3 fives as 2 fives and five ones. Five ones and 1 one give six ones. And 3 subtracted from six is 3.
SECOND STEP:
Regroup 4 twenty-fives as 3 twenty-fives and five fives. Five fives + 2 fives = seven fives. And 4 fives subtracted from seven fives is 3 fives.
THIRD STEP:
1 twenty-five subtracted from 3 twenty-fives is 2 twenty-fives.
Thus, $431_{\text{five}} - 143_{\text{five}} = 233_{\text{five}}$.

Example 3

$$\begin{array}{r} 24 \\ \cancel{3}\cancel{0}2_{\text{five}} \\ -133_{\text{five}} \\ \hline 114_{\text{five}} \end{array}$$

FIRST STEP:
Regroup 3 twenty-fives as 2 twenty-fives and 4 fives and five ones.
SECOND STEP:
Five ones + 2 ones = seven ones. And 3 subtracted from seven is 4.
THIRD STEP:
3 fives subtracted from 4 fives is 1 five.
FOURTH STEP:
1 twenty-five subtracted from 2 twenty-fives is 1 twenty-five.
Thus, $302_{\text{five}} - 133_{\text{five}} = 114_{\text{five}}$.

Exercises 1.6

Show regrouping when subtracting in base ten:

1. $\begin{array}{r} 47 \\ -29 \\ \hline \end{array}$
2. $\begin{array}{r} 423 \\ -145 \\ \hline \end{array}$
3. $\begin{array}{r} 307 \\ -128 \\ \hline \end{array}$

Subtract in base five:

4. 44_{five}
   -23_{five}

5. 41_{five}
   -12_{five}

6. 432_{five}
   -143_{five}

7. 203_{five}
   -124_{five}

8. 2314_{five}
   -1323_{five}

9. 4001_{five}
   -3112_{five}

★10. Use the multiplication table from exercise 12 for Section 1.5 to explain how this table can be used to read the basic division facts in base five.

★11. Choose an example of a division of a four-digit number by a two-digit number to illustrate division in base five.

12. (Refer to Section 1.5, exercise 11.) Your great-great-granduncle Scrooge was exceedingly clever. After making sure you could add in varying bases, he now wants to see if you can subtract. A second door to the safe must also be opened. The combination is given by the bases of the differences below.

   a.   43
       -24
       ——
        14

   b. 887
      -118
      ——
       768

   c. 2476
      -1378
      ——
       1087

1.7 BINARY NUMERATION SYSTEM

REVIEW

- $5^{-1} = \dfrac{1}{5^1}$ or $\dfrac{1}{5}$
- $5^{-2} = \dfrac{1}{5^2}$ or $\dfrac{1}{25}$
- $2^{-4} = \dfrac{1}{2^4}$ or $\dfrac{1}{16}$
- When dividing a number by 2, the only possible remainders are 0 and 1

OBJECTIVES

- Display the structure of the binary numeration system
- List consecutive whole numbers in base two
- Write a base-ten numeral for a number given in the binary notation
- Convert from base ten to base two

The *binary numeration system* is a base-two system. It uses only two digits: 0 and 1. This small number of digits makes the binary system very convenient for use with electronic computers.

The structure of this system follows exactly the same principle as the base-ten and base-five numeration systems. Except now the place values are powers of two. Figure 1.5 shows the places values.

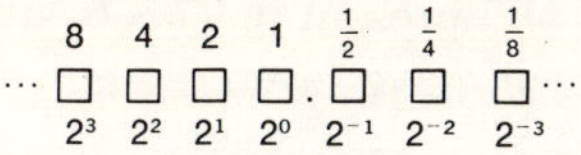

1.7 Binary Numeration System

```
  8   4   2   1  1/2 1/4 1/8
···□   □   □   □ . □   □   □ ···
 2³  2²  2¹  2⁰ 2⁻¹ 2⁻² 2⁻³
```

FIGURE 1.5

The first sixteen numbers in the binary system are listed in Table 1.1.

Table 1.1 *The First Sixteen Numbers in the Decimal and Binary Systems*

Base Ten	Base Two
1	1
2	10
3	11
4	100
5	101
6	110
7	111
8	1000
9	1001
10	1010
11	1011
12	1100
13	1101
14	1110
15	1111
16	10000

To find a base-ten name for a number given in the binary notation, we simply expand by powers of two and add. Here is an example:

$$\begin{aligned} 110101_{two} &= 1 \times 2^5 + 1 \times 2^4 + 0 \times 2^3 + 1 \times 2^2 + 0 \times 2^1 + 1 \times 2^0 \\ &= 32 + 16 + 4 + 1 \\ &= 53 \end{aligned}$$

To convert from base ten to base two, we can subtract the powers of two from the given number starting with the highest power. In the following example we find the binary name for 97.

1 Mathematics for Fun and Numeration Systems

The highest power of 2 present in 97 is 6, since $2^6 = 64$. We know now that the binary numeral will have seven digits.

$$\begin{array}{r} 97 \\ -64 \\ \hline 33 \end{array}$$ We can fill in the seventh digit:

$\underline{1}\ \_\ \_\ \_\ \_\ \_\ \_$

The highest power of two present in 33 is 5, since $2^5 = 32$.

$$\begin{array}{r} 33 \\ -32 \\ \hline 1 \end{array}$$ $\underline{1}\ \underline{1}\ \_\ \_\ \_\ \_\ \_$

Since only 1 is left, we can conclude that there are no fourth, third, second, and first powers of two. The only power of two to be recorded is the zero power, since $2^0 = 1$.

$\underline{1}\ \underline{1}\ \underline{0}\ \underline{0}\ \underline{0}\ \underline{0}\ \underline{1}$

We conclude that $97 = 1100001_{two}$.

There is a short way to find the binary numeral for a number given in base ten, based on successive divisions by two. This process for the number 97 is illustrated below.

```
2)97         EXPLANATION:
2)48   1     Divide 97 by 2. Record the quotient 48 and the re-
2)24   0     mainder 1. Now divide 48 by 2, record the quotient 24
2)12   0     and the remainder 0. Continue this procedure. The next
2)6    0     to the last step is 1 divided by 2, which is 0 and the re-
2)3    0     mainder 1.
2)1    1
  0    1
```

To get the binary numeral for 97, we write the digits from the bottom up: $1100001_{two} = 97$. This method can be used with any numeral in base ten.

Exercises 1.7

1. What digits are used in the binary numeration system?
2. Why is this system called binary? Look up the meaning of the word *binary* in a dictionary.

1.8 Adding and Subtracting in Base Two

Give the values of the following:

3. 2^4
4. 2^5
5. 2^6
6. 2^7
7. 2^0
8. 2^{-1}
9. 2^{-3}
10. 2^{-5}

Give base-ten numerals for the following:

11. 101_{two}
12. 1101_{two}
13. 11001_{two}
14. 111001_{two}
15. 101011_{two}
16. 110010110_{two}

Give binary numerals for the following numbers written in base ten:

17. 13
18. 20
19. 33
20. 64
21. 127
22. 128
23. 130
24. 150
25. 256

26. Computers operate in the binary system. If the letters of our alphabet are numbered sequentially from 1 to 26 (in base ten), what would the computer be saying if it printed out: 1000/101/1100/1100/1111?

27. Old great-great-granduncle Scrooge isn't through with you yet! You're into the safe and have your hands on the fortune when the doors automatically lock behind you. The combination for exiting is in code again, this time in the binary system. Scrooge has left another message: "If in the binary system you have a doubt, better not here or you'll never get out." Convert the following binary codes to base ten to discover the combination.

   a. 101   b. 110   c. 1001

1.8 ADDING AND SUBTRACTING IN BASE TWO

REVIEW

- $110011_{two} = 1 \times 2^5 + 1 \times 2^4$
  $+ 1 \times 2^1$
  $+ 1 \times 2^0$

OBJECTIVES

- Add in base two
- Subtract in base two
- Consider the advantages of the binary system over the decimal system

The principles used when adding in base ten and in base five are also used in base two.

An example will illustrate:

```
        (3) (2) (1)
          1   0   1
    +         1   1
    ─────────────────
      1   0   0   0
```

STEPS:
(1) 1 plus 1 gives 1 two and 0 ones = 10. Write 0, remember 1 two.
(2) 1 two + 1 two gives 1 four and 0 twos. Write 0, remember 1 four.
(3) 1 four + 1 four gives 1 eight and 0 fours. Write 0, remember 1 eight.
(4) 1 eight + 0 gives 1 eight. Write 1.

The corresponding problem in base ten would be: $\begin{array}{r} 5 \\ +3 \\ \hline 8 \end{array}$

Adding several numbers given in binary notation is a bit tricky. Because adding sixty-fours and thirty-twos is both clumsy and confusing, we add each column of figures as ones, using a circled number to indicate what we are carrying to the next column. Study the following example and the explanation to its right.

```
    (8) (7) (6) (5) (4) (3) (2) (1)
                1   0   1   1   0   1
                    1   1   0   1   1
                    1   0   1   1   0
                1   0   0   1   1   1
    +                   1   1   1   0
    ─────────────────────────────────
      1   0   0   1   0   0   1   1
```

STEPS:
(1) 1 plus 1 gives 10, plus 1 gives 11. Write 1, remember ①.
(2) ① plus 1 gives 10, plus 1 gives 11, plus 1 gives 100, plus 1 gives 101. Write 1, remember ⑩.
(3) ⑩ plus 1 gives 11, plus 1 gives 100, plus 1 gives 101, plus 1 gives 110. Write 0, remember ⑪.
(4) ⑪ plus 1 gives 100, plus 1 gives 101, plus 1 gives 110. Write 0, remember ⑪.
(5) ⑪ plus 1 gives 100, plus 1 gives 101. Write 1, remember ⑩.
(6) ⑩ plus 1 gives 11, plus 1 gives 100. Write 0, remember ⑩.
(7) Write 0, remember 1.
(8) Write 1.

1.8 Adding and Subtracting in Base Two

Another way to show this problem would be to write 1 in the next column to the left for each $1 + 1$ of the preceding column. This is done above the first line below.

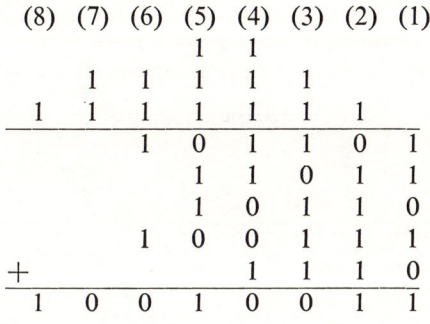

STEPS:
(1) $1 + 1 + 1 = (1 + 1) + 1$, write one 1 in the second column to the left. Since there is one more 1, write 1 for the sum below.
(2) $(1 + 1) + (1 + 1) + 1$, write two 1s in column (3). Since there is one more 1 left over, write 1 for the sum below.
And so forth.

The corresponding addition in base ten would be:

```
 45
 27
 22
 39
 14
---
147
```

Next, let's take a look at subtraction in base two.

$$10_{two}$$
$$- \ 1$$

Since 1 cannot be subtracted from 0, we consider the 1 two as two ones:

$$\overset{2}{\cancel{1}}0_{two}$$
$$\underline{- \ 1}$$
$$ \ 1$$

Another case to examine is one in which two zeros occur:

$$100_{two}$$
$$\underline{- \ \ 1}$$

1 Mathematics for Fun and Numeration Systems

As before, we cannot subtract 1 from 0. We take the 1 four and think of it as one two and two ones:

$$\begin{array}{r} 12 \\ \cancel{1}00_{two} \\ -1 \\ \hline 11_{two} \end{array}$$

And one more case to get you going on subtraction:

$$\begin{array}{r} 1000_{two} \\ -1 \\ \hline \end{array}$$

Think of the 1 eight as one four and one two and two ones—that's eight, right?

$$\begin{array}{r} 112 \\ \cancel{1}000_{two} \\ -1 \\ \hline 111_{two} \end{array}$$

Perhaps you have noticed that whenever there are zeros, following the regrouping we get ones. In base ten we get nines when we subtract from zeros. So, base ten—nines, base two—ones. In base five—fours.

As a bonus, here is a problem in subtraction worked out for you. Study it to be sure you understand what is being done.

$$\begin{array}{r} 1001011_{two} \\ -110010_{two} \\ \hline 11001_{two} \end{array}$$

Once you catch on, there is not much to it. It can be very automatic—just like the computer. No wonder the distinguished mathematician and philosopher Gottfried Wilhelm von Leibniz (1646-1716) seriously urged that we discard the decimal system in favor of the binary. Having only two digits makes things quite simple, except for the length of the numerals.

Exercises 1.8

Add:

1. $\begin{array}{r} 101_{two} \\ +10_{two} \end{array}$
2. $\begin{array}{r} 101_{two} \\ +11_{two} \end{array}$
3. $\begin{array}{r} 111_{two} \\ +11_{two} \end{array}$

1.8 Adding and Subtracting in Base Two

4. 1101_{two}
    $+\ 101_{two}$

5. 110111_{two}
    $+\ \ 1001_{two}$

6. 1100111_{two}
    $+1111111_{two}$

7. 1101_{two}
    $\ 111_{two}$
    $+\ \ 10_{two}$

8. 11011_{two}
    $\ 1101_{two}$
    $+\ 1111_{two}$

9. 111101_{two}
    100011_{two}
    $+110101_{two}$

Subtract:

10. 101_{two}
    $-\ 11_{two}$

11. 1000_{two}
    $-\ \ \ \ 1$

12. 10011_{two}
    $-\ 1101_{two}$

13. 110011_{two}
    $-\ \ 1111_{two}$

14. 1010101_{two}
    $-\ 101101_{two}$

15. 100000011_{two}
    $-\ \ \ \ 111111_{two}$

16. Think of some reasons why a mathematician would urge the adoption of the binary system to replace the decimal system.

CHAPTER 1 TEST

1. What mathematical concepts and skills do you find most useful for your everyday living?
2. Name some professionals you know who use a great deal of mathematics in their jobs.
3. Write 1,000,000 as a power of 10.
4. Assume that the sum of the measures of the three angles in a triangle is 180°. Divide a quadrilateral, a pentagon, a hexagon, and so on into triangles by drawing all the diagonals from one vertex. Find a formula for the sum of all angles of a polygon with n sides in terms of n.
5. Show 483,026 as a sum of products of powers of ten.
6. Build a sequence of powers of 6 to show that $6^0 = 1$.
7. Expand 3142.034_{five} by powers of five.
8. Find the base-ten numeral for 4032_{five}.

Add:

9. 34_{five}
    $+22_{five}$

10. 3012_{five}
    $+3342_{five}$

11. 4402_{five}
    $+4443_{five}$

Subtract:

12. 41_{five}
    -23_{five}

13. 302_{five}
    $-\ 24_{five}$

14. 30134_{five}
    -12144_{five}

1 Mathematics for Fun and Numeration Systems

15. In the war between Mathematica and Ignoramia, many spies were employed. All the spies from Mathematica used a binary code similar to the one in Section 1.7, exercise 26. Because of the Ignoramian spies' lack of knowledge, they could not decipher the coded messages of the Mathematicans and consequently lost the war. What did this message say?

10011/101/1110/100/1000/101/1100/10000/10111/101/10111/1001/1100/1100/1/10100/10100/1/11/1011

Give base-ten numerals for the following:

16. 111_{two} **17.** 10010_{two} **18.** 110011011_{two}

Give binary numerals for the following written in base ten:

19. 39   **20.** 68   **21.** 99   **22.** 109

Add:

23. $\quad 1101_{two}$
$\quad\ +\ \ 111_{two}$

24. $\quad 111001_{two}$
$\quad +\ \ 10101_{two}$

25. $\quad 110011_{two}$
$\quad +100101_{two}$

Subtract:

26. $\quad 1001_{two}$
$\quad -\ \ \ \ 11_{two}$

27. $\quad 1001101_{two}$
$\quad -\ 110011_{two}$

28. $\quad 1100110_{two}$
$\quad -1000001_{two}$

2

Sets

2.1 SOME FAMILIAR SETS

REVIEW

None

OBJECTIVES

- Provide examples of uses of sets in everyday life
- Decide what natural numbers are
- Decide on the meaning of the phrase *well-defined set*
- Agree on the way of designating sets
- Supply verbal descriptions of given sets
- Use the notation $\in$ or $\notin$ to denote whether something belongs or does not belong to a given set
- Establish a one-to-one correspondence between two given finite sets
- Use the notation for the number of members in a set

2 Sets

- Establish the meaning of finite and infinite sets
- Know the meaning of the empty set, $\emptyset$
- Use three dots correctly
- Discover patterns in listings of members in sets
- Find the number of members in a given set

We often use the word *set* in regular speech. Consider the following phrases:

the set of golf clubs
the set of dishes
the set of executives
the jet set

All of these phrases are heard frequently in everyday conversations.

By a *set* we mean simply a collection of things. Each of the following phrases describes a set pretty well:

the set of all colleges in the United States
the set of all students in this college
the set of all persons between the ages of 17 and 22 in the United States
the set of all continents

In describing a set, we must be able to tell whether any given object belongs to that particular set. Consider, for example, the following description:

the set of all natural numbers which are less than 7

Before we can decide what numbers are in this set, an agreement must be made concerning what is meant by *natural numbers*. The natural numbers are the ordinary counting numbers:

1, 2, 3, 4, ...

The three dots mean that the pattern shown by the first four numbers con-

2.1 Some Familiar Sets

tinues. Having clarified what a natural number is, we can now decide what numbers are in the set described above. They are:

1, 2, 3, 4, 5, 6

Given any natural number, we can tell whether it belongs to this set. Therefore, we say that the set is *well-defined*. Another example of a well-defined set is:

the set of all students in this class less than 19 years old

With a list of the students in this class and their ages, we can determine whether or not a certain student is a member of the set.

To know that a reference is being made to a *set*, the names of things which belong to the set will be enclosed in braces. The set of all natural numbers which are less than 7 would be shown as:

$\{1, 2, 3, 4, 5, 6\}$

or

$\{\text{the first six natural numbers}\}$

or

$\{\text{all natural numbers which are less than or equal to 6}\}$

Each object which belongs to a set is called a *member* or an *element* of the set. For example, the number 4 is a member of the set above. Choosing, in the present discussion, to name the set by the letter A, it can be written:

$A = \{1, 2, 3, 4, 5, 6\}$

The symbol $\in$ means *is a member of* or *belongs to*. To say that the number 4 is a member of A, we write:

$4 \in A$ *Read:* 4 is a member of A

The number 15 is not a member of A. This is written:

$15 \notin A$

2 Sets

The symbol $\notin$ means *is not a member of* or *does not belong to*.
Given two sets

$$B = \{5, 1, 7\} \quad \text{and} \quad C = \{\tfrac{2}{5}, \tfrac{1}{7}, \tfrac{2}{3}\}$$

a *one-to-one correspondence* can be established between them by matching elements in the following manner:

$$B = \{5, 1, 7\}$$
$$\updownarrow \updownarrow \updownarrow$$
$$C = \{\tfrac{2}{5}, \tfrac{1}{7}, \tfrac{2}{3}\}$$

The 5 is matched with $\tfrac{2}{5}$, 1 with $\tfrac{1}{7}$, and 7 with $\tfrac{2}{3}$. For each element in B there is exactly one element in C and for each element in C there is exactly one element in B. To say that set B has three elements, we write $n(B) = 3$. Thus $n(B)$ means *the number of elements in the set B.*

There are other ways of establishing a one-to-one correspondence between B and C. For example:

$$B = \{5, 1, 7\}$$
$$C = \{\tfrac{2}{5}, \tfrac{1}{7}, \tfrac{2}{3}\}$$

In this case, 5 is matched with $\tfrac{2}{3}$, 1 with $\tfrac{2}{5}$, and 7 with $\tfrac{1}{7}$.

It is easy to surmise now that two *finite sets*—sets with a definite number of members—can be put into a one-to-one correspondence if they have the same number of members.

We know that the smallest natural number is 1. So, the phrase the set of all natural numbers less than 1 describes a set which has no members. Such a set is called the *empty* or *null set*. Another description of the empty set is:

the set of all natural numbers between 5 and 6

The empty set is denoted by the symbol $\emptyset$. It is clear that $n(\emptyset) = 0$.

Finite sets were mentioned above, and some examples of such sets have been given. But not all sets are finite; some sets are *infinite sets*. You are familiar with some of them, for example, the set of all natural numbers, N.

$$N = \{1, 2, 3, \ldots\}$$

The three dots indicate that the listing goes on and on. The set is infinite because we can't write the last element.

2.1 Some Familiar Sets

In dealing with infinite sets, it is necessary to make a judgment as to how many elements need to be listed before it is clear what infinite set is intended. For example, writing {2, 4, ...} may not be sufficient. The following three different sets have this beginning.

{2, 4, 8, 16, 32, ...}
{2, 4, 6, 8, ...}
{2, 4, 7, 11, 16, ...}

In the first set, the next three members would be 64, 128, and 256. Each subsequent member is obtained by doubling the preceding element.

In the second set, the next three members would be 10, 12, and 14. This is the set of all even natural numbers.

In the third set, the next three members would be 22, 29, and 37. To obtain the next member, a number one greater than was added to the preceding member is added.

Three dots are also used to avoid the listing of a lot of numbers in a finite set. For example, the set of all natural numbers from 1 through 100 can be shown as:

{1, 2, 3, ..., 98, 99, 100}

Exercises 2.1

1. Give three examples of the use of sets in everyday life.
2. Give three phrases describing various collections of students in your college.
3. Give two different verbal descriptions of the set $M = \{2, 4, 6, 8\}$.
4. Display all possible ways of establishing a one-to-one correspondence between the following sets:

    $A = \{1, 2, 3\}$ and $B = \{7, 8, 9\}$

5. Two sets of people:

    $D = \{3, 5, 9\}$ and $F = \{2, 1, 0\}$

    are separated from each other by a river. They must reach each other by nightfall. All the members of set D have lines that they will throw to the members of set F. In this way, they hope to form a one-to-one correspondence between the sets and thereby transfer across the river. Unfortunately, none of the members can decide which way to make the correspondence. Can you help them by drawing all the ways of making a one-to-one correspondence?

2 Sets

Which of the following sentences describe the empty set?

6. The set of all even natural numbers which are less than 2
7. The set of all odd natural numbers which are less than 2
8. The set of all odd natural numbers which are less than 15 and greater than 13
9. The set of all months that have names beginning with the letter T
10. The set of all days that have names beginning with the letter W
11. Why is the following set not well-defined?

   the set of 36 even numbers

12. Is the following set well-defined?

   the set of all odd natural numbers which are greater than 41

For the set $D = \{1, 3, 5, \ldots\}$, which of the following are true and which are false?

13. $1 \in D$
14. $35 \in D$
15. $26{,}000 \in D$
16. $1.5 \in D$
17. $3 \notin D$
18. $\frac{3}{5} \in D$
19. $57{,}009 \in D$
20. $12.8 \notin D$

21. You are applying for a credit card at a local department store. The manager, who is a mathematician, asks you some questions to determine your knowledge of math. If you give correct answers you will receive your credit card. If your answers are incorrect, the manager will probably deny your application. He tells you:

   "Our company is described by the set $\{2, 4, 6, 8, 10, \ldots\}$."

   a. "Jones, who works in C division, has the number 22. Is he a member of our set?"
   b. "We are thinking of hiring some new people with the numbers 101, 200, and 3601. Should we?"

For the set $F = \{5, 10, 15, \ldots\}$, which of the following are true and which are false?

22. $100 \in F$
23. $501 \in F$
24. $2000 \in F$
25. $5004 \in F$

26. There is a famous sequence in mathematics called the Fibonacci sequence. It can be described by the set

   $A = \{1, 1, 2, 3, 5, 8, 13, 21, \ldots\}$

   Can you discover the pattern of this sequence? Determine if:
   a. $34 \in A$ b. $55 \in A$

How many members does each of the following sets have?

27. The set of all even natural numbers which are less than 100
28. The set of all odd natural numbers which are less than 150
29. The set of all even natural numbers which are greater than 10 and less than 120

2.2 Equivalent Sets

30. The set of all odd natural numbers which are greater than 49 and less than 201

How many members are there in each of the following sets?

31. $\{5, 10, 15, \ldots, 125, 130\}$
32. $\{\frac{1}{3}, \frac{1}{5}, \frac{1}{7}, \ldots, \frac{1}{99}, \frac{1}{101}\}$
33. $\{\frac{2}{3}, \frac{5}{6}, \frac{8}{9}, \ldots, \frac{62}{63}, \frac{65}{66}\}$
34. $\{\frac{1}{2}, \frac{2}{4}, \frac{3}{6}, \ldots, \frac{59}{118}, \frac{60}{120}\}$

For each of the following infinite sets supply the three next members and describe the pattern:

35. $\{1, 3, 5, \ldots\}$
36. $\{1, 1, 2, 3, 5, 8, \ldots\}$
37. $\{1, 4, 9, 16, \ldots\}$
38. $\{\frac{1}{2}, \frac{2}{3}, \frac{3}{4}, \ldots\}$
39. $\{1, 8, 27, 64, \ldots\}$
40. $\{\frac{1}{2}, \frac{1}{4}, \frac{1}{8}, \ldots\}$
41. $\{3, 9, 27, \ldots\}$
42. $\{\frac{1}{2}, \frac{3}{4}, \frac{5}{6}, \ldots\}$
43. $\{2, 6, 24, 120, \ldots\}$
44. $\{.1, .02, .003, \ldots\}$

2.2 EQUIVALENT SETS

REVIEW

- The elements of a finite set can be counted
- The elements of an infinite set go on forever
- To establish a one-to-one correspondence between sets A and B means to assign exactly one element in B to each element of A and vice versa

OBJECTIVES

- Know what it means for two finite sets to be equivalent
- Know what it means for two infinite sets to be equivalent
- Given two sets, tell whether they are equivalent or not
- Describe the pattern, given the correspondence between two infinite sets

For the sets

$$A = \{1, 2, 3\} \quad \text{and} \quad B = \{6, 7, 8\}$$

$n(A) = 3$ and $n(B) = 3$. Therefore $n(A) = n(B)$; that is, the sets A and B have the same number of elements. For this reason, sets A and B are said to be *equivalent sets*. In general, any two finite sets which have the same number of elements are equivalent sets. (Do not confuse this with equal sets, a concept that will be discussed in the next section.)

Whether two finite sets are equivalent can be decided either by counting their members or by establishing a one-to-one correspondence between the sets.

The matter of equivalence of two infinite sets is more complicated. It is no longer sensible to talk about the number of elements in an infinite set, since they cannot be counted—there is no end to them. But it is still sensible to talk about a one-to-one correspondence. To see this, consider two infinite sets: N is the set of all natural numbers, and E is the set of all even natural numbers.

$N = \{1, 2, 3, 4, 5, \ldots\}$
$E = \{2, 4, 6, 8, 10, \ldots\}$

A one-to-one correspondence can be established between N and E in the following manner:

$N = \{1, 2, 3, 4, 5, \ldots\}$
$\quad\quad\updownarrow \updownarrow \updownarrow \updownarrow \updownarrow$
$E = \{2, 4, 6, 8, 10, \ldots\}$

To see that there is exactly one member of E for each member of N and there is exactly one member of N for each member of E, observe that this correspondence is arranged in a pattern which continues on forever. To every member of N we assign a member of E which is twice it. So, given 30 in N, we know that 60 in E is its match. Similarly, given 60 in E, we know that 30 in N is its match. This relationship is true for *any* number. In general, given any number k in N, $2k$ in E is matched with it. Given any number m in E, $m/2$ in N is matched with it.

In case we have not succeeded in establishing a one-to-one correspondence between two infinite sets, one of two things may be true.

1. There *is* a one-to-one correspondence, but we are not clever enough to find it.

2. There *is no* one-to-one correspondence between these two sets.

In the first case, the two sets are equivalent. In the second case, they are not equivalent. Pairs of infinite sets which are not equivalent exist, but we will not be considering such sets.

Exercises 2.2

Which of the following pairs of sets are equivalent and which are not?

1. $A = \{9, 1, 12\}$
   $B = \{\text{Ed, Bob, Jerry, Ken}\}$
2. $C = \{\text{New York, California, Ohio}\}$
   $D = \{\text{Jersey City, Sacramento, Columbus}\}$

2.2 Equivalent Sets

3. $E = \{\text{kite, golf ball}\}$
   $F = \{1, 2\}$
4. $G = \{4, 8, 12, \ldots\}$
   $H = \{15, 10, 5\}$
5. Is it possible for two sets, one of which is finite and the other infinite, to be equivalent?
6. Can the empty set be equivalent to a set which is nonempty?
7. As chief director of the new Robot Assembly Project, you are faced with the task of repairing one robot with the parts from another. For this repair to be acceptable, the parts must be equivalent. All interchangeable parts are given by the sets:

   Robot $A = \{Z, W, Y\}$ Robot $B = \{A, G, D\}$

   Are they equivalent? Can you make the repairs?

For each pair of the following sets, a one-to-one correspondence is shown. Describe the one-to-one correspondence between the elements of A and B:

Example $A = \{2, 4, 6, 8, \ldots\}$
$B = \{1, 3, 5, 7, \ldots\}$

Description Each element of A is matched with an element of B which is 1 less; that is, k in A is matched with $k - 1$ in B.

8. $A = \{1, 2, 3, 4, \ldots\}$
   $B = \{1, 4, 9, 16, \ldots\}$
9. $A = \{1, 2, 3, 4, \ldots\}$
   $B = \{0, 3, 8, 15, \ldots\}$
10. $A = \{1, 2, 3, 4, \ldots\}$
    $B = \{1, \frac{1}{2}, \frac{1}{3}, \frac{1}{4}, \ldots\}$
11. $A = \{\frac{1}{2}, \frac{1}{3}, \frac{1}{4}, \frac{1}{5}, \ldots\}$
    $B = \{\frac{1}{3}, \frac{1}{4}, \frac{1}{5}, \frac{1}{6}, \ldots\}$
12. $A = \{1, 2, 3, 4, \ldots\}$
    $B = \{1, 8, 27, 64, \ldots\}$
13. $A = \{1, 2, 3, 4, \ldots\}$
    $B = \{1, 16, 81, 256, \ldots\}$
14. $A = \{\frac{1}{2}, \frac{1}{3}, \frac{1}{4}, \frac{1}{5}, \ldots\}$
    $B = \{\frac{1}{4}, \frac{1}{9}, \frac{1}{16}, \frac{1}{25}, \ldots\}$
15. $A = \{1, 2, 3, 4, \ldots\}$
    $B = \{3, 5, 7, 9, \ldots\}$
16. $A = \{\frac{1}{2}, \frac{1}{4}, \frac{1}{8}, \frac{1}{16}, \ldots\}$
    $B = \{\frac{3}{4}, \frac{3}{8}, \frac{3}{16}, \frac{3}{32}, \ldots\}$
17. $A = \{1, 2, 3, 4, 5, \ldots\}$
    $B = \{2, 8, 18, 32, 50, \ldots\}$
18. $A = \{1, 2, 3, 4, 5, \ldots\}$
    $B = \{2, 9, 28, 65, 126, \ldots\}$
19. $A = \{1, 2, 3, 4, 5, \ldots\}$
    $B = \{0, 3, 8, 15, 24, \ldots\}$
20. Discuss the validity of the following statement:

    {people on planet Earth} is equivalent to {all stars in universe}

2 Sets

★21. Present an argument showing that the set of natural numbers can be put into a one-to-one correspondence with a set obtained by deleting from the set of natural numbers any finite number of natural numbers beginning with 1 and taken consecutively.

22. You have just completed a survey. You have correlated the results in two sets:

$$A_{\text{opinions}} = \{\text{Mr. Smith, Mr. Jones, Mr. Black}\}$$
$$B_{\text{opinions}} = \{\text{Mrs. Smith, Mrs. Jones, Mrs. Black}\}$$

These sets are obviously equivalent, but what is the nature of the correspondence?

2.3 SUBSETS AND EQUALITY

REVIEW

- Capital letters, such as A, B, and X, are used to designate sets
- $\sqrt{36}$ is read "square root of thirty-six"; it is equal to 6, since $6 \times 6 = 36$
- $10_{\text{two}} = \text{two}$
- $10_{\text{three}} = \text{three}$
- $10_{\text{four}} = \text{four}$
- $\emptyset$ is the symbol for the empty set; $n(\emptyset) = 0$

OBJECTIVES

- Know what *universal set* means
- Know what a *subset* means
- Form subsets from a given set
- Use the symbols $\subseteq$, $\subset$, $\nsubseteq$, and $\not\subset$ correctly
- Recognize when two names name the same set (concept of equality of sets)

If we want to form smaller sets, we have to agree upon what the entire set is from which we are forming smaller sets. Such a set is called a *universal set* or simply a *universe*. For example, in elementary school the set of *whole numbers* is a commonly used universe:

$$W = \{0, 1, 2, 3, \ldots\}$$

Other sets can be formed using some of the elements of W, such as:

$$E = \{0, 2, 4, 6, \ldots\}, \text{ the set of even numbers}$$
$$O = \{1, 3, 5, 7, \ldots\}, \text{ the set of odd numbers}$$
$$P = \{2, 3, 5, 7, \ldots\}, \text{ the set of prime numbers}$$

2.3 Subsets and Equality

Each of these sets is a *subset* of the set of whole numbers, W, because every element of each of these sets is also an element of the set of whole numbers. We write:

$E \subseteq W$
$O \subseteq W$
$P \subseteq W$

As you have no doubt guessed, the symbol $\subseteq$ is read: "is a subset of." By the way, do you see that every set is a subset of itself? That is,

$A \subseteq A$ for every set A

This should be obvious, since every element of a set belongs to that set.

If we consider the universal set to be

the set of all presidents of the United States

then some examples of subsets of this set are:

{George Washington, Millard Fillmore, Chester Alan Arthur, John Fitzgerald Kennedy}
{Richard Milhous Nixon, Lyndon Baines Johnson}
{Calvin Coolidge}

We can see that $A = \{1, 3, 5\}$ is not a subset of $B = \{1, 2, 3, 4\}$, because there is an element in A which is not in B. Which number is that? To say that A is not a subset of B, we write

$A \nsubseteq B$

So, to show that some set is not a subset of another set, all we have to do is display one element in the first set that is not in the second set.

Since we cannot display an element in the empty set, we must conclude that the empty set is a subset of every set:

$\emptyset \subseteq A$ for every set A

We say that two sets are different if one of them has at least one element that is not in the other. We write $A \neq B$ to say that sets A and B are different or *unequal*.

People often say that two sets are *equal*. All they mean by that is that we have two names for the same set—we really don't have two sets. For example, if $A = \{1, 2, 3\}$ and $B = \{3, 1, 2\}$, then we can say that $A = B$.

Sets A and B are the same set, for it does not matter in which order we list the elements.

It is interesting to contemplate the statement:

if $X \subseteq Y$ and $Y \subseteq X$, then $X = Y$

What can you say about it?

Another way to show the same set is to name its elements differently. For example:

$$A = \{4, 6, 7\} \quad \text{and} \quad B = \{6+1, \sqrt{36}, 2^2\}$$

Sets A and B are the same since $6 + 1 = 7$, $\sqrt{36} = 6$, and $2^2 = 4$. So, the two sets have exactly the same elements.

Did you notice that $\subseteq$ is a combination of two symbols, $\subset$ and $=$? This symbol is logical, since if $A \subseteq B$, then it allows the possibility that $A = B$. The symbol $\subset$ is also used—it means "is a *proper* subset of." For example, $A \subset B$ if $A = \{1, 3, 5\}$ and $B = \{1, 3, 5, 7\}$, because B has at least one element which is not in A.

Exercises 2.3

1. Consider the set of all students in this mathematics course to be the universal set. Describe three different subsets of this universal set.

Consider the set of all whole numbers, $W = \{0, 1, 2, 3, \ldots\}$. Form another set S which consists of the square of each whole number:

2. Is S an infinite set?
3. Is $S \subset W$ true? Why?
4. Is $W \subset S$ true? Why?
5. Establish a one-to-one correspondence between S and W.

Let the universal set U be the set of those states of the United States which are located west of the Mississippi River:

6. If P is the set of the states (chosen from U) touching the Pacific Ocean, label each of the following as either true or false.
   a. $P \subseteq U$ b. $P \subset U$ c. $P = U$
7. If C is the set of the states (chosen from U) touching the border of Canada, label each of the following as either true or false.
   a. $C \subseteq U$ b. $C \subset U$ c. $C = U$ d. $C \subset P$
8. If T is the set of the states (chosen from U) directly adjoining the Mississippi River, label each of the following as either true or false.
   a. $T \subset U$ b. $T \subset C$ c. $T \subset P$

2.3 Subsets and Equality

Group the following sets into groups of equal sets. If a base of a numeral is not given, assume that it is ten:

9. $\{1, 2, 3, 4, \ldots\}$
10. $\{2, 4, 6, 8, \ldots\}$
11. $\{1, 3, 5, 7, \ldots\}$
12. The set of all natural numbers
13. $\{\frac{1}{2}, \frac{1}{4}\}$
14. $\{10_{\text{two}}, 10_{\text{three}}, 10_{\text{four}}\}$
15. $\{.25, .5\}$
16. $\{4, 3, 2\}$
17. $\{10_{\text{two}}, 11_{\text{two}}, 100_{\text{two}}\}$
18. $\{\frac{1}{4} + \frac{1}{4}, \frac{1}{8} + \frac{1}{8}\}$
19. The set of all even natural numbers
20. The set of all odd natural numbers
21. The set of all natural numbers which are divisible by 2
22. The set of all natural numbers which are not divisible by 2
23. You have just written a 400-page book entitled *Complex Differentials*. Your publisher informs you that the pages making up subset $J = \{98, 99, 100, \ldots, 121, 122\}$ must be rewritten. How many pages will you have to rewrite?

Form all possible subsets of each of the following sets. In each case tell how many subsets there are. [*Hint: Do not forget that* $\emptyset$ (*the empty set*) *is a subset of every set and that every set is a subset of itself.*]

24. $\{5\}$
25. $\{4, 7\}$
26. $\{1, 3, 6\}$

Given the three sets,

$$A = \{0, 2, 4\} \quad B = \{1, 3\} \quad C = \{0, 3, 6, 9\}$$

label each of the following statements as either true or false:

27. $B \subseteq C$
28. $C \subseteq C$
29. $\emptyset \subset B$
30. $\{0\} \subseteq A$
31. $A \not\subset C$
32. $A \neq B$
★33. $\{1\} \not\subset C$
★34. $\{0, 4\} \not\subset C$
35. Can a finite set be a subset of an infinite set? Why?
36. Can an infinite set be a subset of a finite set? Why?
37. Can a finite set be equal to an infinite set? Why?
38. Why is $X \subset X$ not true for every set X?
★39. How many elements does $\{0\}$ have?
★40. How many elements does $\emptyset$ have?
★41. Explain why the statement $\{0\} \neq \emptyset$ is true.
★42. How many elements does $\{\emptyset\}$ have?

2.4 OPERATIONS ON SETS

REVIEW

None

OBJECTIVES

- Given two sets, tell whether they are disjoint or not
- Give examples of binary operations
- Give examples of singulary operations
- Find the union of two given sets
- Find the intersection of two given sets
- Given a universal set and a subset of it, find the complement of that subset

Given any two sets, they are related to each other in one of two ways. Either they have no elements in common, or they have one or more elements in common. Two sets which have no elements in common are called *disjoint sets*.

Consider the following two sets, A and B:

$$A = \{2, 4, 6\} \qquad B = \{11, 13, 16\}$$

A and B are disjoint sets because they have no elements in common. On the other hand, given $C = \{1, 3, 4\}$ and $D = \{3, 4, 5, 10\}$, C and D are not disjoint since they have at least one element in common. What are the elements which belong to both C and D?

To understand operations on sets, it is helpful to recall some basic things about operations on numbers. Operations performed on pairs of numbers are called *binary operations*. Addition, for example, is such an operation—it associates one number with a pair of numbers. For the pair (3, 4), this can be shown as follows:

$$(3, 4) \xrightarrow{+} 7$$

There are other binary operations, such as subtraction, multiplication, and division.

There are also *singulary operations* on numbers, that is, operations performed on *one* number. Squaring, for example, is such an operation. For the number 6, this operation can be shown as follows:

$$6 \xrightarrow{sq} 36$$

2.4 Operations on Sets

The situation with sets is similar: there are binary and there are singular operations on sets. First, we will consider a binary operation on sets. Given two sets, $A = \{5, 10, 15\}$ and $B = \{10, 11, 12, 13, 14, 15\}$, this operation produces a set, $C = \{5, 10, 11, 12, 13, 14, 15\}$. This operation is performed as follows: if an element is in *one* of the two sets, it is in the resulting set; also if an element is in *both* sets, it is in the resulting set. This operation is called *union*. To indicate a union of two sets, we use the symbol $\cup$. For example, you should read $X \cup Y$ as "the union of X and Y."

A second operation on sets associates with two given sets a set consisting of only those elements which are in both sets. For example, given $X = \{1, 4, 7\}$ and $Y = \{4, 7, 10, 12\}$, this operation associates $Z = \{4, 7\}$ with X and Y. This operation is called *intersection*. To indicate an intersection of two sets, we use the symbol $\cap$. For example, you should read $X \cap Y$ as "the intersection of X and Y."

Brief reflection on the meaning of intersection will suggest that the intersection of any two disjoint sets is the empty set.

The third operation to be considered is a singular operation. For this operation to make sense, it is necessary to specify the universal set. Suppose $U = \{1, 2, 3, 4, 5, 6\}$. Consider $A = \{2, 4\}$. It is true that $A \subset U$. Now consider the set B which consists of all elements of U which are not in A; that is, $B = \{1, 3, 5, 6\}$. B is called the *complement* of A in U. To designate the complement of set B, we write a bar above B, like this: $\bar{B}$. You should read $\bar{B}$ as "complement of set B." Sometimes the symbol B' is used to denote the complement of B.

Exercises 2.4

For each given pair of sets, tell whether they are disjoint sets or not:

1. $A = \{0, 1, 2\}, B = \{2\}$
2. $A = \{0, 2\}, B = \{0, 5\}$
3. $A = \{1, 2, 4, 6, \ldots\}, B = \{1, 3, 5, 7, \ldots\}$
4. $A = \{1, 2, 3, 4, \ldots\}, B = \{0\}$
5. $A = \{1, 3\}, B = \{1, 3\}$
6. $A = \varnothing, B = \{1\}$
7. $A = \varnothing, B = \{0\}$
8. $A = \varnothing, B = \varnothing$
9. $A = \{\frac{1}{2}, \frac{3}{4}\}, B = \{\frac{2}{4}\}$
10. $A = \{\frac{1}{3}, \frac{1}{4}, \frac{1}{5}\}, B = \{0.2\}$
11. $A = \{\text{New York, Florida, California}\}, B = \{\text{Los Angeles}\}$
12. $A = \{\text{San Francisco, Houston, Detroit}\}$
    $B = \{\text{New York, Miami, San Diego, San Francisco}\}$
13. $A = \{\text{all cities in the U.S. with a population of over 1 million}\}$
    $B = \{\text{Chicago}\}$
14. $A = \{\text{all states in the U.S. with a population under 10 million}\}$
    $B = \{\text{California, Maine}\}$

15. Medical science has developed a computerized technique that analyzes bones and prints out data in the form of sets. This procedure eliminates the need for expensive X-ray equipment. If the sets are disjoint, the bone is broken. If the sets are not disjoint, the bone is intact. Analyze the following data and see if your foot, upon which you recently dropped seven textbooks, needs a cast.
    a. $A = \{7, 9, 21, 60\}$; $B = \{11, 10, 4, 3\}$
    b. $A = \{2, 3, 4, 5, 6\}$; $B = \{100, 1000, 90, 2\}$
    c. $A = \{0, 100, 99, 98\}$; $B = \{1000, 999, 998, 0\}$
16. Suppose $5 \in X$ and $5 \in Y$, where X and Y are some sets. Is it true that $X \subseteq Y$? That $Y \subseteq X$? That X and Y are not disjoint?
17. Suppose sets M and P have exactly three members in common. Give two conclusions you are justified in drawing from this fact.

For each pair of the following sets, give (a) their union and (b) their intersection:

18. $A = \{1, 3\}$, $B = \{4\}$
19. $A = \{$all natural numbers$\}$, $B = \{$all even natural numbers$\}$
20. $A = \{$all whole numbers$\}$, $B = \{$all odd numbers$\}$
21. $A = \{5, 10, 15, \ldots\}$, $B = \{10, 20, 30\}$
22. $A = \{$all prime natural numbers$\}$, $B = \{$all even natural numbers$\}$
23. $A = \{100, 200, 300, \ldots\}$, $B = \{1000, 2000, 3000\}$
24. $A = \{$all male students in this school$\}$,
    $B = \{$all female students in this school$\}$
25. $A = \{$all members of the U.S. Congress$\}$,
    $B = \{$all members of the U.S. Senate$\}$

Explain why each of the following statements is true for every set A:

26. $A \cup A = A$ 27. $A \cap A = A$ 28. $A \cup \emptyset = A$ 29. $A \cap \emptyset = \emptyset$
30. Is $\emptyset \cap \emptyset = \emptyset$ true? Why?   31. Is $\emptyset \cup \emptyset = \emptyset$ true? Why?

Let the universal set $U = \{0, 2, 4, 6, 8\}$. Give the complement of each of the following sets in U:

32. $A = \{4, 8\}$
33. $B = \{0\}$
34. $C = \{0, 4, 8, 2\}$
35. $D = \{0, 2, 4, 6, 8\}$

Is it true that for every universal set U:

36. $\bar{U} = \emptyset$? Why?   37. $\bar{\emptyset} = U$? Why?
★38. Can the union of two infinite sets be a finite set?
★39. Given an infinite universal set, can it have a subset whose complement is a finite set?

2.5 SETS OF POINTS

REVIEW

- A *point* is a basic building block in geometry

OBJECTIVES

- Describe segments using set notation
- Demonstrate all possible intersections of segments
- Demonstrate how segments can form polygons
- Describe a line
- Consider rays as subsets of lines
- Show how two rays can form an angle
- Distinguish between the noun *intersection* and the verb *intersect*
- Consider intersections of rays
- Consider intersections of lines

The language of sets finds much use in geometry. The reason for this is very simple: every geometric figure is a *set* of points. For example, Figure 2.1 shows a very simple set, a *segment*. We use the symbol $\overline{AB}$ to designate it. When you see a bar above two capital letters, you should say "segment $\overline{AB}$." The points A and B are *endpoints* of $\overline{AB}$.

Many interesting geometric figures can be built from segments, such as various polygons: triangles, quadrilaterals, pentagons, and so on. But we will leave all of that to geometry.

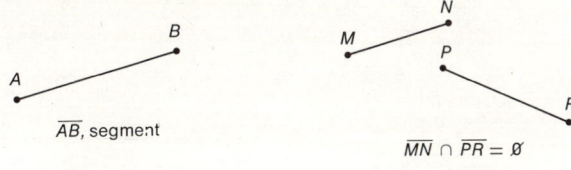

FIGURE 2.1   **FIGURE 2.2**

We do want you to observe, however, that two segments can have different intersections. It can be the empty set, as shown in Figure 2.2. Here $\overline{MN}$ and $\overline{PR}$ do not intersect. Note the important distinction between the noun *intersection* and the verb *intersect*. There always exists an intersection of two sets, even though it may be the empty set. When the intersection of two sets is the empty set, however, we say that the segments do not intersect.

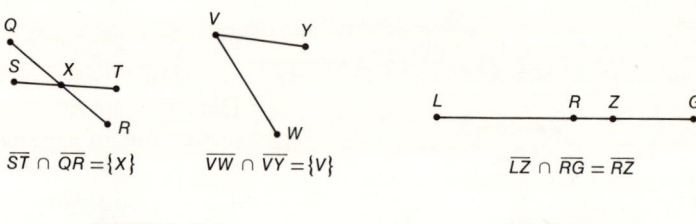

FIGURE 2.3   FIGURE 2.4

Two segments can intersect in exactly one point. This can happen in essentially two ways, which are shown in Figure 2.3. Since the intersection of two sets is always a set, we must put X, the name of the point of intersection, in braces.

The intersection of two segments can also be a segment, as is shown in Figure 2.4. $\overline{RZ}$ does not need to be written within braces, because it is already a name of a set.

Another set of points which is very basic in geometry is a *line*. Figure 2.5 shows a line, which can be named in two ways, line ℓ or $\overleftrightarrow{CD}$. The second name simply uses any two points on the line. The arrow above the names of the two points suggests that a line continues infinitely in two directions.

You probably have already thought that a line can have many subsets that are segments. A line can also have subsets that extend infinitely in one direction—called *rays*. In Figure 2.6, we can identify the following rays: $\overrightarrow{EF}$ and $\overrightarrow{EG}$. The ray $\overrightarrow{EF}$ consists of the point E and all the points of the line to the right of it. Point E is the endpoint of $\overrightarrow{EF}$. It is also the endpoint of $\overrightarrow{EG}$. Note that we use a one-way arrow to indicate a ray and we list the endpoint of a ray first in the name of the ray.

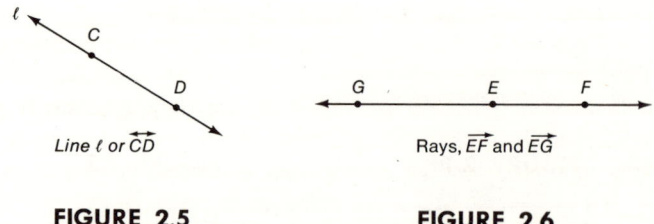

FIGURE 2.5   FIGURE 2.6

One of the most commonly used unions of two rays is an angle. An *angle* is a union of two noncollinear rays that have a common endpoint. Figure 2.7 shows angle BAC. (The symbol $\angle$ is used to denote an angle.)

2.5 Sets of Points

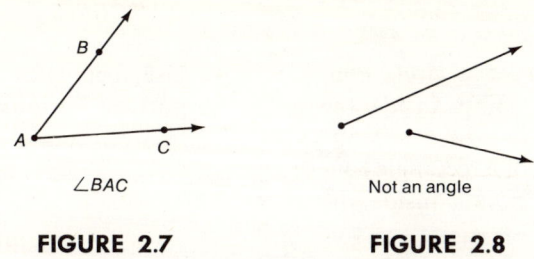

FIGURE 2.7   **FIGURE 2.8**

You should be aware that not every union of two rays is an angle, as Figure 2.8 proves.

Exercises 2.5

1. Give an example illustrating the use of the noun *intersection* and the verb *intersect*.
2. Describe how a union of three segments could form a triangle.
3. Describe how a union of four segments could form a rectangle.
4. Show a way two segments can intersect in one point, different from the two ways shown in Figure 2.3.
5. Draw a picture to show the intersection of two rays, which are subsets of the same line, to be
   a. a segment   b. a ray   c. a point   d. the empty set
6. Draw a picture to show two rays intersecting in one point, but not forming an angle.
7. If two lines intersect, what is their intersection?
★ 8. If two lines are in the same plane and they do not intersect, what kind of lines are they?
★ 9. If two lines in space do not intersect, are they necessarily parallel?
10. As ticket collector for the railroad, your job is to check passengers' tickets against your list to see that everyone is sitting in the proper seat. $\overline{AB}$ represents the train. Tickets are stamped:
   a. $\overline{AD} \cap \overline{CB} = \overline{CD}$ b. $\overline{AC}$ c. $\overline{AB} \cap \overline{DB} = \overline{DB}$
   Where should each ticket holder be seated?

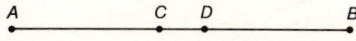

2.6 VENN DIAGRAMS

REVIEW

- The interior of a circle consists of all points inside the circle
- The interior of a rectangle consists of all points inside the rectangle

OBJECTIVES

- Use rectangles and circles to portray various relationships between sets geometrically

A geometric way to think about sets and relationships between them was introduced by the eighteenth-century Swiss mathematician Leonhard Euler. He visualized sets as circles. For example, Figure 2.9 shows sets X and Y as circles, which are appropriately called *Euler circles*. Explain how Figure 2.9 suggests that the sets are disjoint.

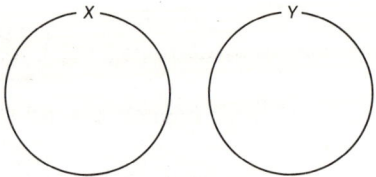

FIGURE 2.9

John Venn, a British logician of the nineteenth century, introduced further refinements to this idea of using pictorial representations of sets. Today we refer to these pictorial representations as *Venn diagrams*. Figure 2.10 shows the universal set U pictured as a rectangle. Sets A and B, pictured by circles, are shown to be subsets of U. How do you conclude from the picture that they indeed are subsets of U? The Venn diagram also suggests that sets A and B are not disjoint. Explain that.

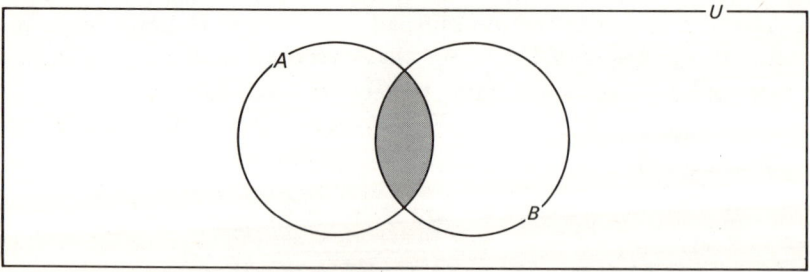

FIGURE 2.10

2.6 Venn Diagrams

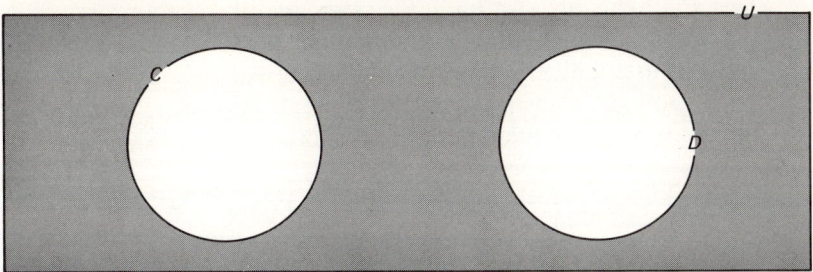

FIGURE 2.11

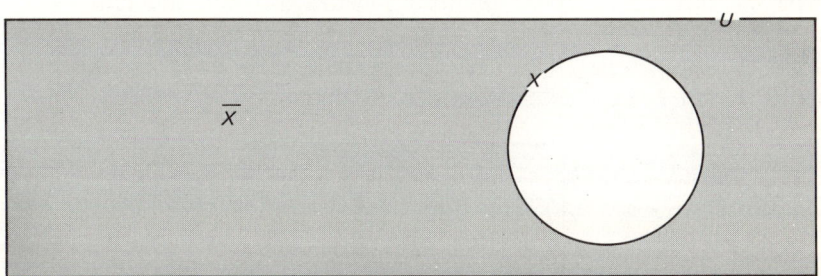

FIGURE 2.12

To show two disjoint sets as in Figure 2.9 and also to indicate that they are subsets of the universe U, we can draw a Venn diagram like Figure 2.11.

Complements of sets can also be pictured by Venn diagrams. Using U again as the universal set, $\bar{X}$, the complement of X in U, can be pictured as in Figure 2.12. The shaded portion represents $\bar{X}$, where X is pictured by a circle including its interior.

Exercises 2.6

Draw a rectangle to represent a universal set U. Then draw three circles within U to picture sets A, B, and C so that:

1. $A \cap B = \emptyset$, but $A \cap C \neq \emptyset$ and $B \cap C \neq \emptyset$
2. $A \cap B \neq \emptyset$, $A \cap C \neq \emptyset$, and $B \cap C \neq \emptyset$
3. $A \cap B = \emptyset$, $A \cap C = \emptyset$, and $B \cap C = \emptyset$
4. $A \cap B = A$ and $C \cap A = C$

2 Sets

5. Two spotlights are cast upon the stage, and the curtain rises on Mario Vibratorio, the well-known singing star. Due to bad planning, the spotlights at first light the stage as in diagram A.
Then spotlight x swings across to shine on Mario.
Finally spotlight y manages to join spotlight x.

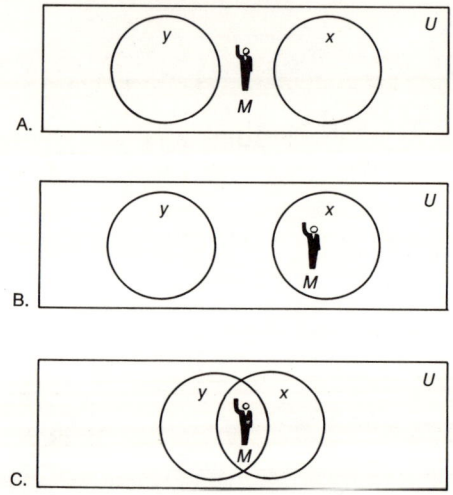

A.

B.

C.

After the show, Mario demands an explanation of the goof-up. Mathematics comes to your aid once again. You try to soothe Mario by telling him that what happened was merely a demonstration of Venn diagrams illustrating set theory. He demands a more detailed answer. Explain what you mean.

CHAPTER 2 TEST

1. Explain why the phrase

　　the set of whole numbers between 7 and 8

describes the empty set.

For the set $T = \{3, 6, 9, 12, \ldots\}$, which of the following are true and which are false?

2. $45 \in T$　　**3.** $17 \in T$　　**4.** $345 \notin T$　　**5.** $222 \notin T$

6. It is 1984. Everyone has a number so that Big Brother will know where you are and where you have been. The well-known criminal 36 (alias Easy Lifter) has been arrested on charges of stealing from the garbage cans of the rich. His alibi is, "At 8:30 PM I was in Set J with numbers

2.6 Venn Diagrams

99, 28, and 42!" You immediately call Big Brother to check the validity of 36's statement. The computer prints out:

$$J_{8:30 \text{ PM}} = \{99, 28, 42\}$$

Could 36 be telling the truth?

For each of the following sets, supply the next three members:

7. $\{1, 2, 4, 7, \ldots\}$ 8. $\{2, 3, 5, 9, \ldots\}$
9. If two finite sets are equivalent, what can you say about the number of members in each set?
10. Display a one-to-one correspondence between the set of all even whole numbers and the set of all odd whole numbers. Give the formula connecting the corresponding numbers in terms of n, where n is the nth whole number.
11. Give all subsets of $A = \{0, 2, 4\}$.
12. $A = \{11_{\text{two}}, 111_{\text{two}}, 1111_{\text{two}}\}$ and $B = \{15, 7, 3\}$. Explain why $A = B$ is true.

Given the three sets:

$$A = \{0, 2, 4\} \qquad B = \{0, 5\} \qquad C = \{6\}$$

which of the following are true and which are false?

13. $A \subseteq B$ 14. $B \not\subseteq A$ 15. $A \cap B = \emptyset$
16. $B \cap C = \emptyset$ 17. $B \cup C = \{0, 5, 6\}$ 18. $A \cup B = \{2, 4, 5\}$
19. In a recent football competition, the results were analyzed by a computer. It printed out a set of the number of games each team had won. The five teams are A, B, C, D, and E. Due to a power brown-out, most of the data were lost. You have only several statements from which to build the final league table. Can you do it from the following data? (In each case, the number of elements within the braces is the number of games the team won.)
$A = \{10, 3, 5, 4, 6\}$ $B = \{1, 5, 7, 9, 11, 16, 20, 25\}$
$n(A) = n(D)$ $n(C) = n(E)$ $E = \{2, 3\}$

20. What is the complement of the set of all even numbers in the set of all whole numbers?
21. Draw a picture of two triangles intersecting in exactly three points.
22. Draw a picture of a situation in which the intersection of two rays is a segment.
23. Draw pictures of two different situations in which the union of two rays is a line.
24. Draw a Venn diagram showing three sets such that no two sets are disjoint.

3

Some Remarkable Features of Numbers

3.1 EVEN, ODD, PRIME, COMPOSITE NUMBERS

REVIEW

- Squaring a number means multiplying it by itself:
  $6^2 = 6 \cdot 6$
- In $3 \cdot 5 = 15$, 3 and 5 are factors of 15

OBJECTIVES

- Tell whether a number is divisible by a given number
- Define an even number
- Define an odd number
- Prove that the square of an even number is even
- Prove that the square of an odd number is odd
- Partition the set of natural numbers into three subsets
- Define a prime number
- Define a composite number
- Partition the set of natural numbers into two subsets on the basis of divisibility by 3

The set of natural numbers is $N = \{1, 2, 3, \ldots\}$; they are also called the counting numbers. All sorts of subsets of N can be formed, some with very interesting properties.

3.1 Even, Odd, Prime, Composite Numbers

Before we look at subsets of N, let's clarify the concept of *divisibility*. We say that one number is divisible by another if the quotient is a natural number. For example, 12 is divisible by 3, because $12 \div 3 = 4$ and 4 is a natural number. But 12 is not divisible by 5, because $12 \div 5 = \frac{12}{5}$, which is not a natural number.

Every *even* number is divisible by 2:

$$E = \{2, 4, 6, \ldots\}$$

Every natural number which is not divisible by 2 is an *odd* number:

$$O = \{1, 3, 5, \ldots\}$$

You should be able to explain the following two statements:

$$E \cup O = N \qquad E \cap O = \emptyset$$

Note that each of the sets—natural numbers, even numbers, and odd numbers—is an infinite set. Each set has a smallest number, but none has a largest number.

We have said that an even number is divisible by 2. There is another way to look at even numbers: every even number can be given as a product of 2 and some natural number. Here are some examples:

$$16 = 2 \cdot 8 \qquad 34 = 2 \cdot 17 \qquad 58 = 2 \cdot 29$$

We shall assume this to be true of every even number. Doing so formally will give us the right to use this assumption in case we want to *prove* some other things about even numbers.

One of the operations we perform quite often is squaring a number, which is the same as multiplying the number by itself. For example:

$$4^2 = 16 \qquad 8^2 = 64 \qquad 7^2 = 49 \qquad 11^2 = 121$$

Perhaps you have already thought to yourself that the square of an even number is even and the square of an odd number is odd. If you did, you made a good observation.

Making such observations shows good mathematical sense. After making them though, we are still not certain that we are right. After all, there are too many even numbers to try to verify that we are right. In cases like this we resort to a *proof*, a logical sequence of statements that verifies our claim. We reason out that something is true for all numbers. As an example, let's prove that the square of every even natural number is even.

3 Some Remarkable Features of Numbers

We will state this in a fancy, but brief, form.

$\forall_n$ if n is even, n^2 is even

The symbol simply means "for every." This very handy symbol says a lot in very little space—it tells us that there are no exceptions, not even one.

Before trying to prove that something is true, we should be pretty convinced that it is true. One way to try to convince ourselves is through a few examples:

$2^2 = 4 \quad 4^2 = 16 \quad 6^2 = 36 \quad 12^2 = 144$

Now do you half-way believe that the square of an even number is even?

One more thing about proofs: the statements we prove are called *theorems*. Now we are ready to state our first theorem and give its proof.

Theorem 3.1 $\forall_n$ if n is even, then n^2 is even

Proof Since n is even, $n = 2k$. (Remember we assumed that every even number can be written as a product of 2 and some natural number.) Now we square the terms on each side of the equals sign:

$$\begin{aligned} n^2 &= (2k)^2 \\ &= (2k)(2k) \\ &= 4k^2 \\ &= 2(2k^2) \end{aligned}$$

Since k is a natural number, $2k^2$ is also a natural number. That's easy to see. Thus, n^2, which is the square of any natural number, is a product of 2 and some natural number. But that means that n^2 is an even number, which is what we wanted to prove.

Now we know that there is no exception: the square of every even number is an even number.

Not to ignore the odd numbers, observe the following examples:

$7 = 2 \cdot 3 + 1$
$9 = 2 \cdot 4 + 1$
$11 = 2 \cdot 5 + 1$
$23 = 2 \cdot 11 + 1$

3.1 Even, Odd, Prime, Composite Numbers

It looks like every odd number can be shown as a product of 2 and some whole number, plus 1. We shall assume this to be true of every odd natural number. Stated in general terms:

$\forall n$, if n is odd, then $n = 2k + 1$

We have proved that the square of every even number is even. What about the square of every odd number? Let's try a few:

$3^2 = 9 \quad 9^2 = 81 \quad 13^2 = 169 \quad 15^2 = 225$

By the way, did you ever multiply a number ending in 5 by itself the quick way? Follow the steps for 35:

Ignore the 5, that leaves 3.
Multiply 3 by the next number, which is 4: $3 \cdot 4 = 12$
Now write 25 to the right of 12, that's 1225.
So, $35^2 = 1225$
Try this on some other numbers ending in 5, even three-digit, if you like.

Now back to the question of the square of every odd number. Let's prove that it is odd.

Theorem 3.2 $\quad \forall n$, if n is odd, then n^2 is odd

Proof Since n is odd, $n = 2k + 1$. Now we have:

$n^2 = (2k + 1)^2$
$ = 4k^2 + 4k + 1$
$ = 2(2k^2 + 2k) + 1$

We have expressed n^2, the square of any odd number, as a product of 2 and some whole number, plus 1. That means only one thing: n^2 is an odd number. And that's what we were supposed to prove.

We want to discuss one more thing about natural numbers. The numbers by which a natural number is divisible are called its *divisors*. Table 3.1 lists all the divisors of some natural numbers.

3 Some Remarkable Features of Numbers

Table 3.1 *The Divisors of the First Ten Natural Numbers*

Number	Set of Divisors
1	{1}
2	{1, 2}
3	{1, 3}
4	{1, 2, 4}
5	{1, 5}
6	{1, 2, 3, 6}
7	{1, 7}
8	{1, 2, 4, 8}
9	{1, 3, 9}
10	{1, 2, 5, 10}

Of the numbers listed at the left, 1 is the only number which has exactly one divisor. The numbers 2, 3, 5, and 7 have exactly two divisors each. Each of the numbers 4, 6, 8, 9, and 10 has more than two divisors.

From the data and discussion in Table 3.1, it appears that we can split the set of all natural numbers into three subsets according to the number of divisors a natural number has. To be fancy, we say that the set of natural numbers is *partitioned* into three subsets:

$A = \{1\}$, the set of all natural numbers having exactly one divisor

$B = \{2, 3, 5, 7, \ldots\}$, the set of all natural numbers having exactly two divisors; such numbers are called *prime* numbers

$C = \{4, 6, 8, 9, 10, \ldots\}$, the set of all natural numbers having more than two divisors; such numbers are called *composite* numbers

The divisors of a composite number are either prime or composite. If a divisor is composite, it can be written as a product of two or more divisors, which in turn are either prime or composite. Continuing this procedure, we see that every composite number can be written as a product of primes. (We will prove this formally in Section 3.8.) For example:

$$40 = 4 \cdot 10$$
$$= (2 \cdot 2) \cdot (5 \cdot 2)$$

Since 2 and 5 prime, 40 has been written as a product of primes.

Exercises 3.1

1. Show 2684 as a product of 2 and a natural number.
2. Show 3997 as a product of 2 and a natural number, plus 1.

3.1 Even, Odd, Prime, Composite Numbers

3. Fred, the union leader, addressed his members: "We have 1763 members. United we stand, divided we fall." He made this statement because he thought that 1763 could only be divided by itself and 1. Was he right? Show why or why not.
4. Why is 1298^2 an even number?
5. Why is 3093^2 an odd number?
6. You are going to tile your penthouse with plush velvet tiles. You know that you will need 224^2 tiles. The tile suppliers tell you that each tile in a box of 200 tiles costs $10. However, because a box of tiles has to be broken to provide fewer than 200 tiles, those tiles cost $11 each. How much will it cost to tile your floor?
7. Is 41 a divisor of 287? Why?
8. Is 11 a divisor of 253? Why?
9. Is 11 a divisor of 259? Why?
10. Which natural number has only one divisor?
11. At least how many divisors does a composite number have?
12. How many prime numbers are even?
13. The criterion

    is divisible by 3

    partitions the set of natural numbers into two subsets: the set of all natural numbers which are divisible by 3 and the set of all natural numbers which are not divisible by 3. List the first five natural numbers in each of these two sets.
14. Is it true that every third number in the sequence 1, 2, 3, ... is divisible by 3? List the first four such numbers.
15. Is it true that every fourth number of the sequence 1, 2, 3, ... is divisible by 4? List the first four such numbers.
16. Is it true that every nth number in the sequence 1, 2, 3, ... is divisible by n? List the first four such numbers.

Give an example of a composite number which has exactly:

17. 2 prime factors
18. 5 prime factors
19. Write 102 as a product of prime numbers.
20. At the recent undertakers' convention in Miami, lunch was being prepared for 230 undertakers. The chef had ordered 1020 New York steaks so that each undertaker could have at least 4 steaks. Seventy-six hungry undertakers asked if they could have 5 steaks. This request caused panic in the kitchen until the chef determined how many extra steaks had been ordered. Would there be enough steaks for the 76 hungry undertakers? How many undertakers could have 5 steaks?

3.2 PRIMES—HOW MANY AND HOW TO FIND THEM

REVIEW

- A prime number has exactly two divisors
- A composite number has more than two divisors

OBJECTIVES

- Raise and answer the question of how many primes there are
- Prove that there is an infinite number of primes
- Show that the formulas $n^2 + n + 41$ and $n^2 - 79n + 1601$ yield prime numbers for some values of n and composite numbers for some values of n

Have you ever wondered how many prime numbers there are? To gain some insight into answering this question, let's form products of the consecutive primes, then add 1 to each such product.

$2 \times 3 + 1 = 7$
$2 \times 3 \times 5 + 1 = 31$
$2 \times 3 \times 5 \times 7 + 1 = 211$
$2 \times 3 \times 5 \times 7 \times 11 + 1 = 2311$
$2 \times 3 \times 5 \times 7 \times 11 \times 13 + 1 = 30,031.$

We could go on getting more numbers this way, but the five numbers above are enough to illustrate the point. The following statements are true about these five numbers.

7 is a prime number
31 is a prime number
211 is a prime number
2311 is a prime number
30,031 is *not* a prime number because $30,031 = 59 \times 509$

We can conclude that the procedure above will yield prime numbers in some instances and composite numbers in others. But the following two observations can be made about each *composite* number found in this sequence.

1. This composite number is greater than any one of the primes used in forming the product.

2. None of the primes used in forming the product is a factor of this composite number.

3.2 Primes—How Many and How to Find Them

To verify these two observations for 30,031, notice that

1. 30,031 is greater than each of 2, 3, 5, 7, 11, and 13.
2. None of 2, 3, 5, 7, 11, and 13 is a factor of 30,031.

We are now ready to tackle the question of how many primes there are. We will *prove* that there is an infinite number of primes. By the way, this question dates back to the time of the Greeks—Euclid answered the question about 300 B.C.!

Theorem 3.3   The set of prime numbers is infinite.

Proof* To prove that there is an *infinite* number of primes, we assume that there is a *finite* number of primes (hoping, of course, to prove this assumption false).

Suppose there are n primes. Assume that p_n is the greatest existing prime. Take the product of all primes arranged in ascending order and add 1 to it:

$$p_1 p_2 p_3 \cdots p_n + 1$$

Call the resulting number k. It was shown that $k = p_1 p_2 p_3 \cdots p_n + 1$ is either prime or composite. These two cases will be examined.

CASE 1 k is prime. Since k is larger than p_n, the assumption that p_n is the greatest existing prime is false. Since the assumption that p_n is the greatest existing prime is a direct consequence of the assumption that there is a finite number of primes, the latter assumption is false. Thus, there is an infinite number of primes.

CASE 2   The second alternative is that k is composite. The number k is larger than p_n, where p_n is the greatest assumed prime. k has at least one prime number as its factor, because every natural number different from 1 is either prime or a product of primes. But none of $p_1, p_2, \ldots, p_n$ is a factor of k. Therefore, the prime which is a factor of k is greater than p_n. So, we have established the existence of a prime which is greater than the greatest assumed prime. Thus, the number of primes is not finite.

Since the number of primes is infinite, it is not possible to know them all. But, at any time in history there is a greatest prime discovered to that

---

*This proof is optional. Its omission will have no bearing on subsequent study.

point. For example, in 1963 Donald B. Gillies established $2^{11213} - 1$ as prime. He accomplished this at the University of Illinois using an electronic computer known by the name of Illiac.

The task of testing a large number to establish whether it is prime or composite involves trying the consecutive prime numbers to see whether any of them are divisors of the given number. There is no magic formula to manufacture primes, contrary to some claims of amateur mathematicians made in the past. For example, one mathematician claimed that substitution of natural numbers for n in

$$n^2 + n + 41$$

would produce primes. And, indeed, it does for all natural numbers from 1 through 40 (test some of them). But substituting 41 for n yields

$$41^2 + 41 + 41$$

which is equal to 41×43. This is obviously not a prime number.

Another formula proposed for producing primes was

$$n^2 - 79n + 1601$$

Substituting natural numbers 1 through 79 for n, prime numbers are obtained (test some of them). But 80 for n yields 1681, which is not a prime number because $1681 = 41 \times 41$. Thus, there is no royal road to the discovery of primes!

Exercises 3.2

1. Compute the product of all consecutive prime numbers through 17, and add 1 to the product. Test to see whether the result is prime or not.
2. We formed products of consecutive prime numbers and added 1 to these products. When the result was a composite number, we claimed that none of the primes used in forming the product was a factor of the composite number. How would you argue that this is true?
3. Find a number which is not prime by replacing n in $n^2 + n + 41$ by a natural number different from 41.
4. You have just memorized the formula $n^2 - 79n + 1601$, which produces prime numbers for n less than 80. Your car insurance is due and you need money. You go to the student union, where no one knows a number larger than 79, and bet everyone $10 that you can produce a

3.3 The Amazing Primes

prime number for any number they mention. Find a prime number for the following n:

a. 30   b. 79   c. 6

Use the formula $n^2 - 79n + 1601$ and substitute the following numbers for n to obtain prime numbers:

5. 10   6. 20   7. 50

3.3 THE AMAZING PRIMES

REVIEW

- There is an infinite number of prime numbers

OBJECTIVES

- Use Eratosthenes' sieve to identify all prime numbers between 1 and 100
- Continue the sieve to identify all prime numbers between 100 and 200
- Show prime numbers as multiples of 6, plus or minus 1
- Identify some twin primes
- Show some even numbers greater than 2 as sums of two primes
- Identify some reversible primes
- Identify some symmetrical primes

Prime numbers have attracted much attention through the years. With each passing year a little more becomes known about them. As far back as 230 B.C. there was a crude method in use for determining primes. This method was called a *sieve* because of its nature, and it was named after Eratosthenes who devised it.

To illustrate this method, we shall give steps to shake all the primes between 1 and 100 from the sieve. We describe the steps in the display below.

1. List all the natural numbers 1 through 100.
2. Cross out 1, since it is not a prime.
3. Circle 2, since it is prime.
4. Cross out each multiple of 2.
5. Circle 3, since it is prime.
6. Cross out each multiple of 3; some of these are already crossed out as multiples of 2.
7. Circle 5, since it is prime.
8. Cross out all multiples of 5 still left.
9. Circle 7, since it is prime.
10. Cross out all multiples of 7 still left.
11. Circle 11, since it is prime.

You should notice that all multiples of 11 have already been crossed out. And you should see that only prime numbers between 1 and 100 are left, which we have also circled.

~~1~~	②	③	~~4~~	⑤	~~6~~	⑦	~~8~~	~~9~~	~~10~~
⑪	~~12~~	⑬	~~14~~	~~15~~	~~16~~	⑰	~~18~~	⑲	~~20~~
~~21~~	~~22~~	㉓	~~24~~	~~25~~	~~26~~	~~27~~	~~28~~	㉙	~~30~~
㉛	~~32~~	~~33~~	~~34~~	~~35~~	~~36~~	㊲	~~38~~	~~39~~	~~40~~
㊶	~~42~~	㊸	~~44~~	~~45~~	~~46~~	㊼	~~48~~	~~49~~	~~50~~
~~51~~	~~52~~	(53)	~~54~~	~~55~~	~~56~~	~~57~~	~~58~~	(59)	~~60~~
(61)	~~62~~	~~63~~	~~64~~	~~65~~	~~66~~	(67)	~~68~~	~~69~~	~~70~~
(71)	~~72~~	(73)	~~74~~	~~75~~	~~76~~	~~77~~	~~78~~	(79)	~~80~~
~~81~~	~~82~~	(83)	~~84~~	~~85~~	~~86~~	~~87~~	~~88~~	(89)	~~90~~
(91)	~~92~~	~~93~~	~~94~~	~~95~~	~~96~~	(97)	~~98~~	~~99~~	~~100~~

The frequency of occurrence of primes has been a subject of study. It was found that as numbers become greater and greater, the spacing between prime numbers grows greater but in an irregular fashion.

Taking a closer look at primes reveals that each prime greater than 3 seems to be either 1 less or 1 more than a multiple of 6, as the following list shows:

3.3 The Amazing Primes

$5 = (6 \times 1) - 1$
$7 = (6 \times 1) + 1$
$11 = (6 \times 2) - 1$
$13 = (6 \times 2) + 1$
$17 = (6 \times 3) - 1$
$19 = (6 \times 3) + 1$
$23 = (6 \times 4) - 1$
$29 = (6 \times 5) - 1$
$31 = (6 \times 5) + 1$
$37 = (6 \times 6) + 1$
and so on

An examination of a table of primes also reveals that some pairs of primes consist of numbers differing by 2. Some of these pairs are:

3, 5   5, 7   11, 13   17, 19

These are called *twin primes*. Although it seems reasonable to expect that there would be an infinite number of pairs of twin primes, this remains an unproved conjecture. No one has yet succeeded in proving it! Some of the greater known twin primes are:

209,267   and   209,269

Another unproved conjecture is that every even number greater than 2 can be shown as a sum of two primes (not necessarily different). The following are some examples:

$4 = 2 + 2$
$6 = 3 + 3$
$8 = 3 + 5$
$10 = 5 + 5 = 3 + 7$
$12 = 5 + 7$
$14 = 7 + 7 = 3 + 11$
$16 = 3 + 13 = 5 + 11$
and so on

No one has proved that this is possible to do for every even number greater than 2.

During the course of listing, some results were announced which were in error. For example, Euler once announced that 1,000,009 was prime, but he later discovered that it was a product of two primes: 293 and 3413. Fermat was once asked in a letter if 101,007,901,169 was prime. He shot back that it was the product of two primes: 899,423 and 112,303.

3   Some Remarkable Features of Numbers

Today, the availability of high-speed electronic computers makes it much easier to investigate prime numbers. Jerome Niebaum of the University of Omaha investigated the so-called *reversible primes* using the Control Data 3400 computer at the Vogelback Computing Center at Northwestern University. Reversible primes are those primes which yield primes after the digits are reversed. For example, the following are reversible primes:

37; 73
79; 97
143; 341
1009; 9001

Of course, each one-digit prime is a reversible prime. There are four of these: 2, 3, 5, and 7. They are called reversible primes of order one. Niebaum found that there are 9 reversible primes of order two. They are: 11, 13, 17, 31, 37, 71, 73, 79, and 97. There are 43 reversible primes of order three. The first five of these are: 101, 107, 113, 131, and 149. He found that there are 203 reversible primes of order four. For example, 1009 and 9001 are such primes. It can be proved that in order for a prime to be reversible, it must begin and end with a 1, 3, 7, or 9.

In his investigation, Niebaum also explored the so-called *symmetrical primes*. Some of these are:

11, 101, 131, 151, 181

He discovered, for example, the following unique sequence of symmetrical primes:

131
10301
1003001
100030001

Curiously, he demonstrated the danger of leaping to conclusions: 10000300001 is not a prime—it is divisible by 19! Furthermore, he verified that:

1000003000001 is divisible by 29
100000030000001 is divisible by 139
10000000300000001 is divisible by 61
1000000003000000001 is divisible by 59

Thus, the primes are quite amazing indeed!

3.3 The Amazing Primes

Exercises 3.3

Show that each of the following prime numbers is 1 less or 1 more than some multiple of 6:

1. 41
2. 61
3. 71
4. 107
5. 181
6. 233
7. 709
8. 827
9. You are the chief codebreaker for the Greek secret service. You must decipher the secret sign of the enemy underworld. Upon capturing an enemy spy, you find the following matrix of numbers. Because you are both a mathematician and a Greek, you remember Eratosthenes. Can you discover what the secret sign looks like?

2	3	5	7
12	18	11	20
4	13	48	40
17	19	23	29

10. Continue the Eratosthenes' sieve to obtain all prime numbers between 100 and 200.
11. Find all twin primes beyond 17 and 19, which are less than 100.

Show each of the following even numbers as a sum of two primes:

12. 24
13. 50
14. 100
15. 224
16. 240
17. It is claimed that from 2 on there is at least one prime number between any natural number and its double. Verify this for natural numbers 2 through 10.
18. Why is 23 not a reversible prime?
19. You have just been on a trip through time and space, passing through many higher dimensions. Unfortunately, while passing through the time warp, all of your calculations were reversed. Of the following primes, which are still primes after this reversal?
    a. 1   b. 71   c. 43   d. 41   e. 97
★20. Prove that a natural number cannot be a reversible prime if it begins or ends with 0, 2, 4, 5, 6, or 8.

3.4 SOME FANTASTIC NUMBERS

REVIEW

- The number 1 is a divisor of every number
- Every number is a divisor of itself

OBJECTIVES

- Define proper divisors of a number
- Partition the set of natural numbers into 1, deficient numbers, perfect numbers, and abundant numbers
- Explore perfect numbers
- Define amicable numbers
- Give examples of amicable numbers

In this chapter we are looking at many aspects of numbers. Let's consider all the divisors of a given number, except the number itself. Such divisors are called *proper divisors*.

Example 1 All the proper divisors of 8 are 1, 2, and 4. The sum of these divisors is $1 + 2 + 4$ or 7.

In this case, the sum of the proper divisors of a number is less than the number:

$7 < 8$

Could the sum of the proper divisors of a number be greater than the number? Consider the number 12. Its proper divisors are 1, 2, 3, 4, and 6. Their sum is:

$1 + 2 + 3 + 4 + 6 = 16$ and $16 > 12$

Could the sum of the proper divisors of a number be equal to the number? Consider the number 6. Its proper divisors are 1, 2, and 3:

$1 + 2 + 3 = 6$

Thus, we can partition the set of natural numbers into the following subsets:

1. Number 1, which has no proper divisors.
2. Numbers for which the sum of the proper divisors is less than the number. These are called *deficient numbers*. (Prime numbers are deficient.)

3.4 Some Fantastic Numbers

3. Numbers for which the sum of the proper divisors is equal to the number. These are called *perfect* numbers.
4. Numbers for which the sum of the proper divisors is greater than the number. These are called *abundant* numbers.

Perfect numbers are not very easy to find. Explorations of these numbers have been going on for hundreds of years. The first four perfect numbers have been known since the first century. They are:

6, 28, 496, and 8128

But the discovery of the next perfect number took some ten centuries. It is:

33,550,336

Today only eighteen perfect numbers are known. The greatest known perfect number has 2663 digits.

Euclid has shown that if the natural number n is such that $2^{n+1} - 1$ is prime, then $2^n(2^{n+1} - 1)$ is a perfect number. Of course, every number of this form is even, since it has a factor 2. No one knows, however, whether there is an odd perfect number!

Another interesting relation between pairs of numbers has been found. For example, the sum of the proper divisors of 220 is equal to 284 and the sum of the proper divisors of 284 is equal to 220. For this reason, 220 and 284 constitute a pair of *amicable numbers*. Other amicable pairs of numbers are

| 1184 | and | 1210 |
| 17,296 | and | 18,416 |

Over 400 pairs of amicable numbers are known.

Exercises 3.4

Give the proper divisors of each of the following numbers:

1. 27   2. 38   3. 51   4. 64

For each of the following numbers, tell whether it is deficient, perfect, or abundant:

5. 16   6. 46   7. 31   8. 28   9. 100
10. Prove that every prime number is deficient.
11. Verify that 496 is a perfect number.

3   Some Remarkable Features of Numbers   64

12. It's spring again, and the math department is alive with the sound of numbers in love (this sound, to those who listen for it, is a friendly buzz). Can you tell which numbers are in love? (*Hint:* Determine the amicable numbers.)
    a. 220, 284   b. 17,296, 18,416   c. 20, 24
13. It is the year 1984 (again). Big Brother is still watching. People are given numbers according to their status. The classification is as follows: (1) important people are given perfect numbers, (2) politicians are given prime numbers, (3) average citizens are given abundant numbers, (4) criminals are given deficient numbers. To whom will the following numbers be assigned?
    a. 496   b. 18   c. 9   d. 11   e. 24
14. Prove that 220 and 284 are amicable numbers.

3.5   GREATEST COMMON DIVISOR

REVIEW

- If $a \div b$ is a natural number, then b is a divisor of a
- A common divisor of two numbers is a number which divides each of the two numbers. For example, 4 is a common divisor of 20 and 36

OBJECTIVES

- Write a statement relating dividend, divisor, quotient, and remainder
- Define the greatest common divisor of two numbers
- Use the Euclidean technique to determine the greatest common divisor of two numbers
- Determine the GCD of some special pairs of numbers

It is instructive to observe the relations between numbers when performing division. For example, when dividing 25 by 7, the quotient 3 and the remainder 4 are obtained. This can be shown as:

$$25 \div 7 = 3, \ r4$$

dividend   divisor   quotient   remainder

Another way to show this relation is the following:

$$25 = (7 \times 3) + 4$$

3.5 Greatest Common Divisor

To show this relation in general, we introduce the following abbreviations:

> d for dividend
> v for divisor
> q for quotient
> r for remainder

The relationship illustrated above can now be generalized as follows:

$$d = vq + r$$

In case r is 0, the above reduces to

$$d = vq$$

as in the case of $32 \div 8 = 4$. It is true that $32 = 8 \cdot 4$.

The above relationship is useful when determining the *greatest common divisor* of two numbers, which is the greatest natural number that divides each of the two given numbers. For example, 6 is the greatest common divisor of 18 and 24 because 6 divides 18 and 24, and it is the greatest number which does that. This is written as:

> GCD (18, 24) = 6   (*Read:* the greatest common divisor of 18 and 24 is 6)

If the greatest common divisor of two numbers is 1, it is said that the two numbers are *relatively prime*. For example, 8 and 21 are relatively prime because:

> GCD (8, 21) = 1

The most primitive way of determining the GCD of two numbers is to list the set of divisors of each number and to choose the greatest number appearing in both sets. For example, for 18 and 24, the work would take the following form:

> set of divisors of 18: $\{1, 2, 3, ⓖ, 9, 18\}$
> set of divisors of 24: $\{1, 2, 3, 4, ⓖ, 8, 12, 24\}$

Observe that 6 is the greatest divisor appearing in both sets; therefore:

> GCD (18, 24) = 6

Euclid developed a technique for finding the GCD based on the division relation, which was explored above. Observe one more fact about

3 Some Remarkable Features of Numbers

this relation, using the original example, $25 = (7 \times 3) + 4$. From this it follows that

$$25 - (7 \times 3) = 4$$

Now consider GCD (25, 7). It is true that any divisor of 25 and 7 is also a divisor of $25 - (7 \times 3)$. This can be proved as follows:
Suppose some number k is a divisor of 25 and 7. Observe that

$$\frac{25 - (7 \times 3)}{k} = \frac{25}{k} - \frac{7 \times 3}{k}$$
$$= \frac{25}{k} - \frac{7}{k} \times 3$$

Since k is a divisor of 25 and 7, $25/k$ and $7/k$ are natural numbers, and therefore $(7/k) \times 3$ is a natural number. Since $25 > 7 \times 3$, $(25/k) - (7/k) \times 3$ is a natural number. Thus, k is a divisor of $25 - (7 \times 3)$.
Since, $25 - (7 \times 3) = 4$, it follows that the divisor of 25 and 7 is also a divisor of 4. Generally:

Any divisor of both d and v in $d - vq = r$ is also a divisor of r.

To continue this process, it can be stated that GCD (25, 7) is also GCD (7, 4). Now

$$7 = 4 \times 1 + 3$$
$$7 - 4 \times 1 = 3$$

Therefore, GCD (7, 4) is also GCD (4, 3). Next:

$$4 = 3 \times 1 + 1$$
$$4 - 3 \times 1 = 1$$

Therefore, GCD (4, 3) = GCD (3, 1), which, of course, is 1. The above development illustrates a technique for determining the GCD of any pair of numbers. This technique is further illustrated in determining GCD (284, 128).

1. Divide: $284 \div 128 = 2$, $r28$
   So: $284 = (128 \times 2) + 28$
   Therefore, GCD (284, 128) = GCD (128, 28).
2. Divide: $128 \div 28 = 4$, $r16$
   So: $128 = (28 \times 4) + 16$
   Therefore, GCD (128, 28) = GCD (28, 16).

3.5 Greatest Common Divisor

3. Divide: $28 \div 16 = 1, r12$
   So: $28 = (16 \times 1) + 12$
   Therefore, GCD (28, 16) = GCD (16, 12).
4. Divide: $16 \div 12 = 1, r4$
   So: $16 = (12 \times 1) + 4$
   Therefore, GCD (16, 12) = GCD (12, 4).

It is easy to see that GCD (12, 4) = 4. Therefore, GCD (284, 128) = 4.
One obvious application of the GCD is when simplifying fractional numerals. For example, knowing that GCD (284, 128) = 4, the following simplification can be accomplished:

$$\frac{128}{284} = \frac{32 \times 4}{71 \times 4} = \frac{32}{71}$$

Note that 32 and 71 are relatively prime, and therefore no further simplification is possible. In general, if GCD $(x, y) = a$, then x/a and y/a are relatively prime.

Exercises 3.5

1. Prove that the GCD of any two prime numbers is 1.
2. Prove that GCD $(n, 1)$ for any natural number n is 1.
3. What is GCD $(n, 2)$ for any odd number n?
4. What is GCD $(n, 2)$ for any even number n?
5. Using the relation $48 = 30 + 18$, give an argument to show that any divisor of 48 and 30 must also be a divisor of 18.

Determine the GCD of each of the following pairs of numbers:

6. (35, 66)   7. (260, 611)   8. (806, 1116)
9. An ambitious mole who has forsaken country life for the big city comes face to face with a grill full of holes. The mole, who is a mathematical genius, has to analyze the situation to determine whether or not he will fit through the holes. He doesn't want to get stuck midway through the unfamiliar territory. Each hole is labeled as a pair because the holes change size as they pass through the grill. The mole has to calculate the greatest common divisor of the two holes to see if he will fit through. If the mole is a size 9, which hole will he fit through?
      a. (18, 12)   b. (9, 27)   c. (32, 24)
10. One often hears the expression "He's in the prime of his life." You have two brothers: Mark, age 21, and Jim, age 31. Are they in the "relative prime" of their lives?

3.6 LEAST COMMON MULTIPLE

REVIEW

- If x is divisible by y, then the quotient $x \div y$ is a natural number
- The greatest common divisor, abbreviated GCD, of two numbers is the greatest natural number that divides each of the two numbers

OBJECTIVES

- Find multiples of a given number
- Find common multiples of two given numbers
- Find the least common multiple, abbreviated LCM, of two given numbers
- Show the GCD of two numbers as a difference of some multiples of the two numbers

To find multiples of a given number in a systematic way, we multiply the given number by 1, 2, 3, and so on. For example, the following are multiples of 4:

4, 8, 12, 16, 20, ...

There is no largest number in this set; however, there is a smallest number. In this case it is 4.

Given two numbers, they have many *common* multiples. For example, the common multiples of 3 and 8 are 24, 48, 72, and so on. They are multiples of 3 and 8 because each of them is divisible by both 3 and 8. The least of these common multiples is 24. To indicate the least common multiple of 3 and 8, we write:

LCM (3, 8) = 24

There is an interesting relationship between the greatest common divisor and the least common multiple of two numbers. It is suggested by the five examples presented in Table 3.2.

Table 3.2 *Relationship between GCD and LCM*

Pair	Product	GCD	Product ÷ GCD	LCM
(2, 5)	10	1	10	10
(3, 6)	18	3	6	6
(4, 6)	24	2	12	12
(10, 8)	80	2	40	40
(12, 1)	12	1	12	12

3.6 Least Common Multiple

From Table 3.2 it seems to be true that dividing the product of two numbers by their greatest common divisor yields the least common multiple.

We shall now make an observation about the GCD of two numbers which will be needed in order to prove a rather important theorem. To illustrate this observation, consider the GCD of 18 and 84. It is 6. We see that 6 can be represented as a difference of some multiples of 18 and 84. Here it is:

$$6 = 5 \cdot 18 - 1 \cdot 84$$

We claim that this is true for every pair of numbers: the GCD of any pair of numbers can be represented as a difference of some multiples of the two numbers. Why this is so should become clear after studying the method by which we arrive at this difference. This is done for the numbers 18 and 84.

In determining GCD (18, 84) the steps would be the following:

$$\text{GCD } (18, 84) = \text{GCD } (12, 18) = \text{GCD } (6, 12)$$

The corresponding division relations are

$$84 = \text{⑱} \times 4 + \text{⑫}$$
$$18 = \text{⑫} \times 1 + \text{⑥}$$

Solving these two equations for the two remainders, 6 and 12, yields the following:

$$6 = 1 \times \text{⑱} - 1 \times \text{⑫}$$
$$12 = 1 \times \text{⑭} - 4 \times \text{⑱}$$

The first equation expresses the GCD, 6, in terms of 18 and 12. But the aim is to express it in terms of 18 and 84. Since 12 was expressed in terms of these two numbers, 12 can be replaced in the second equation by this expression for 12. From the above:

$$6 = 1 \times 18 - 1 \times 12$$

Replacing 12 by $1 \times 84 - 4 \times 18$:

$$6 = 1 \times 18 - 1 \times (1 \times 84 - 4 \times 18)$$

Simplifying

$$6 = 1 \times 18 + 4 \times 18 - 1 \times 84$$

Simplifying further:

$$6 = 5 \times 18 - 1 \times 84$$

Thus, the GCD (18, 84) has been expressed as a difference of the multiples of 18 and 84. To verify that $6 = 5 \times 18 - 1 \times 84$ is true:

$$5 \times 18 - 1 \times 84 = 90 - 84$$
$$= 6$$

Exercises 3.6

Give the LCM and two more multiples of each of the following pairs of numbers:

1. (6, 15) **2.** (21, 4) **3.** (9, 33) **4.** (11, 13)

Give the GCD of each of the following pairs of numbers, and express each GCD as a difference of some multiples of the two numbers:

5. (72, 36) **6.** (125, 80) **7.** (210, 90)

8. You are Michelangelo's assistant. Your job is to order primer paint for the ceiling job he is about to do. Michelangelo says, "Luigi, we're working on a tight budget. I want you to order exactly the right amount of paint—not a drop more or a drop less!" You measure a portion of the ceiling and find that it is 10 ft by 15 ft. The paint manufacturer tells you that one can of paint covers 10 sq ft of the LCM of the two dimensions. How many cans will you need for this portion of the ceiling?

3.7 DIVIDING THE PRODUCT OF TWO NUMBERS BY A PRIME NUMBER

REVIEW

- A prime number is greater than 1 and has exactly two factors
- The multiplication property for equations is:
  if $a = b - c$, then
  $ax = bx - cx$

OBJECTIVES

- Explore divisibility of a product of two numbers and of each of the two numbers by a prime number
- Prove a theorem concerning the divisibility described in the first objective

3.7 Dividing the Product of Two Numbers by a Prime Number

We want to divide products of two numbers by a prime number and see what we can learn from this.

Example 1 Consider the product 6·5. It is divisible by 3 (3 is a prime number):

$$30 \div 3 = 10$$

Note that 6, which is one of the factors in 6·5 is also divisible by 3:

$$6 \div 3 = 2$$

Example 2 Consider the product 10·15. It is divisible by 5 (is 5 a prime number?):

$$150 \div 5 = 30$$

Note that each of the factors in 10·15 is also divisible by 5:

$$10 \div 5 = 2$$
$$15 \div 5 = 3$$

Example 3 Consider the product 10·6 and the prime number 7. This product is not divisible by 7:

$$60 \div 7 = 8, r4$$

Neither of the two factors is divisible by 7:

$$10 \div 7 = 1, r3$$
$$6 \div 7 = 0, r6$$

These examples suggest a theorem which we now state and prove.

Theorem 3.4 If a prime number divides a product of two natural numbers, then it divides at least one of them.

That the theorem is not true for nonprime divisors can be seen from the following example. The number 4 divides the product of 2 and 6:

$$\frac{2 \times 6}{4} = \frac{12}{4} = 3$$

but 4 divides neither 2 nor 6.

Proof Let n_1 and n_2 be two natural numbers. Let p be a prime number which divides the product $n_1 \cdot n_2$. Assume that p does not divide n_1. We shall prove that p must then divide n_2. Since p is a prime number, its only divisors are 1 and p. It was assumed that p does not divide n_1; therefore, GCD $(p, n_1) = 1$. Therefore, 1 can be expressed as a difference of some multiples of p and n_1:

$$1 = ap - bn_1$$

(The correct relation might be $1 = bn_1 - ap$, but this would not change the argument.)

Multiply each member of the equation by n_2:

$$n_2 = apn_2 - bn_1n_2$$

It is known that p divides apn_2 and p divides bn_1n_2 (because, by hypothesis, it divides n_1n_2). Thus, p is a factor of $apn_2 - bn_1n_2$. Since $apn_2 - bn_1n_2 = n_2$, p is a factor of n_2; thus p divides n_2.

Exercises 3.7

1. Consider the product $8 \cdot 6$. Find the two prime numbers which divide this product. Examine each of these primes to see whether it divides the factors 8 and 6.
2. Consider the product $4 \cdot 5$. Find all prime numbers which divide this product and examine the divisibility of each of the factors 4 and 5 in relation to these prime numbers.
3. Examine the divisibility of the product $4 \cdot 8$ and each of its factors in relation to the prime number 5.
4. Give an example of Theorem 3.4 and verify the truth of the theorem for this example.
5. Give an example in which a composite number divides the product of two factors, but it does not divide either of the factors.
6. Eight women have nine hats each. Some of the women become women's libbers and decide never to wear hats again. This leaves a prime set of women who still wear hats. The number of women who still wear hats is the smallest prime that divides the product of all the women and their hats. How many women still wear hats? How many hats do these women have among them, including the hats left behind by the women's libbers? How many libbers are there?

3.8 FUNDAMENTAL THEOREM OF ARITHMETIC

REVIEW

- Given two different prime numbers, one number cannot divide the other

OBJECTIVES

- Factor a composite number into prime factors
- Use factor trees in factoring composite numbers into prime numbers
- Prove the Fundamental Theorem of Arithmetic
- Use prime factorization in finding the greatest common divisor of two numbers

In Section 3.1 we defined a composite number as a natural number with more than two factors. We also considered the question of factoring a composite number into factors which are prime numbers. Let's look at some examples.

Example 1 $78 = 2 \cdot 39$
$ = 2 \cdot 3 \cdot 13$

So, 78 has been factored into prime numbers 2, 3, and 13.

Example 2 To factor 52, we display the method in a different form, called a *factor tree*.

So, 52 has two prime factors, 2 and 13, and 2 is used twice in the product.

Example 3 Here are three different factor trees for factoring 24.

 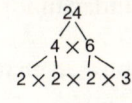 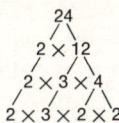

Although each tree is different, the final factorizations are the same, except for the order of factors.

These examples suggest that a composite number can be factored into prime factors in exactly one way, except for the order of factors.

Theorem 3.5   *Fundamental Theorem of Arithmetic.* Every composite number can be factored uniquely, except for the order of factors, into prime factors.

Proof* A composite number has at least one factor which is different from 1 and the number itself. If this factor is not prime, then it is composite and, therefore, it can be factored further. The process is continued until every factor is prime. This proves that every composite number can be factored into prime factors in some way. Now it must be proved that such a factorization is unique.

Let us assume that the prime factors have been arranged in order, from the least to the greatest and that this factorization is:

$$p_1 p_2 \cdots p_n$$

Furthermore, let us assume that there are two such factorizations, the second being:

$$q_1 q_2 \cdots q_k$$

Since each factorization is equal to the same number, the following is true:

$$p_1 p_2 \cdots p_n = q_1 q_2 \cdots q_k$$

Since p_1 divides the given number, p_1 must divide one of the factors $q_1, q_2, \ldots, q_k$. Since the factors are prime, p_1 is equal to that factor. Since the factors are arranged in ascending order, p_1 cannot be less than q_1. But, by the same argument, q_1 must divide one of the factors $p_1, p_2, \ldots, p_n$ and cannot be less than p_1. If follows that $p_1 = q_1$.

Having established that $p_1 = q_1$, divide each side of

$$p_1 p_2 \cdots p_n = q_1 q_2 \cdots q_k$$

by p_1, obtaining:

$$p_2 p_3 \cdots p_n = q_2 q_3 \cdots q_k$$

---

*This proof is optional. Its omission will have no bearing on subsequent study.

3.8 Fundamental Theorem of Arithmetic

Now repeat the argument showing that

$$p_2 = q_2$$

and so on. After n times, it will be established that

$$p_1 = q_1, p_2 = q_2, \ldots, p_n = q_k$$

thus proving that the prime factorization is unique.

In Section 3.5 we showed two methods for finding the greatest common divisor. Now that we can factor each number into its prime factors, we can use this technique to develop a third method for finding the GCD.

Example 1 Find GCD (8, 12). First, factor each number into primes:

$$8 = \boxed{2 \cdot 2} \cdot 2$$
$$12 = \boxed{2 \cdot 2} \cdot 3$$

Now take the product of all common factors of both numbers; they are circled above: $2 \cdot 2$. This is the greatest common divisor, which is 4.

Example 2 Find GCD (16, 25). Factor each number:

$$16 = 2 \cdot 2 \cdot 2$$
$$25 = 5 \cdot 5$$

Since they have no common prime factors, the numbers are relatively prime. That is, their GCD is 1.

Example 3 Find GCD (36, 48). Factor each number:

$$36 = 2 \cdot 2 \cdot 3 \cdot \boxed{3}$$
$$48 = 2 \cdot 2 \cdot 2 \cdot 2 \cdot \boxed{3}$$

Take the product of all common factors (circled):

$$2 \cdot 2 \cdot 3 = 12$$

Thus, GCD (36, 48) = 12.

Exercises 3.8

1. Show two different factor trees for the factorization of 12.
2. Show as many factor trees for the factorization of 48 as you can.
3. A squirrel begins to climb a strange new tree called a numeration tree. He wants to know how high he must climb to reach the top. Each branch of the tree has certain numbers that give the height of the tree

in feet. On the first branch, the squirrel reads 3 × 2 × 6. He calculates the height and then decides to check the numbers on the next branch to make sure he was right. On the second branch he finds 3 × 2 × 2 × 3. How high is the tree?

Factor each of the following composite numbers into prime factors, arranging them from the least to the greatest:

4. 144   **5.** 210   **6.** 365   **7.** 172

8. You are the pharaoh's chief architect. You are designing the great pyramids, and you must provide your mathematically minded pharaoh with a numerical configuration of your plans. If the pyramid is 96 ft high and the pharaoh wants it to have 5 floors, how many configurations have you to choose from? (*Hint:* Draw number trees for 96.)

Use prime factorization to find the greatest common divisor of each of the following pairs of numbers:

9. (12, 52)   **10.** (33, 77)   **11.** (24, 60)   **12.** (56, 84)

13. In the last section, there were eight women with nine hats. Now the women want to sell some of their hats and buy sweaters instead, but they want to end up with the same number of articles. Find the GCD of the above two numbers, which will tell you how many sweaters they can buy if they sell three hats.

3.9 DIVISIBILITY RULES

REVIEW

- The number x is divisible by y, if $x \div y$ is a natural number

OBJECTIVES

- To tell whether a number is divisible by each of the following numbers, using an appropriate rule:

  2, 5, 10, 4, 3, 9, 11, 6, 12, 37, and 7.

- Derive the rules for divisibility by 8 and by 99

Can you tell at a glance whether one number is divisible by another? There are some convenient and quick ways of determining divisibility for some numbers.

3.9 Divisibility Rules

Divisibility by 2, 5, and 10

Divisibility by 2, 5, and 10 is very easy to determine.

A number is divisible by 2 if its ones digit is 0, 2, 4, 6, or 8.
A number is divisible by 5 if its ones digit is 0 or 5.
A number is divisible by 10 if its ones digit is 0.

The rules are derived by considering the structure of a base-ten numeral. Examine this illustration:

$$2436 = \underline{2 \times 1000 + 4 \times 100 + 3 \times 10} + \underset{\text{digit}}{\overset{\text{Ones}}{6}}$$

Since 10 is divisible by 2, 5, and 10, so are 100 and 1000. The long underlined portion of the numeral above names a number that is divisible by 2, 5, and 10—and this will be true with *every* number, not just with the number 2436 used in illustrating the structure. Hence, if the value of the ones digit is divisible by 2, so is the number, and conversely. This follows in the case of 5 and 10.

Divisibility by 4

A number is divisible by 4 if and only if the last two digits in its base-ten numeral (tens and ones) form a number divisible by 4.

The number 100—hence 1000, and so on—is divisible by 4, and any number can be expanded as shown in this illustration:

$$2436 = 2 \times 1000 + 4 \times 100 + \overset{\text{Value of the last two digits}}{\overline{3 \times 10 + 6}}$$

Divisibility by 3 and by 9

A number is divisible by 3 (or 9) if and only if the *sum* of the digit values in its base-ten numeral is divisible by 3 (or 9).

The why of this rule depends on the fact that every power of ten is one more than an "all-nines" number:

$$10 = 9 + 1 \qquad 100 = 99 + 1 \qquad 1000 = 999 + 1 \qquad \text{and so on}$$

3 Some Remarkable Features of Numbers

Using 2436 as an illustration once more, we write:

$$2436 = 2 \times 1000 + 4 \times 100 + 3 \times 10 + 6$$
$$= 2 \times (999 + 1) + 4 \times (99 + 1) + 3 \times (9 + 1) + 6$$

Sum of digit values

$$= \underline{2 \times 999 + 4 \times 99 + 3 \times 9} + \overbrace{2 + 4 + 3 + 6}$$

Since any "all-nines" number is divisible by 3 and by 9, the first underlined portion of the numeral above names a number that is divisible by 3 and by 9, and, again, this will be so for every number. The remaining portion is just the sum of the digit values of the numeral.

In the case of the number 2436, the sum of the digit values is $2 + 4 + 3 + 6 = 15$. Since 15 is divisible by 3, so is 2436. Since 15 is not divisible by 9, neither is 2436. Incidentally, the sum of the digit values may itself be tested for divisibility by adding *its* digit values. Thus, for 15, we have $1 + 5 = 6$, divisible by 3 but not by 9.

As a second illustration, take the number 88,769. Adding digit values: $8 + 8 + 7 + 6 + 9 = 38$. Adding again: $3 + 8 = 11$. Adding again: $1 + 1 = 2$. Since this is divisible by neither 3 nor 9, the number 88,769 is itself divisible by neither 3 nor 9.

Divisibility by 11

A power of ten is either 1 greater or 1 less than some multiple of 11. Observe the following:

$$10 = 11 - 1$$
$$100 = 99 + 1 \;(9 \times 11 + 1)$$
$$1000 = 1001 - 1 \;(91 \times 11 - 1)$$
$$10{,}000 = 9999 + 1 \;(909 \times 11 + 1)$$
and so on

Hence, if the digit values of a numeral are alternately added and subtracted, beginning with the ones digit and proceeding from right to left, the resulting "alternating-digit sum" may be tested for divisibility by 11. For 2436, we have $6 - 3 + 4 - 2 = 5$. Hence, 2436 is not divisible by 11. For 74,712, we have $2 - 1 + 7 - 4 + 7 = 11$. Since 11 is divisible by 11, so is 74,712.

To explain this rule, consider the following analysis.

3.9 Divisibility Rules

$$74{,}712 = 7 \times 10{,}000 + 4 \times 1000 + 7 \times 100 + 1 \times 10 + 2$$
$$= 7 \times (9999 + 1) + 4 \times (1001 - 1) + 7 \times (99 + 1)$$
$$+ 1 \times (11 - 1) + 2$$
$$= 7 \times 9999 + 7 + 4 \times 1001 - 4 + 7 \times 99 + 7 + 1 \times 11$$
$$- 1 + 2$$
$$= \underbrace{7 \times 9999 + 4 \times 1001 + 7 \times 99}_{\uparrow} + 7 - 4 + 7 - 1 + 2$$

Divisible by 11 since
9999, 1001, and 99 are
each divisible by 11

Thus for 74,712 to be divisible by 11, $7 - 4 + 7 - 1 + 2$ must be divisible by 11. This analysis can be generalized to any number.

Exercises 3.9

Test each of the following numbers for divisibility by 2, 5, and 4:

1. 23
2. 30
3. 42
4. 60
5. 72
6. 132
7. 477
8. 561
9. 700
10. 855
11. 1296
12. 5291
13. 20,700
14. 47,232
15. 60,006
16. 80,806
17. 222,222
18. 353,343
19. 591,926
20. 938,190
21. Test the numbers in exercises 1–20 for divisibility by 3.
22. Test the numbers in exercises 1–20 for divisibility by 9.
23. Test the numbers in exercises 1–20 for divisibility by 11.
24. Show that a number is divisible by 6 if and only if it is divisible by both 2 and 3. What numbers in exercises 1–20 are divisible by 6?
25. Show that a number is divisible by 12 if and only if it is divisible by both 3 and 4. What numbers of exercises 1–20 are divisible by 12?
26. You are chief coach for this year's Super Bowl. It is your job to see that the 11 players on the field always have a vitamin pill to suck. If each player takes a pill each time he goes onto the field, how many times can the players be changed if you have 231 pills for your squad?
★27. Present an argument that if x and y are each divisible by z, so is $x + y$.
★28. Why is 0 divisible by any nonzero whole number?

3 Some Remarkable Features of Numbers

★29. Present an argument that each whole number whose numeral in base ten consists of an even number of nines ($\underbrace{999\ldots 9}_{\text{Even number of nines}}$) is divisible by 11.

★30. Present an argument that every number in the sequence 11, 1001, 100,001, 10,000,001, ... is divisible by 11.

★31. Verify that $37 \times 27 = 999$ and $37 \times 27{,}027 = 999{,}999$. Then show that a number is divisible by 37 if the sum of the numbers formed by successive triples of digits, from right to left, is divisible by 37. Thus, for 435,823 we find $823 + 435 = 1258$, and for 1258 we find $258 + 1 = 259 = 7 \times 37$. Hence, 435,823 is divisible by 37.

Test the following numbers for divisibility by 37:

★32. 5291 ★33. 222,222
★34. 353,343 ★35. 591,926

36. Invent a number and test it for divisibility by 2, 5, 10, 4, 3, 9, 11, and 7. How many divisors does it have? Can you improve on this? If you cannot, tell why.

37. You have 207 party cakes. You are an egalitarian. You want each guest to have exactly the same number of cakes. If you know you must invite more than 4 guests, exactly how many people can you invite?

The rule for divisibility by 7 is illustrated with the number 25,554,844. Divide the numeral into groups of three digits each, starting from the right: 844; 554; 25. (The last group may have one, two, or three digits.)

a. Form the sum of the odd-place groups.

$$\begin{array}{r} 844 \\ +\ 25 \\ \hline 869 \end{array}$$

b. Form the sum of the even-place groups (only one group here).

554

c. Subtract the number in b from that in a.

$$\begin{array}{r} 869 \\ -554 \\ \hline 315 \end{array}$$

d. If 315 is divisible by 7, so is the original number. Since $315 = 7 \times 45$, it is divisible by 7. Thus, 25,554,844 is divisible by 7. Using this rule, test each of the following numbers for divisibility by 7.

★38. 74,572,526 ★39. 1,230,659,365

3.10 Casting Out Nines

We may test 315 for divisibility by 7 as follows. Multiply 5 by 2 (10) and subtract the product from 31; thus $31 - 10 = 21$. Since 21 is divisible by 7, so is 315. Using this rule, test each of the following numbers for divisibility by 7:

★40. 861   ★41. 903   ★42. 569

The rule in exercises 40–42 can be extended to larger numbers. This is illustrated on 84,455:

$$
\begin{array}{r}
84{,}455 \\
2 \times 5 = \underline{10} \\
8435 \\
2 \times 5 = \underline{10} \\
833 \\
2 \times 3 = \underline{6} \\
77 \\
2 \times 7 = \underline{14} \\
-7
\end{array}
$$

Therefore, 84,455 is divisible by 7. Using this rule, test each of the following numbers for divisibility by 7:

★43. 86,107
★44. 23,157
★45. 306,501
★46. Derive a rule for divisibility by 8.
★47. Derive a rule for divisibility by 99.

3.10 CASTING OUT NINES

REVIEW

- Illustration of the use of distributive property:

  $4 \times 1000 = 4 \times (999 + 1)$
  $ = 4 \times 999 + 4 \times 1$
  $ = 4 \times 999 + 4$

- If $732 \div 26 = 28$, $r4$
  then $26 \times 28 + 4 = 732$

OBJECTIVES

- Define nines excess of a number
- Find nines excesses of numbers
- Prove a theorem about the computations with nines excesses
- Check the results of computations using nines excesses
- Check the results of computations using elevens excesses

3 Some Remarkable Features of Numbers

It was once practice in our schools to have students check numerical computations by *casting out the nines*. The *nines excess* of a number is the remainder obtained when the number is divided by 9.

Example 1 The nines excess of 5 is 5.
Example 2 The nines excess of 9 is 0.
Example 3 The nines excess of 12 is 3.
Example 4 The nines excess of 43 is 7, since $43 = 4 \cdot 9 + 7$.
Example 5 The nines excess of 100 is 1, since $100 = 11 \cdot 9 + 1$.

Actually, it is not necessary to divide a number by 9 to find out what the nines excess is, because the nines excess of a number is equal to the nines excess of the sum of the digit values in its base-ten numeral.

Example 1 The sum of the digit values of 4873 is:

$$4 + 8 + 7 + 3 \quad \text{or} \quad 22$$

In turn, the sum of the digit values of 22 is:

$$2 + 2 \quad \text{or} \quad 4$$

Conclusion: the nines excess of 4873 is 4.

It is not too difficult to rationalize this rule. We illustrate it using 4873.

$$4873 = 4 \times 1000 + 8 \times 100 + 7 \times 10 + 3$$

$$ = 4 \times 999 + 8 \times 99 + 7 \times 9 + \underbrace{4 + 8 + 7 + 3}_{\text{Sum of digit values}}$$

$$ = (\text{a multiple of } 9) + (4 + 8 + 7 + 3)$$

Hence, when 9 is divided into 4873, it divides the number named in the underlining, so that the remainder must equal the remainder obtained from dividing 9 into the sum of the digit values. This argument is obviously of general type, applicable to any number.

We shall now prove a theorem that justifies checking numerical computations by casting out the nines.

Theorem 3.6   If addition, subtraction, or multiplication is carried out with the nines excesses in place of the original numbers, the nines excess of this result will equal the nines excess of the result of the original computation.

3.10 Casting Out Nines

Proof Suppose two numbers a and b are added. Let their nines excesses be denoted by E_a and E_b. Then by definition of nines excess:

$$a = \text{(a multiple of 9)} + E_a$$
$$b = \text{(a multiple of 9)} + E_b$$

Since the sum of two multiples of 9 is again a multiple of 9, we find:

$$a + b = \text{(a multiple of 9)} + (E_a + E_b)$$

Hence, $a + b$ and $E_a + E_b$ have the same nines excess. Similarly, $a - b$ and $E_a - E_b$ have the same nines excess. (In this case, $E_a - E_b$ may be negative. Thus, if $a = 50$ and $b = 17$, $a - b = 33$ with excess 6, and $E_a - E_b = 5 - 8 = -3$. Add 9 to the -3 to bring it into the usual 0-to-8 nines-excess range: $-3 + 9 = 6$.)

If the numbers a and b are multiplied, we have

$$\begin{aligned} ab &= (9m + E_a)(9n + E_b) \\ &= (9 \times 9mn) + 9nE_a + 9mE_b + E_aE_b \\ &= 9 \times (9mn + nE_a + mE_b) + E_aE_b \\ &= \text{(a multiple of 9)} + E_aE_b \end{aligned}$$

In the above, m and n denote whole numbers, the quotients obtained when 9 is divided into a and b, respectively. Hence the nines excess of ab is equal to that of E_aE_b.

Any computation involving addition, subtraction, and multiplication may be made in steps each involving the performance of one operation on two numbers. The results shown above for two numbers can therefore be extended to apply to computation involving several numbers and operations.

Checking by casting out the nines consists of repeating calculations with nines excesses in place of original numbers. The following shows an addition check:

	Excesses
2436	6
218	2
+1602	0
4256 ✔	8

3 Some Remarkable Features of Numbers

The following is a subtraction check:

```
              Excesses
   2407          4
 −  982          1
   1525 Wrong:   3
   1425 ✓
```

At one time in school use, a special framework was used for the multiplication check:

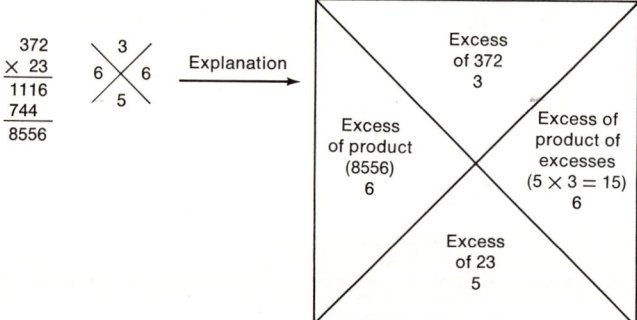

In case multiplication does not check as a whole, the trouble can be localized by checking the computation of each partial product by excesses.

Division can be checked by forming the corresponding basic division relation and checking it:

```
       28
   26)732      Relation: 26 × 28 + 4 = 732 → 3
      52       Excesses:  8 ×  1 + 4 =  12 → 3 ✓
     212
     208
       4
```

Casting out the nines fails to catch one very common numerical error, the transposition of two digits. If an answer ought to be 2936 and is accidentally given as 2396, the false answer will pass the nines check because the sum of the digit values—the nines excess—is unchanged. *Casting out elevens* will catch a transposition error.

The elevens check takes the same form as the nines check. The rule for finding the elevens excess of a number (remainder on division by eleven) is to form the "alternating sum" of the digit values: Add the value of the ones

3.10 Casting Out Nines

digit, subtract the value of the tens digit, add the value of the hundreds digit, subtract the value of the thousands digit, and so forth. For the number 180,739, for example, we find $9 - 3 + 7 - 0 + 8 - 1 = 20$ whose elevens excess is 9. Hence, the elevens excess of 180,739 is 9.

Exercises 3.10

Perform each operation; then check by casting out nines:

1. 5682
    +4371

2. 1096
    − 961

3. 365
    × 93

4. 24)3695

5. Check the computations in exercises 1–4 by casting out elevens.
6. You have the terrible disease known as "nonaphobia," a terrible fear of the number 9. Because you cannot stand the digit, you cast it out wherever possible. On this premise, check the following:

   a. 236
       +331
       ‾‾‾‾
       567

   b. 431
       −280
       ‾‾‾‾
       151

CHAPTER 3 TEST

1. What kind of a number is the square of an even number?
2. What kind of a number is the square of an odd number?
3. You may have heard the old western song "Even is even and odd is odd, but the squares are just the same." Is it a true statement?
4. How many divisors does a prime number have? A composite number?
5. How many prime numbers are there?

Show the following primes as products of 6 and some number, plus or minus 1:

6. 67   7. 89   8. 143   9. 341

10. You perhaps have heard of the movie *The Prime of Miss Jean Brodie.* By counting the letters in "Miss Jean Brodie" and subtracting 1, determine just what her prime is.

Define each of the following:

11. deficient number   12. perfect number
13. abundant number   14. amicable numbers

3   Some Remarkable Features of Numbers

15. Determine GCD (48, 108).
16. Find LCM (4, 14) and two more multiples of the two numbers.
17. Find GCD (30, 45), and express it as a difference of some multiples of the two numbers.
18. If a prime number divides a product of two natural numbers, then what is true about divisibility of these two natural numbers in relation to this prime number?
19. Give an example in which a composite number divides the product of two factors but does not divide either of the factors (different from the example in exercise 5 of Section 3.7, please).
20. State the Fundamental Theorem of Arithmetic.
21. Show as many factor trees for the factorization of 56 as you can.

Factor each of the following numbers into prime factors, arranging them from the least to the greatest:

22. 102   23. 78   24. 245
25. You are Mother Nature and have decided to build some number trees upon their bases. The first bases you prepare are:
    a. $2 \times 2 \times 2 \times 3$ b. $3 \times 3 \times 3 \times 2$
    Finish the trees.
26. What is an easy way of telling whether a number is divisible by 2?

State the divisibility rule for divisibility of a number by:

27. 4   28. 3   29. 9

Perform each operation; then check by casting out nines and elevens:

30. 3892   31.   3294
    +7065   − 397

32.   473   33. 47)4036
    × 28

4

Whole Numbers, Integers, and Rationals

4.1 THE SYSTEM OF WHOLE NUMBERS

REVIEW

- A prime number is a natural number greater than 1 which has exactly two divisors
- A composite number is a natural number which has more than two divisors

OBJECTIVES

- Define a binary operation
- Know the meaning of the solution set of an equation
- State six properties of addition and multiplication
- Give examples proving lack of closure of the set of whole numbers under subtraction and division
- Prove lack of closure of various number sets under various operations

The most fundamental of all mathematical processes is counting. For this, we use the natural numbers, $N = \{1, 2, 3, \ldots\}$. Tossing zero into the set of natural numbers results in the set of *whole numbers*, $W = \{0, 1, 2, 3, \ldots\}$. Both N and W are infinite sets. Each has a least member, but neither has a greatest member.

4 Whole Numbers, Integers, and Rationals

The usefulness of whole numbers is increased by performing operations on them. The four basic operations are addition, subtraction, multiplication, and division. Each of these operations is a *binary operation*, since each is performed on *two* numbers. Knowing the operations enables us to do some algebra, such as solving equations.

To solve an equation such as:

$$x + 5 = 9$$

means to determine the replacement for x which will result in a true statement. In this case, 4 is this replacement, since $4 + 5 = 9$ is true. We say that $\{4\}$ is the *solution set* of the equation $x + 5 = 9$.

Of the four operations we mentioned, addition and multiplication are the most basic. To consider the set of whole numbers, along with these two operations, we observe certain *properties* these operations have. For easy reference, these properties are listed below.

1. *Closure property:*

The sum and the product of any two whole numbers are whole numbers.

2. *Commutative property:*

Two whole numbers can be added or multiplied in any order:
$a + b = b + a$
$a \cdot b = b \cdot a$

3. *Associative property:*

When using three numbers, they can be regrouped in addition and multiplication:
$(a + b) + c = a + (b + c)$
$(a \cdot b) \cdot c = a \cdot (b \cdot c)$

4. *Distributive property:*

Multiplication can be distributed over addition:
$a \cdot (b + c) = (a \cdot b) + (a \cdot c)$

5. Property of *additive identity:*

The number 0 is the additive identity since for every whole number a:
$a + 0 = a$

6. Property of *multiplicative identity:*

The number 1 is the multiplicative identity since for every whole number a:
$a \cdot 1 = a$

4.1 The System of Whole Numbers

To prove that the set of whole numbers is not closed under subtraction, we only have to show one example (a *counter example*) in which the difference of two whole numbers is not a whole number. For example, $3 - 5$ proves the lack of closure under subtraction, since -2 is not a whole number.

Exercises 4.1

Solve each of the following equations:

1. $x + 3 = 24$
2. $21 - x = 14$
3. $x \cdot 12 = 144$
4. $x - 9 = 27$
5. $5x = 65$
6. $x \div 11 = 11$
7. Superman, who is faster than a speeding bullet, is fired upon by the archvillain Alphonso! Superman sees the bullet coming and determines its speed to be 70 ft/sec. Also being a mathematician, he calculates the time it will take the bullet to reach him so he will know how soon he must fly away. The formula is:

$$t = \frac{V}{10} + \frac{1}{8}$$

   where t is in sec if velocity is in ft/sec. How long does Superman have before he must depart?
8. Prove that the set of odd numbers is not closed under addition by citing one counterexample. (*Hint:* For the set of odd numbers to be closed under addition, the sum of every pair of odd numbers must be an odd number.)
9. You wish to become a resident at a very exclusive adult-living complex known as The Set of Whole Numbers. After applying, the manager tells you, "Sorry, we are closed under subtraction." At this point, you remember your math course and say, "But the set of whole numbers is not closed under subtraction. Look at this example!" What example would you show him?
10. Prove that the set of whole numbers is not closed under subtraction by citing one counterexample.
11. Prove that the set of whole numbers is not closed under division.
12. Prove that the set of even numbers is not closed under division.
13. Prove that the set of prime numbers is not closed under addition.
14. Prove that the set of prime numbers is not closed under multiplication.
15. Prove that the set of composite numbers is not closed under addition.
16. Why is there no additive identity in the set of natural numbers?

For each statement, tell which property it illustrates:

17. $5 + 49 = 49 + 5$
18. $(9 \cdot 3) \cdot 5 = 9 \cdot (3 \cdot 5)$
19. $15 \cdot 37$ is a whole number
20. $19 + 0 = 19$

21. $(27 + 36) + 14 = 27 + (36 + 14)$ **22.** $136 \cdot 1 = 136$
23. $76 \cdot 9 = 9 \cdot 76$ **24.** $73 + 0$ is a whole number
25. $2 \times (5 + 3) = (2 \times 5) + (2 \times 3)$
26. Big Brother, whom you have already met, keeps everyone's personal traits on computer records. Someone who illustrates a closure property is to be silenced. Spies have the commutative property. People who easily make friends have associative properties. People who think they are superior have the distributive property. From the following printout, tell in which classification the subjects belong.
   a. $(2 + 3) + 100 = 2 + (3 + 100)$
   b. $3 + 5 = 8$
   c. $8 \cdot (2 + 2) = (8 \cdot 2) + (8 \cdot 2)$
27. Perform the operations in two different orders in $(125 + 137) + 63$ to convince yourself that $125 + (137 + 63)$ is easier to carry out.
28. Do the same for $(78 \cdot 25) \cdot 4$ and $78 \cdot (25 \cdot 4)$. Is the last grouping easier for computations? What property is helpful here?
29. Why is it impossible to verify all instances of the commutative property of addition in the set of whole numbers?

4.2 THE SYSTEM OF INTEGERS, ADDITION

REVIEW

- A *number line* is a line on which numbers are assigned to points
- Closure property of addition of whole numbers:

  the sum of any two whole numbers is a whole number

- Commutative property of addition:

  $a + b = b + a$

- Associative property of addition:

  $(a + b) + c = a + (b + c)$

- Property of additive identity:

  $a + 0 = a$

OBJECTIVES

- Know what the set of integers consists of
- Add integers
- Define an additive inverse of an integer
- State the property of additive inverse
- Define the absolute value of an integer
- Solve equations in the set of integers

4.2 The System of Integers, Addition

The whole numbers are not adequate to cover certain situations. Temperatures do fall below zero sometimes, and no whole number can describe these temperatures. There are equations which cannot be solved if only whole numbers are available. For example, the equation:

$$x + 9 = 5$$

does not have a whole-number solution. For such situations we turn to *integers*.

The set of integers consists of *negative integers*, 0, and *positive integers*.

$$I = \{\ldots, -3, -2, -1, 0, +1, +2, +3, \ldots\}$$

To simplify writing, $+$ will not be written when referring to positive integers. It is important to remember, however, that when writing 5, for example, positive 5 is meant.

To solve the equation $x + 9 = 5$, it is necessary to know how to add integers. There is no problem with the addition of positive integers, since it is like adding whole numbers. Thus,

$$5 + 6 = 11, 9 + 5 = 14, 17 + 13 = 30, \text{ and so on for positive integers.}$$

Sometimes a pictorial model is helpful in attempting to remember how to do certain operations. A *number line* is a very popular model for the purpose of adding integers. Here are some examples of addition of integers done with the help of the number line.

Example 1 Add: $-3 + (-4)$
Start at 0; move 3 units to the left; now 4 units to the left. You are now at the answer: $-3 + (-4) = -7$.

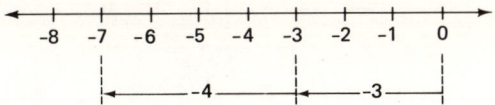

Example 2 Add: $-5 + (-4)$
Here is a picture of this addition. Describe the moves and tell the answer.

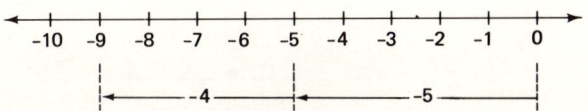

4 Whole Numbers, Integers, and Rationals

These last examples suggest the following generalization:

The sum of two negative integers is a negative integer.

Such diagrams can also be used as models for adding a positive and a negative integer.

Example 3   Add: $3 + (-2)$
Start at 0; move 3 units to the right; now 2 units to the left. You are now at the answer: $3 + (-2) = 1$.

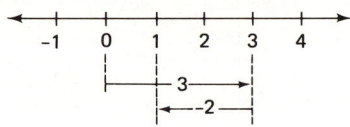

No doubt you noticed that a positive number is represented by a move to the right (positive direction) and a negative number by a move to the left (negative direction).

Example 4   Add: $5 + (-3)$
Here is a picture of this addition. Describe the moves and tell the answer.

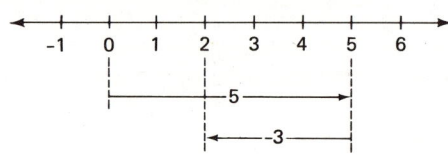

Study the following three pictures and the statements below them, and see whether you can arrive at a method for adding two integers.

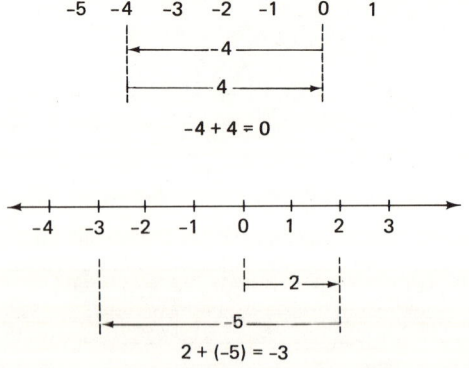

4.2 The System of Integers, Addition

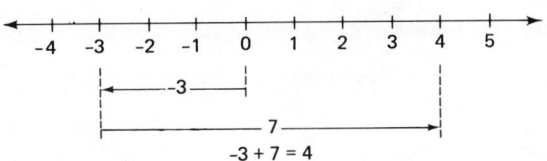

$$-3 + 7 = 4$$

These examples show that the answer to an addition of a positive integer and a negative integer can be a positive integer, or 0, or a negative integer. Let us analyze this closer. When is the answer 0? Here are some examples:

$$5 + (-5) = 0 \qquad -12 + 12 = 0 \qquad 125 + (-125) = 0$$

Two integers whose sum is 0 are called *additive inverses*.

Example 1 -5 is the additive inverse of 5
Example 2 9 is the additive inverse of -9
Example 3 0 is the additive inverse of 0

The observation concerning the sum of two additive inverses is stated as a property.

PROPERTY OF ADDITIVE INVERSE For every integer a, $a + (-a) = 0$.

And now a word of caution! Some people read $-a$ as negative a. This causes a problem because $-a$ can be a positive number! Just replace a by -2:

$$-a = -(-2)$$
$$ = 2$$

So, $-a$ is a *positive* number when a is replaced by -2.

Our suggestion:

Read $-a$ as "additive inverse of a"

When a is replaced by a name of a positive number, $-a$ is negative.
When a is replaced by 0, $-a$ is 0. So, 0 is its own additive inverse. Furthermore, 0 is the only integer for which this is true.
When a is replaced by a name of a negative number, $-a$ is positive.

Our ability to talk about operations with integers and to make rules will be increased if we introduce another concept. It is the concept of the *absolute value* of a number. Its meaning should become clear from the following examples.

4 Whole Numbers, Integers, and Rationals

By the way, read $|-5|$ as "the absolute value of five."

$|-5| = 5 \qquad |10| = 10 \qquad |-3| = 3 \qquad |16| = 16 \qquad |0| = 0$

SUMMARY   The absolute value of a positive integer is the integer.
The absolute value of a negative integer is the additive inverse of the integer; so, it is positive.
The absolute value of 0 is 0.

With just a little bit of thought, you will convince yourself that addition of integers has the closure property, commutativity, associativity, and additive identity. And now we can add another property—the property of additive inverse—which the whole numbers do not have.

Exercises 4.2

1. In the set of integers is there a least integer? A greatest integer?
2. What is the name of the property which tells that the sum of a pair of integers is an integer?

Solve each of the following equations in the set of integers:

3. $x + 12 = 8$ 4. $3 + x = 0$ 5. $x + 12 = 1$
6. $x + 7 = 19$ 7. $x + (-10) = 0$ 8. $26 + x = 5$
9. A bomb has been planted in a large hotel in the heart of the city. Being an experienced bomb expert, you know that the bomb is on a time fuse given by the equation $0 = -123 + t$, where t is in minutes. How long do you have before detonation?

Tell the absolute value of each of the following integers:

10. -8 11. 12 12. 0
13. -25 14. -100
15. We have an inchworm training for the Inchworm Olympics. Since all inchworms are math majors, this worm uses a number line to train on. If his training pattern consists of the following sums, draw his path.
     a. $7 + (-6)$ b. $(-3) + (-1)$ c. $(-2) + 2$

Compute the sums. Draw pictures only if you need to.

16. $5 + (-3)$ 17. $-4 + 1$ 18. $-3 + 0$
19. $6 + 8$ 20. $-13 + 13$ 21. $3 + (-7)$
22. $2 + (-1)$ 23. $0 + 6$ 24. $-6 + 3$

4.3 Subtraction of Integers

Compute the sums:

25. $100 + (-70)$
26. $-75 + 50$
27. $125 + (-125)$
28. $-20 + 45$
29. $36 + (-76)$
30. $16 + (-148)$
31. $29 + (-14)$
32. $45 + (-82)$
33. $-15 + 166$

For each statement, tell which property of addition for the set of integers it illustrates:

34. $-12 + (-17) = -17 + (-12)$
35. $-37 + 37 = 0$
36. $2 + (-8)$ is an integer
37. $(-2 + 6) + (-4) = -2 + [6 + (-4)]$
38. $-7 + 0 = -7$
39. Explain why there is no property of additive inverse in the set of whole numbers.
40. For every integer a, $|a|$ is a nonnegative integer. Give three examples illustrating this, using one positive integer, one negative integer, and 0.
41. Your bank manager says, "You are overdrawn by $300. Therefore, you are absolutely $300 in debt." Being a mathematician, you say, "Excuse me, but I cannot be *absolutely* $300 in debt." Why not?
★42. In the set of integers every equation of the form $x + a = b$, where a and b are integers, has a solution. What is this solution?

4.3 SUBTRACTION OF INTEGERS

REVIEW

- The symbol $>$ means *is greater than*
- The symbol $<$ means *is less than*
- $a - b = c$ means $a = c + b$
- The symbol $\not>$ means *is not greater than*
- The symbol $\not<$ means *is not less than*

OBJECTIVES

- Compare integers
- Subtract integers by adding

4 Whole Numbers, Integers, and Rationals

The number line proved to be a useful device for adding integers. We can also use it to tell which of two given integers is larger. We will follow a very simple rule:

> The number whose point on the number line is farther to the right is the larger.

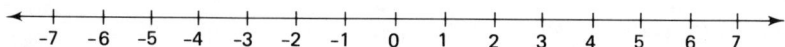

Example 1 $5 > -2$ *Read:* five *is greater than* negative two
This statement is true because the point corresponding to 5 is *to the right* of the point corresponding to -2.

From the above it follows that:

$-2 < 5$ *Read:* negative two *is less than* five

This is true since the point corresponding to -2 is to the left of the point corresponding to 5.

You should be able to tell now why each of the following is true:

$-3 > -7$ $1 > -6$ $7 > -1$ $-1 > -7$
$-4 < 2$ $-1 < 7$ $0 < 5$ $-7 < 0$

Perhaps you have already thought to yourself that every negative number is less than every positive number. The following are also true:

> $a > 0$ for every positive a
> $-a < 0$ for every positive a; note that $-a$ is negative if a is positive

We could use the number line to subtract one integer from another, but there is another way to do it. Simply change every subtraction to an addition. Before doing this, let us reason a bit:

$7 - 2 = n$ means $n + 2 = 7$

Since $5 + 2 = 7$, it follows that $7 - 2 = 5$. But $7 + (-2) = 5$ also. So, instead of subtracting 2, we can add -2.

This suggests a general procedure for changing a subtraction problem into an addition problem: add the additive inverse.

4.3 Subtraction of Integers

Example 1 $5 - (-2) = ?$
Add the additive inverse of -2 instead:

$$5 + 2 = 7$$

So, $5 - (-2) = 7$.
Example 2 $-9 - (-7) = ?$
Add the additive inverse of -7 instead:

$$-9 + 7 = -2$$

So, $-9 - (-7) = -2$.
Example 3 $2 - 9 = ?$
Add the additive inverse of 9 instead:

$$2 + (-9) = -7$$

So, $2 - 9 = -7$.

We have the pattern:

$a - b = a + (-b)$ for all integers a and b
To subtract an integer, add its additive inverse.

Exercises 4.3

1. Answer the riddle: when is subtraction not subtraction? Illustrate your answer with an example.

Subtract by changing to addition:

2. $20 - 7$ **3.** $76 - 14$ **4.** $3 - 35$
5. $5 - 55$ **6.** $23 - (-4)$ **7.** $10 - (-20)$
8. $-3 - 1$ **9.** $-15 - (-3)$ **10.** $-12 - (-4)$
11. $-5 - (-15)$ **12.** $0 - 7$ **13.** $0 - (-67)$

For each statement tell whether it is true or false:

14. $5 > -70$ **15.** $-13 < -1$ **16.** $3 \not> 1$
17. $-2 \not< 0$ **18.** $0 \not> 1$ **19.** $-100 < 0$
20. $|-100| < 0$ **21.** $-5 < -1$ **22.** $|-5| < |-1|$
23. $15 > 3$ **24.** $|15| > |3|$ **25.** $0 > -20$
26. $|0| > |-20|$ **27.** $|5 + (-2)| = |5| + |-2|$
28. $|-3 + (-6)| = |-3| + |-6|$ **29.** $|-9 + 2| = |-9| + |2|$
30. $|3 - (-5)| = |3| - |-5|$ **31.** $|8 - 6| = |8| - |6|$
32. $|-2 - (-10)| = |-2| - |-10|$ **33.** $|3 - 5| > |3| - |5|$
34. $|27 - (-3)| > |27| - |-3|$ **35.** $|-3 - (-5)| > |-3| - |-5|$

4 Whole Numbers, Integers, and Rationals

36. You are the judge in a shot-put competition. Three competitors, A, B, and C, each put their shot. You record the results as follows:

$$A > 70 \text{ ft} \quad B = 35 \text{ ft} \quad C > 40 \text{ ft} \quad A > C \quad B < C$$

Interpret your results on a line.

★37. Examine several instances of $|a - b| = |b - a|$ by replacing a and b by names of various integers. Do you think this is true for all integers a and b?

★38. Give one instance of replacements for a and b in $|a - b| = |a| - |b|$ to prove that it is not true for all integers a and b.

4.4 MULTIPLICATION OF INTEGERS

REVIEW

- Distributive property:
  $a(b + c) = ab + ac$
- Commutative property of multiplication:
  $ab = ba$
- Associative property of multiplication:
  $(ab)c = a(bc)$

OBJECTIVES

- Multiply two integers
- Prove that the product of two negative integers is positive

Multiplying two positive integers is like multiplying two natural numbers. For example, $5 \times 6 = 30$ and $9 \times 7 = 63$:

> The product of two positive integers is a positive integer.
> If $a > 0$ and $b > 0$, then $ab > 0$

Similarly, the integer 0 is just like the whole number 0.

> The product of any integer and 0 is 0.
> $a \cdot 0 = 0$ for every integer 0

Now comes the new part: what should the product be when one or both of the integers are negative? More importantly, how are such decisions made?

One way is to agree that once a particular pattern is established it shall continue. We shall accept that and proceed to observe certain patterns.

4.4 Multiplication of Integers

Let's start with an example:

$$2 \times 4 = 8$$
$$2 \times 3 = 6$$
$$2 \times 2 = 4$$
$$2 \times 1 = 2 \quad \longleftarrow \text{To obtain next number,}$$
$$\text{subtract 2.}$$

2 is repeated   To obtain next number, subtract 1.

The pattern followed in each column is described above. There should be no difficulty to know how to continue with the pattern. Here are the next few steps:

$$2 \times 0 = 0$$
$$2 \times (-1) = -2$$
$$2 \times (-2) = -4$$
$$2 \times (-3) = -6$$
$$\cdot$$
$$\cdot$$
$$\cdot$$

What would be the next step?

Assuming that the pattern continues, the following are true:

$$2 \times (-1) = -2 \quad 2 \times (-2) = -4 \quad 2 \times (-3) = -6 \ldots$$

We have a case of the product of a positive integer (2) and a negative integer (-1) being equal to a negative integer (-2). Similarly, for 2 and $-2[2 \times (-2) = -4]$, 2 and $-3[2 \times (-3) = -6]$, and so on.

Would this be true if the pattern is built using a different positive integer, for example, 5?

$$5 \times 4 = 20$$
$$5 \times 3 = 15$$
$$5 \times 2 = 10$$
$$5 \times 1 = 5$$
$$5 \times 0 = 0$$

Describe the pattern for each of the three columns in this case. Now continue the same pattern:

$$5 \times (-1) = -5$$
$$5 \times (-2) = -10$$
$$5 \times (-3) = -15$$

Again, it is the case that a positive integer (5) multiplied by a negative integer (-1, or -2, or -3) results in a product which is a negative integer.

> The product of a positive and a negative integer is negative.
> If $a > 0$ and $b < 0$, then $ab < 0$

To decide about the product of two negative integers, similar patterns can be constructed and an assumption made that they continue. Keep in mind that now we know what the product of a negative and a positive integer is. Construct a pattern using -2 as one of the factors:

$$-2 \times 4 = -8$$
$$-2 \times 3 = -6$$
$$-2 \times 2 = -4$$
$$-2 \times 1 = -2$$

Describe the pattern found in each column. Continue according to the same patterns:

$$-2 \times 0 = 0$$
$$-2 \times (-1) = 2$$
$$-2 \times (-2) = 4$$
$$-2 \times (-3) = 6$$

Thus, a negative integer (-2) multiplied by a negative integer (-1, or -2, or -3) is equal to a positive integer (2, or 4, or 6).

Would the results be the same if a negative integer different from -2, for example, -5, is used?

$$-5 \times 4 = -20$$
$$-5 \times 3 = -15$$
$$-5 \times 2 = -10$$
$$-5 \times 1 = -5$$

Describe the pattern found in each column. Continue the pattern:

$$-5 \times 0 = 0$$
$$-5 \times (-1) = 5$$
$$-5 \times (-2) = 10$$
$$-5 \times (-3) = 15$$

And again, the product of a negative integer (-5) and a negative integer (-1, or -2, or -3) is a positive integer (5, or 10, or 15).

> The product of two negative integers is positive.
> If $a < 0$ and $b < 0$, then $ab > 0$

4.4 Multiplication of Integers

The rules for multiplication of two integers were arrived at on a purely intuitive basis: a pattern in the multiplication of whole numbers was observed, and the rule was suggested by assuming that this pattern continues.

There is a more sophisticated way to establish that, for example, the product of two negative integers is positive. It is by *proving* first that $-1 \cdot (-1) = 1$ on the basis of the assumption that $1 \cdot (-1) = -1$. Follow the proof below step by step.

$$-1 \cdot [1 + (-1)] = (-1 \cdot 1) + (-1) \cdot (-1) \quad \text{(distributive property)}$$
$$= -1 + (-1) \cdot (-1) \quad (-1 \cdot 1 = -1)$$

It is not known what $(-1) \cdot (-1)$ is equal to. But the following is known:

$$-1 \cdot [1 + (-1)] = -1 \cdot 0 \quad [1 + (-1) = 0 \text{ and for every } a,$$
$$= 0 \quad a \cdot 0 = 0]$$

Thus,
$$-1 + (-1) \cdot (-1) = 0$$
But the only number added to -1 which yields 0 for the sum is 1. Therefore,
$$(-1) \cdot (-1) = 1$$

This result can serve as a basis for finding the product of any two negative integers. Here is one example to illustrate the procedure:

$$(-6) \cdot (-7) = [6 \cdot (-1)] \cdot [7 \cdot (-1)]$$
$$= (6 \cdot 7) \cdot [-1 \cdot (-1)]$$
$$= 42 \cdot 1$$
$$= 42$$

$(6 \cdot 7) \cdot [-1 \cdot (-1)]$ is obtained from $[6 \cdot (-1)] \cdot [7 \cdot (-1)]$ by repeated applications of the commutative and the associative properties of multiplication. The effect of these two properties is that the factors in a product can be rearranged in any order.

The example above can be generalized to show that the product of two negative integers is a positive integer, assuming that $-1 \cdot (-1) = 1$.

Given that $a > 0$ and $b > 0$, it follows that $-a < 0$ and $-b < 0$.

$$-a \cdot (-b) = [a \cdot (-1)] \cdot [b \cdot (-1)]$$
$$= (ab) \cdot [-1 \cdot (-1)]$$
$$= (ab) \cdot 1$$
$$= ab.$$

Since a is positive and b is positive, ab is also positive. Thus, $-a \cdot (-b)$ is positive.

Exercises 4.4

Tell the products:

1. $3 \times (-5)$
2. -11×2
3. 0×8
4. $12 \times (-13)$
5. -2×7
6. 16×0
7. $0 \times (-25)$
8. $-11 \times (-10)$
9. $-6 \times (-5)$
10. -5×0
11. 5×12
12. 0×2678

13. How many times have you heard the phrase "Two wrongs do not make a right"? Say you have two bank accounts. In one, you are $10 in debt. In the other, you are $15 in debt. If these particular banks indicate a debt by a − sign, then do your two "wrongs" (debts) produce a right (credit), at least theoretically?

14. According to the new rules of the NFL, distances to the left of the field center line are to be negative, and distances to the right are to be positive. What is the area of the left-hand portion and the right-hand portion if the new field is 30 yards (positive) across? What is the total area? Can you suggest after computing the total area why the NFL decided to drop the idea?

Tell which of the following are true:

15. If $a > 0$ and $b < 0$, then $ab < 0$.
16. If $-a < 0$ and $-b < 0$, then $ab > 0$.
17. If $-a > 0$ and $-b > 0$, then $ab > 0$.
18. $|a| \cdot |b| > 0$.
19. If $-a > 0$ and $b < 0$, then $ab > 0$.
20. If $a < 0$ and $b > 0$, then $|a| \cdot b > 0$.
21. Using the following four assumptions:
    a. $(-1) \cdot (-1) = 1$
    b. the product of two positive integers is a positive integer
    c. commutative and associative properties of multiplication
    d. $-a = a \cdot (-1)$
    prove that $-5 \cdot (-8) = 40$.
22. Another riddle for you to solve: when can a and b which are less than zero be greater than zero?
★23. Given $a < 0$ and $b < 0$, prove that $ab > 0$.
★24. Given $a < 0$ and $-b < 0$, prove that $ab < 0$.
★25. Given $a < 0$ and $-b < 0$, prove that $a(-b) > 0$.

4.5 RATIONAL NUMBERS

REVIEW

- The solution of the equation $x + 3 = 8$ is 5, since $5 + 3 = 8$ is true
- Properties for the set of whole numbers:
  1. Closure of addition:
     $a + b$ is a whole number for all whole numbers a and b
  2. Closure of multiplication:
     ab is a whole number for all whole numbers a and b
  3. Commutativity of addition:
     $a + b = b + a$ for all whole numbers a and b
  4. Commutativity of multiplication:
     $ab = ba$ for all whole numbers a and b
  5. Associativity of addition:
     $(a + b) + c = a + (b + c)$ for all whole numbers a, b, and c
  6. Associativity of multiplication:
     $(ab)c = a(bc)$ for all whole numbers a, b, and c
  7. Distributivity of multiplication over addition:
     $a(b + c) = (ab) + (ac)$ for all whole numbers a, b, and c
  8. Additive identity:
     $a + 0 = a$ for each whole number a
  9. Multiplicative identity:
     $a \cdot 1 = a$ for each whole number a

OBJECTIVES

- Solve equations which have rational-number solutions
- Tell whether two fractional names name the same rational number
- Add rational numbers
- Multiply rational numbers
- Prove the closure property of rational numbers under addition
- Prove the commutative property of addition of rational numbers
- List the properties of whole numbers
- State the property of multiplicative inverse for rational numbers

In elementary school children study fractions, since they are very useful to know. In algebra, many equations do not have solutions in the set of integers. For example, the equation:

$$-3x = 2$$

does not have an integer for its solution. But it does have a solution in the set of rational numbers. The rational number $-\frac{2}{3}$ is the solution of the equation $-3x = 2$.

Any number which has a name of the form a/b, where a and b are integers ($b \neq 0$) is a *rational number*.

We know that every integer has a name of this form. For example, $4 = \frac{4}{1}$, $-9 = -\frac{9}{1}$, $0 = \frac{0}{1}$. It follows that every integer is a rational number. Therefore, the set of integers is a subset of the set of rational numbers.

Every rational number can be given a lot of names. Here is a beginning of the infinite set of names for the rational number $\frac{2}{3}$.

$$\{\tfrac{2}{3}, \tfrac{4}{6}, \tfrac{6}{9}, \tfrac{8}{12}, \ldots\}$$

You, no doubt, spotted the pattern according to which these names are arranged. Give the next four names which fall into this pattern.

Any two names from the set above name the same number. For example, $\frac{2}{3} = \frac{8}{12}$.

This statement of equality simply tells us that $\frac{2}{3}$ and $\frac{8}{12}$ are two names for the same number. Observe also that $2 \cdot 12 = 3 \cdot 8$. In general:

$$\frac{a}{b} = \frac{c}{d} \text{ means that } ad = bc \ (b \neq 0, d \neq 0)$$

So, we have a way of testing whether two fractional names name the same number.

Using this pattern, we can tell that $\frac{4}{7}$ and $\frac{16}{28}$ are two names for the same number, since $4 \cdot 28 = 7 \cdot 16$. Verify this. We also know that $\frac{2}{3}$ and $\frac{10}{14}$ are not names for the same number, since $2 \cdot 14 \neq 3 \cdot 10$.

Adding Rational Numbers

Example 1
$$\frac{2}{3} + \frac{4}{5} = \frac{2 \times 5 + 3 \times 4}{3 \times 5}$$
$$= \frac{10 + 12}{15}$$
$$= \frac{22}{15}$$

4.5 Rational Numbers

In general:

$$\frac{a}{b} + \frac{c}{d} = \frac{ad + bc}{bd}$$

Multiplying Rational Numbers

Example 1 $\quad \frac{4}{5} \times \frac{2}{3} = \frac{4 \times 2}{5 \times 3} = \frac{8}{15}$

In general:

$$\frac{a}{b} \times \frac{c}{d} = \frac{ac}{bd}$$

Using the properties of integers which were assumed earlier, a number of theorems concerning the operations on rational numbers can be proved. Two theorems are proved here. Reasons supporting the statements made in the proofs are given.

Theorem 4.1   *Closure of the Set of Rational Numbers under Addition.* For all rational numbers a/b and c/d, $a/b + c/d$ is a rational number.

Proof

$\dfrac{a}{b} + \dfrac{c}{d} = \dfrac{ad + bc}{bd}$	Addition of rational numbers
$ad + bc$ is an integer	Closure of integers under multiplication and addition
bd is an integer	Closure of integers under multiplication
$\dfrac{ad + bc}{bd}$ is a rational number	Form of a rational number name
$\dfrac{a}{b} + \dfrac{c}{d}$ is a rational number	Substitution $\left(\dfrac{a}{b} + \dfrac{c}{d} = \dfrac{ad + bc}{bd}\right)$

Thus, it is proved that the set of rational numbers is closed under addition.

4 Whole Numbers, Integers, and Rationals

Theorem 4.2 *Commutativity of Addition of Rational Numbers.* For all rational numbers a/b and c/d, $a/b + c/d = c/d + a/b$.

Proof

$$\frac{a}{b} + \frac{c}{d} = \frac{ad + bc}{bd} \qquad \text{Addition of rational numbers}$$

$$= \frac{bc + ad}{bd} \qquad \text{Commutative property of addition of integers}$$

$$= \frac{cb + da}{bd} \qquad \text{Commutative property of multiplication of integers}$$

$$= \frac{c}{d} + \frac{a}{b} \qquad \text{Addition of rational numbers}$$

Thus, it is proved that the addition of rational numbers is commutative.

It is instructive to observe how additional properties of addition and multiplication are gained as new systems of numbers are introduced. The system of whole numbers possesses the following properties:

1. Closure of addition
2. Closure of multiplication
3. Commutativity of addition
4. Commutativity of multiplication
5. Associativity of addition
6. Associativity of multiplication
7. Distributivity of multiplication over addition
8. Additive identity
9. Multiplicative identity

After introducing the set of integers, a tenth property was gained, namely, the property of additive inverse. It asserted that every integer a has a unique additive inverse $-a$, such that $a + (-a) = 0$.

Having introduced the set of rational numbers, another property of multiplication is gained, which multiplication of integers does not have. Before stating this property, let's look at some examples.

Example 1 Given the rational number $\frac{2}{3}$, is there a rational number such that the product of the two numbers is equal to 1?

Answer Yes, this number is $\frac{3}{2}$ because $\frac{2}{3} \times \frac{3}{2} = 1$.

4.5 Rational Numbers

Example 2 Determine such a number for $\frac{5}{7}$.

Answer It is $\frac{7}{5}$ because $\frac{5}{7} \times \frac{7}{5} = 1$.

Example 3 Determine such a number for .3.

Answer It is $\frac{10}{3}$ because $.3 = \frac{3}{10}$ and $\frac{3}{10} \times \frac{10}{3} = 1$.

PROPERTY OF MULTIPLICATIVE INVERSE   For every nonzero rational number a/b there exists a unique *multiplicative inverse* b/a such that $a/b \cdot b/a = 1$. A multiplicative inverse is also called a *reciprocal*.

This property gives the set of rational numbers the closure property under division. Division of rational numbers has an analogy in the subtraction of integers. The latter was accomplished by resorting to addition. Similarly, division of rational numbers is accomplished by resorting to multiplication. This is illustrated below:

$$\frac{2}{3} \div \frac{5}{7} = \frac{\frac{2}{3}}{\frac{5}{7}} = \frac{\frac{2}{3} \times \frac{7}{5}}{\frac{5}{7} \times \frac{7}{5}} = \frac{\frac{2}{3} \times \frac{7}{5}}{1} = \frac{2}{3} \times \frac{7}{5}$$

The above shows that the quotient

$$\frac{2}{3} \div \frac{5}{7}$$

is the same as the product

$$\frac{2}{3} \times \frac{7}{5}$$

The conversion of division of two rational numbers to multiplication can be generalized to any two rational numbers as shown below:

$$\frac{a}{b} \div \frac{c}{d} = \frac{\frac{a}{b}}{\frac{c}{d}} = \frac{\frac{a}{b} \times \frac{d}{c}}{\frac{c}{d} \times \frac{d}{c}} = \frac{\frac{a}{b} \times \frac{d}{c}}{1} = \frac{a}{b} \times \frac{d}{c}$$

Thus, for all rational numbers a/b and c/d:

$$\frac{a}{b} \div \frac{c}{d} = \frac{a}{b} \times \frac{d}{c}$$

Since division is defined in the set of rational numbers, equations of the form $ax = b$ can now be solved. Here a and b are integers, $a \neq 0$. The

4 Whole Numbers, Integers, and Rationals

solution of the equation $ax = b$ is b/a. This can be verified by substituting b/a for x:

ax	b
$a\left(\dfrac{b}{a}\right)$	b
b	

Exercises 4.5

Solve each equation in the set of rational numbers:

1. $5x = 9$
2. $3x = 7$
3. $2x = 13$
4. $9x = 2$
5. $11x = 3$
6. $15x = 7$
7. Tarzan no longer swings through the trees, because of his ignorance of rational numbers. He had to grab hold of a vine at precisely the right moment, or he would plummet to the ground. All rational vines obey this basic law, "If the length of the vine is known, then its position is also known." Unfortunately, one day Tarzan ran into one of these rational vines, with the result that he spent quite some time in the Treetops Convalescent Hospital. The equation Tarzan *should* have applied to the rational vine is $3Q = 20H$, where Q is the position ($3Q$ = 3-dimensional) and H = height. Instead, Tarzan used his old equation, $3Q = 21H$. If the vine's height was 25 ft, find the difference in the two positions.
8. Make the necessary substitutions of integers for a and b in a/b to obtain the rational number 1. How many such substitutions are there?
9. State the property of the rational number 1 which is analogous to the property of the integer 1 for multiplication.
10. State the property of the rational number 0 which is analogous to the property of the integer 0 for addition.

Tell which statements are true and which are false:

11. $\frac{1}{2} = 3$
12. $\frac{4}{7} = \frac{2}{3}$
13. $\frac{5}{11} = \frac{30}{66}$
14. $\frac{2}{7} = \frac{7}{12}$
15. $\frac{1}{4} = \frac{8}{11}$
16. $\frac{9}{21} = \frac{3}{7}$

Find the sums:

17. $\frac{1}{2} + \frac{2}{3}$
18. $\frac{3}{7} + \frac{1}{3}$
19. $\frac{4}{5} + \frac{2}{3}$
20. $\frac{4}{7} + \frac{3}{2}$
21. $\frac{4}{3} + \frac{7}{2}$
22. $\frac{2}{9} + \frac{4}{5}$

Find the products:

23. $\frac{2}{3} \times \frac{4}{5}$
24. $\frac{1}{6} \times \frac{5}{3}$
25. $\frac{4}{3} \times \frac{1}{5}$
26. $\frac{2}{7} \times \frac{3}{5}$
27. $\frac{2}{5} \times \frac{1}{3}$
28. $\frac{4}{9} \times \frac{5}{3}$

4.6 Ordering Rational Numbers

29. In Big Brother's society of 1984, assuming an alias is virtually impossible because your original number is in storage deep in the computers. Of the following, the first number is the one a person claims is his, and the second is the number in computer storage. Which persons are trying to operate under aliases?
    a. 1/7, 3/21   b. 1/2, 33/64   c. 1/5, 600/3000

Name the property illustrated by each statement:

30. $\frac{4}{5} \times \frac{5}{4} = 1$
31. $\frac{1}{2} \times (\frac{1}{3} + \frac{1}{4}) = (\frac{1}{2} \times \frac{1}{3}) + (\frac{1}{2} \times \frac{1}{4})$
32. $\frac{2}{3} \div \frac{4}{7} = \frac{2}{3} \times \frac{7}{4}$
33. $\frac{2}{3} \div \frac{4}{7}$ is a rational number
34. $\frac{5}{6} \cdot 1 = \frac{5}{6}$
35. $\frac{4}{9} \times \frac{10}{11}$ is a rational number
36. $\frac{1}{2} + \frac{7}{8} = \frac{7}{8} + \frac{1}{2}$
37. $\frac{7}{8} + 0 = \frac{7}{8}$
38. $(\frac{4}{3} \times \frac{1}{2}) \times \frac{3}{5} = \frac{4}{3} \times (\frac{1}{2} \times \frac{3}{5})$

★39. State and prove the closure property of the set of rational numbers under multiplication.

★40. State and prove the associative property of multiplication of rational numbers.

★41. State and prove the associative property of addition of rational numbers.

★42. State and prove the distributive property of multiplication over addition of rational numbers.

43. Another riddle to test your powers of perception: what can you add in but close at the same time?

4.6 ORDERING RATIONAL NUMBERS

REVIEW

- $-3 < 0$, that is, -3 is negative
- $4 > 0$, that is, 4 is positive
- In $\frac{a}{b}$, a is the numerator and b is the denominator

OBJECTIVES

- State the rule for telling whether a rational number is positive or negative
- State the rules for comparing two rational numbers

A rational number involves two integers. Here are examples of combinations of positive and negative integers making up rational numbers:

$$\frac{-2}{-3} \qquad \frac{-2}{3} \qquad \frac{2}{-3} \qquad \frac{2}{3}$$

How do we tell which of these rational numbers are positive and which are negative? There is a simple rule for this.

$$\text{If } ab > 0, \text{ then } a/b > 0$$
$$\text{If } ab < 0, \text{ then } a/b < 0$$

Since $-2 \cdot (-3) = 6$ and $6 > 0$, $\frac{-2}{-3} > 0$ or positive.

Since $-2 \cdot 3 = -6$ and $-6 < 0$, $\frac{-2}{3} < 0$ or negative.

We shall write rational numbers the same way we wrote integers. To designate a positive rational number, we use no sign; for example, $\frac{2}{3}$ is positive. Thus, $\frac{-2}{-3} = \frac{2}{3}$. We use the minus sign to designate a negative rational number. For example, $-\frac{2}{3}$ is negative.

To compare two rational numbers, it is best to have the signs attached to integers. Here are some examples illustrating the procedure for comparing rational numbers:

$$\frac{2}{3} < \frac{1}{2} \text{ because } 2 \cdot 2 < 3 \cdot 1 \ (4 < 3)$$

$$\frac{-2}{3} < \frac{-1}{3} \text{ because } -2 \cdot 3 < 3 \cdot (-1) \ (-6 < -3)$$

$$\frac{-5}{7} < \frac{1}{4} \text{ because } -5 \cdot 4 < 7 \cdot 1 \ (-20 < 7)$$

Notice that in each example above, the denominator was a positive number. Every rational number can be so represented. For a positive number, both the numerator and denominator will be positive. For a negative number, the numerator will be negative and the denominator positive.

$$a/b < c/d \text{ if } ad < bc$$
b and d are positive integers

We know that $a/b < c/d$ means the same as $c/d > a/b$ and $ad < bc$ means the same as $bc > ad$.

$$c/d > a/b \text{ if } bc > ad$$
b and d are positive integers

Exercises 4.6

For each of the following rational numbers, give a name of the form a/b or $-\frac{a}{b}$, where a and b are positive integers:

4.6 Ordering Rational Numbers

1. $\dfrac{-2}{-5}$
2. $\dfrac{1}{-3}$
3. $\dfrac{-(-2)}{3}$
4. $\dfrac{-(-1)}{-5}$
5. $\dfrac{3}{-(-7)}$
6. $\dfrac{-3}{-7}$
7. $\dfrac{4}{-5}$
8. $\dfrac{-(-3)}{-7}$
9. $\dfrac{-2}{-(-5)}$
10. $\dfrac{-(-3)}{-(-4)}$

Tell which of the following are true and which are false:

11. $\dfrac{4}{9} = \dfrac{2}{3}$
12. $\dfrac{-3}{-4} > \dfrac{2}{3}$
13. $\dfrac{-5}{-9} = \dfrac{5}{9}$
14. $\dfrac{9}{18} = \dfrac{1}{2}$
15. $\dfrac{-1}{2} < 0$
16. $\dfrac{1}{2} > \dfrac{1}{3}$
17. $\dfrac{5}{8} > \dfrac{5}{7}$
18. $0 < \dfrac{1}{3}$
19. $\dfrac{4}{7} < \dfrac{-5}{2}$
20. $\dfrac{4}{-9} = \dfrac{-4}{9}$
21. $\dfrac{-5}{-3} = \dfrac{3}{5}$
22. $0 < \dfrac{-1}{3}$

23. To get into the maximum-security top-secret Nuclear Banana Factory, you have to keypunch your ID number onto a computerized card. If the number you punch corresponds to the master number in the computer, you are allowed to enter. Check the following numbers to see if they agree with the master numbers (in parentheses).

   a. $\dfrac{-101}{1000}$ $\left(\dfrac{-1010}{100000}\right)$

   b. $\dfrac{-72}{-73}$ $\left(\dfrac{144}{146}\right)$

   c. $\dfrac{-20}{34}$ $\left(\dfrac{80}{-136}\right)$

Give names for the following using a minus sign in front or no sign at all:

24. $\dfrac{-5}{6}$
25. $\dfrac{-3}{-5}$
26. $\dfrac{3}{-7}$
27. $\dfrac{-4}{-(-9)}$

4.7 DECIMALS

REVIEW

- An illustration of the division procedure:

$$\begin{array}{r} 1937, r16 \\ 26 \overline{)50378} \\ \underline{26} \\ 243 \\ \underline{234} \\ 97 \\ \underline{78} \\ 198 \\ \underline{182} \\ 16 \end{array}$$

OBJECTIVES

- Determine a decimal name for a rational number given in fractional form
- Define terminating decimal numerals
- Define repeating nonterminating decimal numerals
- Give a fractional name for a given decimal, terminating or nonterminating

Given a rational number in the form a/b, it is possible to determine a decimal name for this number. To accomplish this, division is performed as illustrated below.

Example 1 $\frac{1}{4} = .25$

$$\begin{array}{r} .25 \\ 4\overline{)1.00} \\ \underline{8} \\ 20 \\ \underline{20} \\ 0 \end{array}$$

Example 2 $-\frac{1}{5} = -.2$

$$\begin{array}{r} .2 \\ 5\overline{)1.0} \\ \underline{1\,0} \\ 0 \end{array}$$

Decimal numerals, such as .25 and $-.2$, are called *terminating decimal numerals*. These numerals were obtained by dividing one integer by another, and at some point the remainder 0 was reached. However, as the examples below show, not all rational numbers have names which are terminating decimal numerals.

Example 3 $\frac{1}{3} = .333\ldots$

$$\begin{array}{r} .333\ldots \\ 3\overline{)1.000} \\ \underline{9} \\ 10 \\ \underline{9} \\ 10 \\ \underline{9} \\ 1 \end{array}$$

4.7 Decimals

Instead of writing .333..., where the three dots indicate that the pattern (in this case, repetition of 3) continues on forever, $.\overline{3}$ will be written to indicate the same thing. Thus, $\frac{1}{3} = .\overline{3}$.

Example 4 $\quad \frac{1}{6} = .1\overline{6}$

```
        .166...
    6) 1.000
       6
       ─
        40
        36
        ──
         40
         36
         ──
          4
```

Such decimal numerals as $.\overline{3}$ and $.1\overline{6}$ are called *repeating nonterminating decimal numerals*.

When dividing one integer by another, one of two things happens: the quotient appears as either a terminating decimal numeral, or as a repeating nonterminating decimal numeral. Are there any other possibilities? Could it be that a decimal would continue on and on without any repeating pattern? To gain insight into the answer to this question, study the division of 1 by 7 below:

```
       .1 4 2 8 5 7 1 4...
    7) 1.0 0 0 0 0 0 0 0
       7
       ─
      ③0
       2 8
       ───
        ②0
        1 4
        ───
         ⑥0
         5 6
         ───
          ④0
          3 5
          ───
           ⑤0
           4 9
           ───
            ①0
             7
            ───
             ③0
```

It follows that $\frac{1}{7} = .\overline{142857}$. How do we know that this block of six digits will repeat in the same order? Observe that the last remainder, 3, was encountered previously. This means that from now on every remainder, which occurred previously, will keep repeating in the same order, thus forming a block of repeating digits.

4 Whole Numbers, Integers, and Rationals

Thus, a clue to the question raised above is obtained. When dividing an integer by 7, the only possible remainders are 0, 1, 2, 3, 4, 5, 6. If the remainder at any point should be 0, the division would stop, and a terminating decimal numeral would result. Otherwise, a repetition of the same remainder must occur in at most the sixth step, since there are only six possible nonzero remainders. Once the same remainder occurs for the second time, a continuous repetition of previous steps from that point on takes place.

Generally, if some integer is divided by an integer n and if no remainder is 0, then a repetition of a block of at most $n - 1$ digits takes place. The following can now be concluded.

PROPERTY OF DECIMAL NUMERALS FOR RATIONAL NUMBERS Every rational number has a decimal name which is either a terminating decimal numeral or a repeating nonterminating decimal numeral.

Consider now a reverse problem: given a decimal numeral, determine a name of the form a/b for the same number. If the decimal numeral is terminating, this is very easy to do as shown in the following two examples.

$$.37 = \frac{37}{100} \qquad .5063 = \frac{5063}{10000}$$

What about the case of a repeating nonterminating numeral? The examples below illustrate a procedure which leads to the names of the form $\frac{a}{b}$.

Example 1 $.\overline{5} = ?$
Let $x = .\overline{5}$; then $10x = 5.\overline{5}$.

$$\begin{aligned}10x &= 5.\overline{5} \\ x &= .\overline{5} \quad \text{Now subtract} \\ \hline 10x - x &= 5.\overline{5} - .\overline{5} \\ 9x &= 5 \\ x &= \tfrac{5}{9}\end{aligned}$$

Thus, $.\overline{5} = \tfrac{5}{9}$.

Example 2 $9.\overline{5} = ?$
We already know that $.\overline{5} = \tfrac{5}{9}$.

$$9.\overline{5} = 9 + .\overline{5}$$
$$= 9 + \tfrac{5}{9} = 9\tfrac{5}{9} = \tfrac{86}{9}$$

4.7 Decimals

Example 3 $.\overline{42} = ?$
Let $x = .\overline{42}$; then $100x = 42.\overline{42}$.

$$100x = 42.\overline{42}$$
$$x = .\overline{42}$$
$$\overline{99x = 42}$$
$$x = \tfrac{42}{99} \text{ or } \tfrac{14}{33}$$

Thus, $.\overline{42} = \tfrac{42}{99}$ or $\tfrac{14}{33}$.

Example 4 $.6\overline{225} = ?$
Let $x = .6\overline{225}$; then $100x = 62.\overline{25}$.

$$100x = 62.\overline{25}$$
$$x = .6\overline{225}$$
$$\overline{99x = 61.63}$$
$$x = \tfrac{61.63}{99} = \tfrac{6163}{9900}$$

Thus, $.6\overline{225} = \tfrac{6163}{9900}$.

A thoughtful examination of Examples 1–4 gives a clue as to how to decide whether to multiply by 10 or by 100 or by some other power of 10 in order, after subtracting, for a terminating decimal or a whole number to result. You may wish to do some more experimenting on your own to gain proficiency in this procedure.

Exercises 4.7

1. In an earlier section, you met two terrible diseases, longnumberitis and shortnumberitis. A derivative of these diseases (perhaps even more deadly) is fractionitis. Any math teacher afflicted with this disease cannot say any fractions. Convert the following fractions to decimals to help the poor teacher.
    a. $\tfrac{8}{9}$ b. $\tfrac{10}{25}$ c. $\tfrac{2}{10}$

Give a decimal numeral for each of the following rational numbers:

2. $\tfrac{4}{5}$ 3. $\tfrac{4}{25}$ 4. $\tfrac{9}{7}$ 5. $\tfrac{2}{15}$ 6. $\tfrac{3}{11}$
7. $\tfrac{5}{13}$ 8. $3\tfrac{2}{5}$ 9. $\tfrac{3}{8}$ 10. $\tfrac{3}{15}$ (*Hint:* simplify first.)

Using the procedure illustrated in the examples above, give a fractional numeral for each of the following:

11. $.\overline{12}$
12. $.1\overline{067}$
13. $.8\overline{8}$
14. $10.8\overline{8}$
15. $.1\overline{313}$
16. $.376\overline{565}$

17. The direct opposite of fractionitis is decimalitis, a disease which makes it impossible to say a decimal number. Help the afflicted by writing the following decimals as fractions.
   a. 3.142   b. 0.5789   c. $0.4\overline{33}$
★18. Write out an argument similar to the one given above showing that a decimal name for $\frac{1}{11}$, if it is a repeating nonterminating decimal numeral, could not have more than ten digits in the repeating block.

4.8 DENSITY OF RATIONAL NUMBERS

REVIEW

- The set of integers: $\{\ldots, -3, -2, -1, 0, 1, 2, 3, \ldots\}$
- The arithmetic mean of two numbers is found by multiplying $\frac{1}{2}$ by the sum of the two numbers; of three numbers by multiplying $\frac{1}{3}$ by the sum of the three numbers; and, in general, of n numbers by multiplying $1/n$ by the sum of the n numbers.
- The arithmetic mean of two numbers is the number which is half-way between the two numbers.

OBJECTIVES

- Define what is meant by discreteness of the set of integers
- Define density of the set of rational numbers
- Find rational numbers between two given rational numbers

Let us return for a moment to integers. Given an integer, say -5, what is the integer preceding it? It is -6. What is the integer following -5? It is -4. It's no problem to see that there is one integer just before it and one integer just after it. For this reason, we say that the set of integers is *discrete*.

The story about rational numbers is quite different. There is no rational number, for example, which just follows $\frac{1}{2}$. And there is more to it than that. For example, there is no integer between 5 and 6; but there are plenty of rational numbers between $\frac{1}{2}$ and $\frac{1}{3}$. As a matter of fact, there are plenty of rational numbers between any two rational numbers. Let us start with a modest claim though.

4.8 Density of Rational Numbers

DENSITY PROPERTY OF RATIONAL NUMBERS Given two rational numbers, there is a rational number between them.

This claim is very easy to verify. We can even tell one way such a number can be found: find the *arithmetic mean* (average) of the two given numbers.

Example 1 Given $\frac{4}{5}$ and $\frac{7}{9}$, we compute the arithmetic mean of $\frac{4}{5}$ and $\frac{7}{9}$, which is $\frac{1}{2}$ of the sum of the two given numbers:

$$\frac{1}{2}\left(\frac{4}{5} + \frac{7}{9}\right) = \frac{1}{2} \cdot \frac{36 + 35}{45}$$
$$= \frac{1}{2} \cdot \frac{71}{45}$$
$$= \frac{71}{90}$$

To verify that $\frac{71}{90}$ is indeed between $\frac{4}{5}$ and $\frac{7}{9}$, observe first that $\frac{4}{5} > \frac{7}{9}$ because $4 \times 9 > 5 \times 7$ ($36 > 35$). It is to be shown that $\frac{4}{5} > \frac{71}{90}$ and $\frac{71}{90} > \frac{7}{9}$.

$\frac{4}{5} > \frac{71}{90}$ because $4 \times 90 > 5 \times 71$ ($360 > 355$)
$\frac{71}{90} > \frac{7}{9}$ because $71 \times 9 > 90 \times 7$ ($639 > 630$)

Knowing that given any two rational numbers a and b there is a third rational number c between them, enables us to prove that there is an infinite number of rational numbers between any two rational numbers. The argument for this is as follows:

Given two rational numbers a and b such that $a < b$, there is another rational number c such that $a < c < b$. Now we have $a < c$; therefore there is a rational number d such that $a < d < c$. And the argument can continue on and on, leading to the conclusion that between any two rational numbers there is an infinite number of rational numbers.

A picture is a good way to show how the process of squeezing lots of rational numbers between two given rational numbers can go on and on. We show it for numbers $\frac{1}{2}$ and $\frac{1}{3}$.

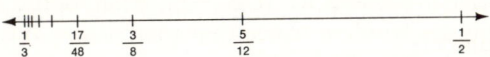

We see that $\frac{5}{12}$ is half-way between $\frac{1}{2}$ and $\frac{1}{3}$. It is the arithmetic mean of these two numbers. Next, $\frac{3}{8}$ is half-way between $\frac{1}{3}$ and $\frac{5}{12}$. Then, $\frac{17}{48}$ is half-way between $\frac{1}{3}$ and $\frac{3}{8}$. At this point, we got tired of computing arithmetic means and just marked some points half-way between the successive pairs of points. Then the intervals get so small that we have to give that up too. At this point, our imagination has to take over and convince us that this process can go on forever.

4 Whole Numbers, Integers, and Rationals

Exercises 4.8

For each of the following integers, give the integer preceding it and the one following it:

1. -8 2. 16 3. -49 4. 0
5. You are about to play a game that the FBI uses to test new agents' powers of perception. You will be given two integers. It is your job to find the integer that they surround.
   a. 1,000,000; 1,000,002   b. $-20, -18$ c. 0, 2

Determine the arithmetic mean of each of the following pairs of rational numbers:

6. $\frac{1}{3}, \frac{1}{4}$ 7. $\frac{3}{5}, \frac{7}{9}$ 8. $\frac{1}{8}, \frac{1}{9}$
9. The following pairs of numbers were heard to say, "Let's stay away from that number. He's mean." Given the two numbers that were speaking, exactly what number was so mean?
   a. 1/80, 1/81   b. 1/3, 1/5   c. 9/11, 10/15
10. In exercises 6–8, order each set of the three numbers (the given pair and their arithmetic mean) by writing a statement of the form: $a < b < c$.

For each of the following pairs of numbers a and b, compute $\frac{2a+b}{3}$:

11. $0, 9$ 12. $5, 6$ 13. $-2, 10$ 14. $8, 10$
15. In exercises 11–14, order each set of the three numbers by writing a statement of the form $x < y < z$.

For each of the following pairs of rational numbers a/b and c/d, compute $\frac{a+c}{b+d}$:

Example $\frac{1}{2}, \frac{1}{5}$

If $\frac{a}{b} = \frac{1}{2}$ and $\frac{c}{d} = \frac{1}{5}$, then $\frac{a+c}{b+d} = \frac{1+1}{2+5} = \frac{2}{7}$.

16. $\frac{3}{4}, \frac{1}{3}$ 17. $\frac{2}{5}, \frac{3}{7}$ 18. $\frac{3}{5}, \frac{1}{9}$ 19. $\frac{4}{3}, \frac{7}{8}$
20. In exercises 16–19, verify that each number you obtained is between the two given numbers.
21. That old mathematician and well-known pirate, Long (Integer) John Numberline, left his treasure buried on a number-line island. You know that its position was half-way between $\frac{5}{7}$ and $\frac{7}{9}$, but only one of the following choices will fit between the two numbers to help you find the treasure.
   a. $\frac{94}{126}$ b. $\frac{50}{63}$ c. $\frac{100}{126}$

CHAPTER 4 TEST

1. Explain the meaning of binary operation.

For each statement, tell what property it illustrates:

2. $-8 + 0 = -8$
3. $\frac{5}{6} \cdot \frac{6}{5} = 1$
4. $7 + (-7) = 0$
5. $(-5)(-7)$ is an integer
6. $7 + 12 = 12 + 7$

Using -3, 5, and -8, in this order, write an illustration of the:

7. Associative property of addition
8. Distributive property of multiplication over addition

Find the answers:

9. $12 + 37$
10. $11 + (-26)$
11. $-9 + (-26)$
12. $-12 + 37$
13. $-3 - (-15)$
14. $3 - (-1)$
15. $-2 - 17$
16. $8 - (-12)$
17. $5 \cdot 60$
18. $2 \cdot (-13)$
19. $-12 \cdot 7$
20. $-4 \cdot (-12)$

21. I withdraw $10 from my bank account and add $35. If after this I have $50, how much did I have to start with?

Tell which of the following are true and which are false:

22. $3 < -15$
23. $-12 > -100$
24. $0 > -16$
25. $-10 < 1$
26. $26 > 1$
27. $0 > 1$
28. $|-2| = |2|$
29. $|0| > |-10|$
30. $|0| > |2|$
31. $|-10| < |-1|$
32. $|3 + (-7)| = |3| + |-7|$
33. $|5 - (-2)| = |5| - |-2|$
34. $|-5 \cdot (-2)| = |-5| \cdot |-2|$

35. The maximum lifetime of a worm is given by the equation $7 + t = 90$, where t is in hours. The oldest worm, who attributed his long life to his feeding grounds beneath a brewery, said he was 85 hours old. Was he telling the truth?

Solve each equation:

36. $x + 5 = 27$
37. $x - 3 = 12$
38. $4 - x = -3$
39. $2x + 3 = 4$
40. $4x = -8$
41. $3x = -5$
42. $-2x + 1 = 1$
43. $3(x + 1) = -4\frac{1}{2}$
44. $x \div 2 = -8$
45. $x \div 3 = -1$
46. $\dfrac{x + 1}{2} = -\dfrac{3}{2}$
47. $\dfrac{2x - 1}{3} = 0$

Compute the answers:

48. $\frac{3}{4} + \frac{2}{5}$
49. $\frac{1}{2} + \frac{3}{7}$
50. $\frac{7}{6} + \frac{1}{5}$
51. $\frac{4}{7} - \frac{1}{2}$
52. $\frac{11}{8} - \frac{4}{3}$
53. $\frac{4}{11} - \frac{3}{10}$
54. $\frac{5}{6} \cdot \frac{1}{7}$
55. $\frac{2}{3} \div \frac{4}{5}$
56. $\frac{4}{7} \div \frac{4}{3}$

Tell which are true and which are false:

57. $\frac{5}{6} = \frac{7}{8}$
58. $-\frac{2}{5} = \frac{4}{-10}$
59. $\frac{1}{2} < -\frac{9}{10}$
60. $\frac{4}{5} > \frac{5}{6}$
61. $\frac{2}{3} < \frac{2}{5}$
62. $-\frac{2}{5} < 0$
63. $-\frac{4}{5} < -\frac{4}{7}$
64. $\frac{1}{100} > 0$
65. $-\frac{1}{2} < -\frac{1}{3}$

Give a decimal numeral for each of the following rational numbers:

66. $\frac{4}{25}$
67. $\frac{5}{6}$
68. $\frac{3}{7}$

69. You go into a small store in the country where the proprietor has not heard about decimals. You wish to buy several candy bars, and the manager asks for 2\frac{3}{8}$. You give him $2.40, but he refuses to accept it on the grounds that it is too little. Is he right?

Give a fractional numeral for each of the following:

70. .37
71. .0039
72. $.\overline{7}$
73. $.\overline{16}$

Compute the arithmetic mean of each of the following pairs of numbers:

74. $\frac{2}{3}, \frac{1}{5}$
75. $\frac{1}{2}, \frac{1}{7}$
76. $\frac{4}{7}, \frac{5}{9}$
77. $\frac{1}{6}, \frac{1}{10}$

78. For exercises 74–77, write statements of the form $a < b < c$ for each triple of numbers.

Tell what is meant by:

79. Discreteness of the set of integers
80. Density of the set of rational numbers
81. International Espionage spies are often very "discrete" about their operations. In the sets below, are the spies discrete?
    a. $A = \{7, 8, 9\}$
    b. $B = \{\frac{1}{2}, \frac{1}{3}, \frac{1}{4}\}$
    c. $C = \{\frac{1}{2}, 1, 2\}$

5

Numbers as Exponents

5.1 WHOLE NUMBER EXPONENTS

REVIEW

- The set of natural numbers: $\{1, 2, 3, 4, \ldots\}$
- The set of whole numbers: $\{0, 1, 2, 3, 4, \ldots\}$
- Example of a *product*: $3 \cdot 5$; in this product, 3 and 5 are the *factors*
- *Real numbers* will mean for now all the numbers you know such as $5, -\frac{1}{2}, \sqrt{2}$

OBJECTIVES

- Recognize exponents, bases, and powers
- Define the whole number powers
- Use the product of powers property

Numbers are used as exponents quite frequently:

5^2 means $5 \cdot 5$ *Read:* the square of five, or five squared, or the second power of five

5^3 means $5 \cdot 5 \cdot 5$ *Read:* the cube of five, or five cubed, or the third power of five

5^4 means $5 \cdot 5 \cdot 5 \cdot 5$ *Read:* five to the fourth power

5 Numbers as Exponents

From here on, we use the same form for naming powers:

5^5 five to the fifth power
5^6 five to the sixth power
5^n five to the nth power

In 4^2, the 2 is the *exponent* and 4 is the *base*, as shown below:

$4^2 \leftarrow$ Exponent
↑
Base

The entire symbol "4^2" is called a *power*. In this case, 4^2 is the second power of 4.

A variable can be used to show various powers. Assuming the set of real numbers as the replacement set, each of the following is true for all replacements of x:

$x^2 = x \cdot x$
$x^3 = x \cdot x \cdot x$
$x^4 = x \cdot x \cdot x \cdot x$
and so on

In general,

$$x^n = \underbrace{x \cdot x \cdot x \cdot \ldots \cdot x}$$

x used n times as a factor

The general statement above applies to n greater than 1, since it would make no sense to talk about a product of one or zero factors. In Chapter 1, in order to reason out what the 1 or 0 power of a number should be, we observed a pattern and decided that it should continue:

$5^4 = 5 \cdot 5 \cdot 5 \cdot 5 = 625$
$5^3 = 5 \cdot 5 \cdot 5 = 125$
$5^2 = 5 \cdot 5 = 25$

Each time we decreased the exponent by 1, we divided the power by 5. So, 5^1 should be 5, since $25 \div 5$ is 5. Similarly, 5^0 should be 1, since $5 \div 5$ is 1.

$5^1 = 5$
$5^0 = 1$

5.1 Whole Number Exponents

In general,

$x^1 = x$ for every number x
$x^0 = 1$ for every nonzero number x

We won't bother to deal with 0^0 — it's too tricky!

In Section 5.4 we will give another justification for the statement $x^0 = 1$.

Powers have certain properties that follow quite naturally from the meaning of exponents. An example will suggest these properties:

$$3^4 \times 3^5 = (3 \times 3 \times 3 \times 3) \times (3 \times 3 \times 3 \times 3 \times 3)$$
$$= 3 \times 3 \times 3 \times 3 \times 3 \times 3 \times 3 \times 3 \times 3$$
$$= 3^9$$

Thus, $3^4 \times 3^5 = 3^9$. Note that the sum of the exponents, $4 + 5$, is 9, and this is the exponent of the product of the two powers.

In general, for each real number x,

$$x^4 \cdot x^5 = (x \cdot x \cdot x \cdot x) \cdot (x \cdot x \cdot x \cdot x \cdot x)$$
$$= x \cdot x \cdot x \cdot x \cdot x \cdot x \cdot x \cdot x \cdot x$$
$$= x^9$$

Generalizing from this example, we can state the following important property:

THE PRODUCT OF POWERS PROPERTY For all natural numbers m and n,

$$x^m \cdot x^n = x^{m+n}$$

The property above is true for all real number replacements of x.

The product of powers property can be extended to more than two powers. This is illustrated for the case of three powers.

$$x^3 \cdot x^5 \cdot x^7 = (x^3 \cdot x^5) \cdot x^7$$
$$= x^8 \cdot x^7$$
$$= x^{15}$$

That is, $x^3 \cdot x^5 \cdot x^7 = x^{3+5+7}$. In general,

$$x^{a_1} \cdot x^{a_2} \cdot \ldots \cdot x^{a_n} = x^{a_1 + a_2 + \cdots + a_n}$$

Read: a_1 as "a subscript one," a_2 as "a subscript two," and a_n as "a subscript n."

5   Numbers as Exponents

It is important to notice that the product of powers property applies to cases of the same base. It does not apply to such cases as $2^3 \times 3^4$, since the bases of these numbers are not the same.

When using negative numbers for bases, an ambiguous situation may arise. For example, what is -3^2 equal to? Is it $(-3)(-3)$, or is it $-(3 \cdot 3)$? That the two are not the same can be seen from the following:

$$(-3)(-3) = 9$$
$$-(3 \cdot 3) = -9$$

To assign a unique meaning to -3^2, an agreement is made that it is equal to $-(3 \cdot 3)$ or -9. In general,

$$-x^n = -(\underbrace{x \cdot x \cdot x \cdot \ldots \cdot x})$$

x used n times as a factor.

Thus, raising to a power comes first; taking the additive inverse comes second.

There may also be a question as to the meaning of an expression where multiplication and raising to a power are involved. For example, what is the meaning of $3 \cdot 4^2$? Is it $3 \cdot (4 \cdot 4)$, or is it $(3 \cdot 4) \cdot (3 \cdot 4)$? That the two are not the same can be seen from the following:

$$3 \cdot (4 \cdot 4) = 3 \cdot 16 = 48$$
$$(3 \cdot 4) \cdot (3 \cdot 4) = 12 \cdot 12 = 144$$

It will be assumed that $3 \cdot 4^2$ means $3 \cdot (4^2)$, which is equal to 48. In general,

$$x \cdot y^n = x \cdot (y^n)$$

Thus, raising to a power comes first; multiplying by another factor comes second.

Exercises 5.1

For each of the following give the simplest name involving no exponents:

Examples $(-3)^2 = (-3) \cdot (-3) = 9$
$-3^2 = -(3 \cdot 3) = -9$
$2 \cdot 5^2 = 2 \cdot 25 = 50$

5.1 Whole Number Exponents

1. 5^3
2. $(-5)^3$
3. -5^3
4. $3 \cdot 6^2$
5. $(3 \cdot 6)^2$
6. -4^3
7. $3 \cdot (-2)^2$
8. $3 \cdot (-2)^3$
9. $10 \times .1^2$
10. $-4 \cdot (\tfrac{1}{2})^2$
11. $(-4 \cdot \tfrac{1}{2})^2$
12. $[-4 \cdot (-\tfrac{1}{2})^2]$
13. We have a square sheet of paper $2'' \times 2''$. Find the area of this piece of paper
    a. initially   b. if its dimensions are doubled
    c. if its dimensions are tripled
14. The same piece of paper is now made into a cube. Repeat exercise 13 for a volume of a cube of side $2''$.
15. Based upon your results from exercises 13 and 14, can you develop a relationship among the dimensions of a box, its area, and its volume?

Find the answers:

16. $(\tfrac{1}{2})^3$
17. $(\tfrac{1}{3})^2$
18. $(.1)^3$
19. $(\tfrac{1}{2})^0$
20. $(-\tfrac{1}{2})^0$
21. $(-\tfrac{1}{2})^2$
22. $(-\tfrac{1}{2})^3$
23. $(-.1)^3$
24. $(-2)^4$
25. $(-2)^5$
26. $(-2)^1$
27. $(-2)^0$
28. 0^5
29. 0^{13}
30. 0^1

For each of the following, give a simpler equivalent expression:

Examples $x^3 x^7 = x^{3+7} = x^{10}$
$xy^2 x^3 y = (x^1 x^3)(y^2 y^1) = x^4 y^3$

31. $x^4 x^8$
32. aa^3
33. $mm^2 n$
34. $ab^2 ab$
35. $x^2 yz^3 z^7$
36. $(a+b)^2(a+b)^3$

37. The space program is about to launch an Earth Resources Satellite weighing 6×10^2 lb. The velocity that the rocket must reach to get into orbit is given by the equation $v = w^2 - 300{,}000$, where w is in lb and v is in ft/sec. Just as you launch the craft, all the computers blow fuses. You must quickly calculate whether or not you must jettison any weight from the craft. If you know the spacecraft is traveling at 7×10^4 ft/sec, will it achieve orbit?
38. If a power has a positive number base and an odd number exponent, is the power a positive or a negative number?
39. If a power has a positive number base and an even number exponent, is the power a positive or a negative number?
40. If a power has a negative number base and an odd number exponent, is the power a positive or a negative number?
41. If a power has a negative number base and an even number exponent, is the power a positive or a negative number?

By merely looking at each of the following, tell whether it names a positive or a negative number. Do not compute the answers.

42. 34^8
43. $(-35)^4$
44. $(-35)^5$
45. $(-1)^{100}$
46. $(-1)^{147}$
47. $(-1)^{10}(-1)^{11}$
48. $(-1)^{12}(-1)^{20}$
49. $(-1)^5(-1)^{15}(-1)^{25}$
50. $(-1)^5(-1)^{15}(-1)^{30}$

51. Biologists have long known of chromosome defects in animals. One rare animal, the hippoglotarhino, has a rare birth defect denoted by a $-x$ sign. If the number of generations is indicated by a power (for example, second generation $= (-x)^2$), tell which of the following generations have birth defects:
   a. 4th generation   b. 11th generation   c. 100th generation

★52. If a real number x, which is between 0 and 1, is raised to a natural number power greater than 1, is the answer greater than, equal to, or less than x?

★53. If a real number x, which is greater than 1, is raised to a natural number power greater than 1, is the answer greater than, equal to, or less than x?

★54. What is 0 raised to any natural number power equal to?

★55. If a real number x, which is between 0 and -1, is raised to an even natural number power, is the answer greater than, equal to, or less than x? Give two examples of this case.

★56. If a real number x, which is between 0 and -1, is raised to an odd natural number power, is the answer greater than, equal to, or less than x? Give two examples of this case.

★57. If a real number x, which is less than -1, is raised to an even natural number power, is the answer greater than, equal to, or less than x? Give two examples of this case.

★58. If a real number x, which is less than -1, is raised to an odd natural number power, is the answer greater than, equal to, or less than x? Give two examples of this case.

★59. Give an argument why $(-x)^2(-x)^6 = x^8$ is true for all real number replacements of x.

★60. Give one specific case to show that the following is not true for all real number replacements of x: $-x^m(-x)^n = x^{m+n}$.

★61. Give an argument why $-x^2(-x^6) = x^8$ is true for all real number replacements of x.

★62. Are there any real number replacements for x and y in $x^m \cdot y^n = y^{m+n}$ which will result in a true statement? Give one case of such replacements.

5.2 THE POWER OF A POWER AND THE POWER OF A PRODUCT

REVIEW

- Commutative property of multiplication:

  $xy = yx$

OBJECTIVES

- Recognize and use the power of a power property
- Prove the theorem:

  $(x^m)^n = (x^n)^m$

- Recognize and use the power of a product property

A few examples will illustrate another property of powers:

$(4^2)^3 = 4^2 \cdot 4^2 \cdot 4^2$
$ = 4^6$
$ = 4^{2 \cdot 3}$

$(7^2)^4 = 7^2 \cdot 7^2 \cdot 7^2 \cdot 7^2$
$ = 7^8$
$ = 7^{2 \cdot 4}$

$(3^5)^2 = 3^5 \cdot 3^5$
$ = 3^{10}$
$ = 3^{5 \cdot 2}$

THE POWER OF A POWER PROPERTY For all whole numbers m and n,

$(x^m)^n = x^{mn}$

A theorem, which is a direct consequence of this property, will now be proved.

Theorem 5.1 For all whole numbers m and n, $(x^m)^n = (x^n)^m$.

Proof

$(x^m)^n = x^{mn}$	Power of a power property
$ = x^{nm}$	Commutative property of multiplication
$ = (x^n)^m$	Power of a power property

In Theorem 5.1, x is any real number. In general, if we don't specify the set for a variable, we mean all real numbers.

5 Numbers as Exponents

Another set of examples reveals the next property of powers:

$$(5 \cdot 6)^2 = (5 \cdot 6) \cdot (5 \cdot 6)$$
$$= 5^2 \cdot 6^2$$

$$(2 \cdot 7)^3 = (2 \cdot 7) \cdot (2 \cdot 7) \cdot (2 \cdot 7)$$
$$= 2^3 \cdot 7^3$$

$$(8 \cdot 3)^4 = (8 \cdot 3) \cdot (8 \cdot 3) \cdot (8 \cdot 3) \cdot (8 \cdot 3)$$
$$= 8^4 \cdot 3^4$$

THE POWER OF A PRODUCT PROPERTY   For each natural number n,

$$(xy)^n = x^n y^n$$

The names given the properties above are suggestive of the nature of the properties. For example, the power of a power suggests $(x^m)^n$, that is, a power raised to a power. Whereas, the power of a product suggests that a product is raised to a power. When analyzing examples, you should think of these names in this way.

Exercises 5.2

Using the power of a power property, give simpler equivalent expressions which have only one exponent:

Examples $(x^3)^4 = x^{3 \cdot 4} = x^{12}$
$[(-3)^2]^3 = (-3)^{2 \cdot 3} = (-3)^6 = 3^6$
$[(-3)^3]^5 = (-3)^{3 \cdot 5} = (-3)^{15}$
$[(-x)^2]^3 = (-x)^{2 \cdot 3} = (-x)^6 = x^6$
$[(-x)^3]^5 = (-x)^{3 \cdot 5} = (-x)^{15}$

1. $(y^2)^6$
2. $[(-a)^2]^4$
3. $[(-m)^3]^6$
4. $[(-n)^5]^7$
5. $[(-2)^4]^2$
6. $[(-5)^6]^3$
7. $[(-5)^5]^3$
8. $[(-4)^1]^3$
9. $[(-4)^1]^4$
10. $[(-.2)^3]^9$
11. $[(-.7)^9]^4$
12. $[(-.8)^3]^1$
13. You can tell by the various diseases that math teachers get that teaching math can be a very hazardous profession. One more disease must be added to the list—long-exponentitis. It prevents the individual from saying exponents larger than 6. Arrange the following numbers so someone with long-exponentitis can say them.
    a. x^{12} b. y^8 c. x^9 d. q^{10}

5.2 The Power of a Power and the Power of a Product

14. You guessed it! The exact opposite of long-exponentitis also exists—short-exponentitis. This is the inability to say an exponent shorter than 6. Rearrange the following expressions so that someone with short-exponentitis can say them.
    a. $(x^2)^3$ b. $(y^5)^5$ c. $(z^2)^5$ d. $(q^3)^4$

By merely looking at each of the following, tell whether it names a positive or a negative number. Do not compute the answers.

15. $[(-1)^2]^3$ 16. $[(-2)^3]^4$ 17. $[(-2)^3]^3$
18. $[(-5)^1]^4$ 19. $[(-5)^1]^5$ 20. $[-(-2)^5]^7$
21. Explain why $-(5)^2 \neq (-5)^2$, but $-(5)^3 = (-5)^3$.
22. Describe all possible cases of $(-x)^n$, where x is a real number and n is a whole number.

For each of the following, compute the answers in each of two ways: (1) Multiply first; then raise to a power. (2) Apply the power of a product property.

    Example $[(-2) \times 4]^3$
    (1) $[(-2) \times 4]^3 = [-8]^3 = -512$
    (2) $[(-2) \times 4]^3 = (-2)^3 \times 4^3 = -8 \times 64 = -512$

23. $(3 \times 2)^2$ 24. $[(-1) \times 3]^2$ 25. $[(-1) \times 3]^3$
26. $[(-1) \times (-2)]^2$ 27. $[(-1) \times (-2)]^3$ 28. $[(-1) \times (-2) \times (-3)]^3$
29. Only mathematicians are allowed into the top-secret Math Research Establishment. The computer in the guardhouse flashes an expression on the screen, and each person who wants to enter must respond with an equivalent expression. Can you gain entrance faced with the following expressions?
    a. $(x^3)^7$ b. x^{81} c. $(y^2)^{10}$

Tell which of the following are true for all real number replacements of the variables:

30. $4x^2 = 16x^2$ 31. $(4x)^2 = 16x^2$ 32. $-4x^2 = 16x^2$
33. $(-4x)^2 = 16x^2$ 34. $-(4x)^2 = -16x^2$ 35. $-(4x)^2 = 16x^2$
36. $(-4x)^3 = -64x^3$ 37. $-(4x)^3 = -64x^3$ 38. $4(xy)^2 = 4x^2y^2$
39. $(4xy)^2 = 16x^2y^2$ 40. $(-4xy)^2 = -16x^2y^2$ 41. $-4(xy)^2 = -16x^2y^2$
42. Give a case of a replacement where $(xy)^m = x^m(-y)^m$.
43. Give a case of a replacement where $(xy)^m = (-x)^m(-y)^m$.

Tell which of the following are true and which are false:

44. $[3(-7)]^2 = 3 \cdot 7^2$ 45. $[(-5)(-4)]^2 = (-5)(-4)^2$
46. $[3(-7)]^3 = 3^3 \cdot 7^3$ 47. $[7(-2)]^3 = 7^3(-2)^3$
48. $[(-3)(-2)]^5 = (-3)^5(-2)^5$ 49. $[(-3)(-2)]^5 = -(3^5 \cdot 2^5)$
50. $[(-3)(-2)]^3 = 3^3 \cdot 2^3$ 51. $[(-3)(-2)]^6 = 3^6 \cdot 2^6$

5 Numbers as Exponents

Simplify, using the appropriate properties of powers:

Examples $[(-a)b]^3 = (-a)^3 b^3 = -a^3 b^3$
$[(-1)a]^2 = (-1)^2 a^2 = 1 \cdot a^2 = a^2$
$[(-1)a]^3 = (-1)^3 a^3 = -1 \cdot a^3 = -a^3$

52. $(2x)^2$
53. $(3xy)^3$
54. $[(-2)x]^2$
55. $[(-2)x]^3$
56. $[(-1)(-z)]^2$
57. $[(-1)(-x)]^3$
58. $(2^3 xy^2)^3$
59. $[xy(-a)]^2$
60. $[(-x)(-y)(-z)]^3$

5.3 THE POWER OF A QUOTIENT AND THE QUOTIENT OF POWERS

REVIEW

- An example of a power of a quotient:

  $(\frac{4}{5})^7$

- An example of a quotient of powers:

  $\dfrac{2^5}{7^3}$

OBJECTIVES

- Recognize and use the power of a quotient property
- Recognize and use the patterns involved in quotients of powers

Using the meaning of a whole number exponent applied to a quotient of two numbers will suggest another property of exponents:

$$\left(\frac{4}{7}\right)^3 = \frac{4}{7} \cdot \frac{4}{7} \cdot \frac{4}{7}$$
$$= \frac{4 \cdot 4 \cdot 4}{7 \cdot 7 \cdot 7}$$
$$= \frac{4^3}{7^3}$$

In general,

$$\left(\frac{x}{y}\right)^n = \underbrace{\frac{x}{y} \cdot \frac{x}{y} \cdot \ldots \cdot \frac{x}{y}}_{n \text{ factors}} = \underbrace{\frac{x \cdot x \cdot \ldots \cdot x}{y \cdot y \cdot \ldots \cdot y}}_{n \text{ factors}} = \frac{x^n}{y^n}$$

5.3 The Power of a Quotient and the Quotient of Powers

This is stated as the next property.

THE POWER OF A QUOTIENT PROPERTY For each whole number n,

$$\left(\frac{x}{y}\right)^n = \frac{x^n}{y^n} \quad [y \neq 0]$$

Since division by 0 is undefined, it is stated that $y \neq 0$.

Note that the above property deals with a power of a quotient. We next consider a quotient of powers:

$$\frac{3^6}{3^4} = \frac{3 \cdot 3 \cdot 3 \cdot 3 \cdot 3 \cdot 3}{3 \cdot 3 \cdot 3 \cdot 3}$$
$$= 3 \cdot 3$$
$$= 3^2$$

Notice that the exponent in the result is 2, which is $6 - 4$. Each of the following examples illustrates the same pattern:

$$\frac{2^7}{2^3} = 2^{7-3} \qquad \frac{7^5}{7^2} = 7^{5-2} \qquad \frac{8^4}{8} = \frac{8^4}{8^1}$$
$$= 2^4 \qquad \qquad = 7^3 \qquad \qquad = 8^{4-1}$$
$$= 8^3$$

The same pattern can be illustrated by the use of a variable. For example:

$$\frac{x^5}{x^2} = \frac{x \cdot x \cdot x \cdot x \cdot x}{x \cdot x} = x \cdot x \cdot x = x^3 \, (x \neq 0)$$

Thus,

$$\frac{x^5}{x^2} = x^{5-2} \, (x \neq 0)$$

The examples above were chosen in such a manner that the exponent in the numerator in each case is greater than the exponent in the denominator. To generalize, it can be stated that

If $b > a$, then $\dfrac{x^b}{x^a} = x^{b-a}$

But what about the cases where

$b = a$ or $b < a$?

5 Numbers as Exponents

These two cases are considered in detail next. Each of the following examples illustrates the case where $b = a$:

$$\frac{3^2}{3^2} \qquad \frac{4^5}{4^5} \qquad \frac{(-5)^2}{(-5)^2}$$

The first example can be computed as follows:

$$\frac{3^2}{3^2} = \frac{3 \times 3}{3 \times 3} = 1$$

Another way to think of each case above is that it is a quotient of a number and itself. Therefore, each quotient is equal to 1. Thus,

$$\frac{4^5}{4^5} = 1 \qquad \frac{(-5)^2}{(-5)^2} = 1$$

To generalize,

If $b = a$, then $\dfrac{x^b}{x^a} = \dfrac{x^a}{x^a} = 1 \quad (x \neq 0)$

Now consider some examples where $b < a$:

$$\frac{3^2}{3^6} = \frac{3 \times 3}{3 \times 3 \times 3 \times 3 \times 3 \times 3} = \frac{1}{3^4}$$

Another way of looking at the above is as follows:

$$\frac{3^2}{3^6} = \frac{1}{3^{6-2}} \quad \text{since} \quad \frac{1}{3^{6-2}} = \frac{1}{3^4}$$

Computing in the same manner, it can be shown that each of the following is true:

$$\frac{5^3}{5^7} = \frac{1}{5^{7-3}} = \frac{1}{5^4}$$

$$\frac{(-4)^5}{(-4)^8} = \frac{1}{(-4)^{8-5}} = \frac{1}{(-4)^3}$$

To generalize,

If $b < a$, then $\dfrac{x^b}{x^a} = \dfrac{1}{x^{a-b}}$

5.3 The Power of a Quotient and the Quotient of Powers

Three cases of the quotient of powers just considered were:

CASE 1 If $b > a$, then $\dfrac{x^b}{x^a} = x^{b-a}$

CASE 2 If $b = a$, then $\dfrac{x^b}{x^a} = 1$

CASE 3 If $b < a$, then $\dfrac{x^b}{x^a} = \dfrac{1}{x^{a-b}}$

The above are true for each real number x ($x \neq 0$) and for all whole numbers a and b, with the restrictions specified in each case.

Exercises 5.3

For each of the following give an equivalent expression using the power of a quotient property. Simplify whenever possible.

1. $\left(\dfrac{x}{y}\right)^3$ 2. $\left(\dfrac{-x}{y}\right)^4$ 3. $\left(\dfrac{-x}{y}\right)^5$

4. $\left(\dfrac{-3a}{s}\right)^4$ 5. $\left(\dfrac{-4m}{n}\right)^3$ 6. $\left[\dfrac{(-x)(-y)}{2}\right]^3$

7. $\left[\dfrac{(-2x)(-y)}{3}\right]^4$ 8. $\left[\dfrac{(-a)(-b)}{(-x)(-y)}\right]^3$ 9. $\left[\dfrac{(-e)(-d)}{(-x)(-y)}\right]^2$

10. In the last section, we met long-exponentitis, an inability to say an exponent larger than 6. Using your knowledge of quotients, help someone afflicted with this disease to say the following.
    a. x^{24}/x^{20} b. x^{225}/x^{220} c. y^{20}/y^{15}
11. With your knowledge of quotients, propose an effective treatment for short-exponentitis for the following numbers:
    a. x^2 b. y^5 c. z^4
12. Explain why $\left(\dfrac{-3}{5}\right)^6 = \left(\dfrac{3}{5}\right)^6$ is true, but $\left(\dfrac{-3}{5}\right)^7 = \left(\dfrac{3}{5}\right)^7$ is false.
13. Specify for what real number replacements of x and y and natural number replacements of n, $\left(\dfrac{-x}{y}\right)^n = \left(\dfrac{x}{y}\right)^n$ is true.
14. Specify for what real number replacements of x and y and whole number replacements of n, $\left(\dfrac{-x}{-y}\right)^n = \left(\dfrac{x}{y}\right)^n$ is true.

Using the patterns for quotients of powers, stated as Cases 1, 2, and 3 in the text, give an equivalent expression for each of the following:

15. $\dfrac{x^4}{x^2}$

16. $\dfrac{a^6}{a}$

17. $\dfrac{x^4 y^7}{x^2 y^2}$

18. $\dfrac{26 x^3 y^2}{13 x^2 y^2}$

19. $\dfrac{-25 m^4 n^3}{5 m^4 n^3}$

20. $\dfrac{a^2 b^3 c^4}{ab^2 c^4}$

21. $\dfrac{(a+b)^5}{(a+b)^2}$

22. $\dfrac{x^3(x+y)^4}{x(x+y)^7}$

23. $\dfrac{(x-y)^2}{x-y}$

24. $\dfrac{-14 a^2 (m+n)^3}{7a(m+n)^5}$

Tell which of the following are true and which are false:

25. $\left(\dfrac{1}{-2}\right)^3 = -\dfrac{1}{8}$

26. $\left(\dfrac{1}{-2}\right)^2 = -\dfrac{1}{4}$

27. $\left(\dfrac{-2}{-3}\right)^3 = \left(-\dfrac{2}{3}\right)^3$

28. $\left(-\dfrac{5}{6}\right) = \left(\dfrac{5}{6}\right)^2$

29. $\left(-\dfrac{5}{6}\right)^3 = \left(\dfrac{5}{6}\right)^3$

30. $\left(\dfrac{-3}{4}\right)^2 = \left(\dfrac{3}{4}\right)^2$

31. $\left(\dfrac{-3}{4}\right)^3 = \left(\dfrac{3}{4}\right)^3$

32. $\left[\dfrac{1}{(-2)(-5)}\right]^3 > 0$

33. $\left[\dfrac{1}{(-2)(-3)(-4)}\right]^3 > 0$

34. $\dfrac{2^3}{2^7} = 2^{7-3}$

35. $\left(\dfrac{0}{4}\right)^3 < 0$

36. $\left(\dfrac{2}{-3}\right)^4 = \left(\dfrac{2}{3}\right)^4$

37. $\left(\dfrac{2}{-3}\right)^5 = \left(\dfrac{2}{3}\right)^5$

38. $\dfrac{1}{2^4} = \dfrac{1}{(-2)^4}$

39. To gain admission to the top-security Quotient Division of the Research Establishment, you must again prove yourself. The computer flashes quotient expressions on the screen, and you must provide equivalent versions. When faced with the following expressions, can you gain entrance?

   a. $\dfrac{x^{-3}}{x^{-2}}$ b. $\dfrac{x^4}{x^2}$ c. $\dfrac{x^9 y^8}{x^3 y^2}$

5.4 INTEGERS AS EXPONENTS

REVIEW

- The set of integers: $\{\ldots, -3, -2, -1, 0, 1, 2, 3, \ldots\}$
- Replacing m in $-m$ by -2 gives the following: $-(-2) = 2$

OBJECTIVES

- Justify $x^0 = 1$ for every non-zero number x in a second way
- State and use the property of integer exponents
- State and use the quotient of powers property

5.4 Integers as Exponents

In considering the quotient of two powers $\frac{x^b}{x^a}$, a separate rule was established for each of the following cases: $b > a$, $b = a$, $b < a$. It would be more convenient to have one rule which would apply to all three of these cases.

If the rule for Case 1

$$\text{If } b > a, \text{ then } \frac{x^b}{x^a} = x^{b-a}$$

is applied to Case 2, where $b = a$, the following result is obtained:

$$\frac{x^b}{x^a} = \frac{x^a}{x^a} = x^{a-a} = x^0 \ (x \neq 0)$$

We have already agreed that for each real number x:

$$x^0 = 1 (x \neq 0)$$

We now have a second reason why this is a logical choice: a number divided by itself is equal to 1.

It follows that, whenever $b > a$ or $b = a$:

$$\frac{x^b}{x^a} = x^{b-a}$$

Thus, Cases 1 and 2 of the quotient of powers are replaced with one case.

We now examine Case 3:

$$\frac{x^b}{x^a}, \text{ where } b < a$$

If the rule for Case 1 is applied to this situation, the following result is obtained:

$$\frac{x^b}{x^a} = x^{b-a}$$

Since in this case $b < a$, $b - a$ does not exist in the set of whole numbers. Considering a and b to be integers, $b < a$ tells that $b - a$ is a negative integer, as shown in the following example:

$$\frac{x^2}{x^5} = x^{2-5} = x^{-3}$$

The last expression contains a negative exponent. The meaning of negative exponents is not yet known to us. We can develop it logically.

5 Numbers as Exponents

Observe:

$$\frac{x^2}{x^5} = x^{2-5}$$
  According to the rule of Case 1
$$= x^{-3}$$

$$\frac{x^2}{x^5} = \frac{1}{x^{5-2}}$$
  According to the rule of Case 3
$$= \frac{1}{x^3}$$

It follows that $x^{-3} = 1/x^3$.

We accept the following property of integer exponents:

$$x^{-m} = \frac{1}{x^m} \quad \text{for each integer } m$$

Now we can state one property to cover all cases of quotients of two powers.

THE QUOTIENT OF POWERS PROPERTY For all integers a and b,

$$\frac{x^b}{x^a} = x^{b-a} \quad (x \neq 0)$$

It is easy to observe that all of the properties that have been developed for whole number exponents also hold for integer exponents. The following examples illustrate this.

Example 1 The product of powers property applied to integer exponents:

$$x^3 \cdot x^{-7} = x^{3+(-7)} = x^{-4}$$

This can be verified by replacing x^{-7} by $1/x^7$.

$$x^3 \cdot \frac{1}{x^7} = \frac{x^3}{x^7} = \frac{1}{x^4}$$

And, by the property of integer exponents, $1/x^4 = x^{-4}$.

Example 2 The power of a power property applied to integer exponents:

$$(x^{-2})^3 = x^{-2 \cdot 3} = x^{-6}$$

Replacing x^{-2} by $1/x^2$, the following is obtained:

5.4 Integers as Exponents

$$(x^{-2})^3 = \left(\frac{1}{x^2}\right)^3 = \frac{1}{(x^2)^3} = \frac{1}{x^6}$$

And, by the property of integer exponents, $1/x^6 = x^{-6}$.

To see that the power of a power property applies to the exponent 0 as well, consider the following example:

$$(x^0)^{-3} = x^{0 \cdot (-3)} = x^0 = 1$$

Using the definition of a negative integer exponent, the same result is obtained:

$$(x^0)^{-3} = \frac{1}{(x^0)^3} = \frac{1}{1^3} = \frac{1}{1} = 1$$

Thus, the new definition of an integer exponent and the quotient of powers property are consistent with all of our previous definitions and properties. Defining division as

$$\frac{x}{y} = x \cdot \frac{1}{y} \quad (y \neq 0)$$

leads to some other patterns involving negative integer exponents. One of these patterns is displayed in the following examples.

$$3 \cdot 4^{-2} = 3 \cdot \frac{1}{4^2} = \frac{3}{4^2}$$

$$5 \cdot 3^{-6} = 5 \cdot \frac{1}{3^6} = \frac{5}{3^6}$$

$$-7 \cdot 6^{-8} = -7 \cdot \frac{1}{6^8} = \frac{-7}{6^8}$$

To generalize:

$$x \cdot y^{-m} = x \cdot \frac{1}{y^m}$$

$$= \frac{x}{y^m} \quad (y \neq 0)$$

Another pattern that can be easily derived is the following:

$$x^{-m} \cdot y^{-m} = \frac{1}{x^m \cdot y^m}$$

5 Numbers as Exponents

Exercises 5.4

For each of the following, give a simplified equivalent expression containing only positive exponents:

$$\text{Examples} \quad \frac{x^{-3}x^7}{x^6} = \frac{x^{-3+7}}{x^6} = \frac{x^4}{x^6} = \frac{x^4}{x^4 \cdot x^2} = \frac{1}{x^2}$$

$$\frac{(x+y)^{-2}(x+y)^{-3}}{(x+y)^{-7}} = \frac{(x+y)^{-2+(-3)}}{(x+y)^{-7}}$$
$$= \frac{(x+y)^{-5}}{(x+y)^{-7}}$$
$$= (x+y)^{-5-(-7)}$$
$$= (x+y)^2$$

1. $\dfrac{x^5 x^2}{x^3}$
2. $\dfrac{xx^8}{x^{11}}$
3. $\dfrac{(a+b)^{10}}{(a+b)^5}$
4. $\dfrac{a^{-2}a^{-4}}{a^{-8}}$
5. $\dfrac{xy^2 x^{-4}}{y^{-2} x^4}$
6. $\dfrac{(a-b)^{-3}}{(a-b)^{-4}}$
7. $\dfrac{m^3 n^{-2}}{mn^{-3} m^2}$
8. $[(x+y)^{-2}]^3$
9. $(x^{-6})^0 (x^2)^{-4}$
10. $\dfrac{-12(a+x)^{-2}}{6(a+x)^6}$
11. $(xy)^{-2} x^2 y^{-4}$
12. $\dfrac{(a+b+c)^{-3}}{(a+b+c)^{-4}}$

13. We have met many diseases during our travels, but perhaps none is so strange as negative-exponentitis. This disease allows only negative exponents to be presented. The unfortunate victim cannot say any positive exponents. Help these poor people by arranging the following exponents as negative exponents.
 a. $1/x^7$
 b. x^7/x^{20}
 c. z^{21}/z^{22}
 d. $(x^{10} \cdot x^3)/(x^7 \cdot x^{10})$
 e. $\dfrac{(x+y)^2 (x+y)^3}{(x+y)^7}$

14. Verify that $x^{-m} = 1/x^m$ applies to the case $m = 0$.
15. There exists a galaxy several million light years away from our own. This distant galaxy is a twin of ours, except that all negative exponents are written $1/x^n$. It is your job to show the inhabitants of this galaxy that $x^{-n} = 1/x^n$. Verify this equality with the following numbers.
 a. $x = 2, n = -4$
 b. $x = -4, n = 2$

Verify that $x^{-m} = 1/x^m$ is true when:

16. x is replaced by 2 and m by -4
17. x is replaced by -4 and m by 2

5.5 Square Root

Tell what replacements of variables in the following will lead to undefined symbols:

18. $\dfrac{x^2}{y^3}$

19. $\dfrac{1}{(a+b)^2}$

20. $\dfrac{x}{(a-b)^3}$

21. $\dfrac{a+b}{a+2b}$

22. $3x^{-2}$

23. $2(x+y)^{-3}$

5.5 SQUARE ROOT

REVIEW

- The symbol $\sqrt{}$ is called a *radical*
- An expression like $\sqrt{x}$ is called a *radical expression*

OBJECTIVES

- Define a principal square root
- State and use the property of the product of square roots
- State and use the property of the square root of the quotient
- Rationalize denominators

The equation $x^2 = 36$ has two solutions, 6 and -6. Both of the following are true:

$$6^2 = 36$$
$$(-6)^2 = 36$$

Therefore, 36 has two square roots, 6 and -6. To distinguish between these two square roots, we shall call the *positive* square root of 36, written $\sqrt{36}$, the *principal square root* of 36.

In general, $\sqrt{x}$, where x is a nonnegative real number, means the nonnegative square root of x, and $\sqrt{x}$ is called the *principal square root* of x.

There are several patterns involving square roots which are illustrated and stated below:

$\sqrt{4 \cdot 9} = \sqrt{36} = 6$, $\sqrt{4} \cdot \sqrt{9} = 2 \cdot 3 = 6$
Therefore, $\sqrt{4 \cdot 9} = \sqrt{4} \cdot \sqrt{9}$

$\sqrt{16 \cdot 4} = \sqrt{64} = 8$, $\sqrt{16} \cdot \sqrt{4} = 4 \cdot 2 = 8$
Therefore, $\sqrt{16 \cdot 4} = \sqrt{16} \cdot \sqrt{4}$.

$\sqrt{25 \cdot 4} = \sqrt{100} = 10$, $\sqrt{25} \cdot \sqrt{4} = 5 \cdot 2 = 10$
Therefore, $\sqrt{25 \cdot 4} = \sqrt{25} \cdot \sqrt{4}$.

5 Numbers as Exponents

These examples illustrate the following:

PRODUCT OF SQUARE ROOTS For all real nonnegative numbers x and y,

$$\sqrt{xy} = \sqrt{x} \cdot \sqrt{y}$$

This pattern is also true for roots other than the square root. For all nonnegative real numbers x and y, and for each natural number $n > 2$:

$$\sqrt[n]{xy} = \sqrt[n]{x} \cdot \sqrt[n]{y}$$

Now study the following examples and observe the pattern illustrated by them:

$$\frac{4}{\sqrt{4}} = \frac{4}{2} = 2, \quad \sqrt{4} = 2 \qquad \text{Therefore,} \quad \frac{4}{\sqrt{4}} = \sqrt{4}.$$

$$\frac{9}{\sqrt{9}} = \frac{9}{3} = 3, \quad \sqrt{9} = 3 \qquad \text{Therefore,} \quad \frac{9}{\sqrt{9}} = \sqrt{9}.$$

$$\frac{81}{\sqrt{81}} = \frac{81}{9} = 9, \quad \sqrt{81} = 9 \qquad \text{Therefore,} \quad \frac{81}{\sqrt{81}} = \sqrt{81}.$$

The pattern displayed above is the following:

SQUARE ROOT IN THE DENOMINATOR For each real positive number x,

$$\frac{x}{\sqrt{x}} = \sqrt{x}.$$

Another pattern is suggested by the following examples.

$$\sqrt{\frac{81}{9}} = \sqrt{9} = 3, \quad \frac{\sqrt{81}}{\sqrt{9}} = \frac{9}{3} = 3$$

Therefore, $\sqrt{\dfrac{81}{9}} = \dfrac{\sqrt{81}}{\sqrt{9}}.$

$$\sqrt{\frac{64}{4}} = \sqrt{16} = 4, \quad \frac{\sqrt{64}}{\sqrt{4}} = \frac{8}{2} = 4$$

Therefore, $\sqrt{\dfrac{64}{4}} = \dfrac{\sqrt{64}}{\sqrt{4}}.$

$$\sqrt{\frac{144}{36}} = \sqrt{4} = 2, \quad \frac{\sqrt{144}}{\sqrt{36}} = \frac{12}{6} = 2$$

Therefore, $\sqrt{\dfrac{144}{36}} = \dfrac{\sqrt{144}}{\sqrt{36}}.$

The following pattern is illustrated by these examples.

5.5 Square Root

SQUARE ROOT OF A QUOTIENT For all real numbers $x \geq 0$ and $y > 0$,

$$\sqrt{\frac{x}{y}} = \frac{\sqrt{x}}{\sqrt{y}}$$

This pattern is true for roots other than the square root.

For all real numbers $x \geq 0$ and $y > 0$, and each natural number $n \geq 2$:

$$\sqrt[n]{\frac{x}{y}} = \frac{\sqrt[n]{x}}{\sqrt[n]{y}}$$

To compute approximations involving square roots, it is convenient to find names for such expressions as $\sqrt{2}/\sqrt{5}$ so that there is no radical sign in the denominator. When this is accomplished, it is said that the denominator has been *rationalized*. To rationalize the denominator in $\sqrt{2}/\sqrt{5}$, the following procedure is used:

$$\frac{\sqrt{2}}{\sqrt{5}} = \frac{\sqrt{2} \cdot \sqrt{5}}{\sqrt{5} \cdot \sqrt{5}} = \frac{\sqrt{10}}{5}$$

Observe that in $\sqrt{10}/5$, the denominator is 5, which is a rational number.

To find a decimal approximation for the expression $\sqrt{2}/\sqrt{5}$, 1.41 is divided by 2.24, since $\sqrt{2} \doteq 1.41$ and $\sqrt{5} \doteq 2.24$. (The symbol $\doteq$ means *is approximately equal to*.) However, beginning with the rationalized expression $\sqrt{10}/5$, the division is performed as follows:

$$\frac{\sqrt{10}}{5} \doteq \frac{3.16}{5} = .632$$

It is obvious that the rationalized expression is easier to work with, since it is considerably easier to divide 3.16 by 5 than to divide 1.41 by 2.24.

The *rationalizing* procedure can be generalized as follows:

For all real numbers $x \geq 0$ and $y > 0$,

$$\frac{\sqrt{x}}{\sqrt{y}} = \frac{\sqrt{x} \cdot \sqrt{y}}{\sqrt{y} \cdot \sqrt{y}} = \frac{\sqrt{xy}}{y}$$

Sometimes expressions of the following forms occur in the denominator:

$$a + \sqrt{b}, \quad a - \sqrt{b}, \quad \sqrt{a} + \sqrt{b}, \quad \sqrt{a} - \sqrt{b}$$

Each of these forms involves a radical. How could such expressions be rationalized? Let's look at some examples.

5 Numbers as Exponents

Example 1 $(4 + \sqrt{3})(4 - \sqrt{3}) = 4 \cdot 4 - 4 \cdot \sqrt{3} + 4 \cdot \sqrt{3} - \sqrt{3} \cdot \sqrt{3}$
$$= 16 - 3$$
$$= 13$$

In general, $(a + \sqrt{b})(a - \sqrt{b}) = a^2 - a\sqrt{b} + a\sqrt{b} - \sqrt{b} \cdot \sqrt{b}$
$$= a^2 - b$$

Using the above procedure, the expression $\dfrac{5}{4 + \sqrt{3}}$ would be rationalized as follows:

$$\frac{5}{4 + \sqrt{3}} = \frac{5(4 - \sqrt{3})}{(4 + \sqrt{3})(4 - \sqrt{3})} = \frac{5(4 - \sqrt{3})}{13}$$

Example 2 $(\sqrt{5} + \sqrt{3})(\sqrt{5} - \sqrt{3}) = \sqrt{5} \cdot \sqrt{5} - \sqrt{3} \cdot \sqrt{5}$
$$+ \sqrt{3} \cdot \sqrt{5} - \sqrt{3} \cdot \sqrt{3}$$
$$= 5 - 3 = 2$$

In general, $(\sqrt{a} + \sqrt{b})(\sqrt{a} - \sqrt{b}) = \sqrt{a}\sqrt{a} - \sqrt{a}\sqrt{b} - \sqrt{a}\sqrt{b}$
$$- \sqrt{b}\sqrt{b}$$
$$= a - b$$

Using the above procedure, the expression $\dfrac{5}{\sqrt{5} + \sqrt{3}}$ would be rationalized as follows:

$$\frac{5}{\sqrt{5} + \sqrt{3}} = \frac{5(\sqrt{5} - \sqrt{3})}{(\sqrt{5} + \sqrt{3})(\sqrt{5} - \sqrt{3})} = \frac{5(\sqrt{5} - \sqrt{3})}{2}$$

Exercises 5.5

Simplify each expression using the patterns shown in the following examples:

Examples $\sqrt{128} = \sqrt{64 \cdot 2} = \sqrt{64} \cdot \sqrt{2} = 8\sqrt{2}$
$\sqrt{243} = \sqrt[4]{81 \cdot 3} = \sqrt[4]{81} \cdot \sqrt[4]{3} = 3\sqrt[4]{3}$
For all $x \geq 0$ and $y \geq 0$,
$\sqrt{100x^4y^6} = \sqrt{100} \cdot \sqrt{x^4} \cdot \sqrt{y^6} = 10x^2y^3$
For all $x > 0$ and $y > 0$,
$\dfrac{xy}{\sqrt{xy}} = \dfrac{xy\sqrt{xy}}{\sqrt{xy}\sqrt{xy}} = \dfrac{xy\sqrt{xy}}{xy} = \sqrt{xy}$
$8\sqrt{3} - 4\sqrt{75} = 8\sqrt{3} - 4\sqrt{25} \cdot \sqrt{3}$
$\qquad = 8\sqrt{3} - 4 \cdot 5\sqrt{3} = 8\sqrt{3} - 20\sqrt{3}$
$\qquad = -12\sqrt{3}$

5.5 Square Root

$$(3\sqrt{5} + \sqrt{5})(3\sqrt{5} - \sqrt{5}) = (3\sqrt{5} + \sqrt{5})\,3\sqrt{5}$$
$$- (3\sqrt{5} + \sqrt{5})\,\sqrt{5}$$
$$= 3\sqrt{5}\cdot 3\sqrt{5} + \sqrt{5}\cdot 3\sqrt{5}$$
$$- 3\sqrt{5}\cdot\sqrt{5} - \sqrt{5}\cdot\sqrt{5}$$
$$= 9\cdot 5 + 3\cdot 5 - 3\cdot 5 - 5$$
$$= 45 + 15 - 15 - 5$$
$$= 40$$

$$\frac{3+\sqrt{5}}{2-\sqrt{3}} = \frac{(3+\sqrt{5})\,(2+\sqrt{3})}{(2-\sqrt{3})\,(2+\sqrt{3})}$$
$$= \frac{(3+\sqrt{5})2 + (3+\sqrt{5})\sqrt{3}}{(2-\sqrt{3})2 + (2-\sqrt{3})\sqrt{3}}$$
$$= \frac{6 + 2\sqrt{5} + 3\sqrt{3} + \sqrt{15}}{4 - 2\sqrt{3} + 2\sqrt{3} - 3}$$
$$= 6 + 2\sqrt{5} + 3\sqrt{3} + \sqrt{15}$$

1. $\sqrt{72}$
2. $\sqrt{63}$
3. $\sqrt{48}$
4. $\sqrt{172}$
5. $\sqrt{a^7}$
6. $\sqrt{a^4 b^5}$
7. $\dfrac{\sqrt{5}}{\sqrt{6}}$
8. $\dfrac{\sqrt{7}}{\sqrt{18}}$
9. $9\sqrt{7} - 5\sqrt{7}$
10. $5\sqrt{2} + 3\sqrt{2}$
11. $\dfrac{2}{\sqrt{12}}$
12. $5\sqrt{3} + 7\sqrt{12}$
13. $9\sqrt{32} - 2\sqrt{18}$
14. $\sqrt{3}(\sqrt{5} + \sqrt{6})$
15. $(2 + \sqrt{5})(2 - \sqrt{5})$
16. $(3 + \sqrt{2})(\sqrt{2} - 3)$
17. $\dfrac{\sqrt[3]{-27}}{\sqrt{x^8}}$
18. $\sqrt{\dfrac{3}{5}}$
19. $\sqrt{\dfrac{4}{10}}$
20. $\sqrt{\dfrac{7}{3}}$
21. $\dfrac{\sqrt{2x^2}}{\sqrt[3]{-8x^3}}$
22. $\sqrt[4]{16x^4}\,\sqrt[4]{32x^8}$
23. $(x + \sqrt{y})(x - \sqrt{y})$
24. $(m + n\sqrt{2})(m - n\sqrt{2})$
25. $\left(x + \dfrac{1}{\sqrt{y}}\right)\left(x - \dfrac{1}{\sqrt{y}}\right)$
26. $\left(\dfrac{1}{\sqrt{a}} - \dfrac{1}{\sqrt{b}}\right)\left(\dfrac{1}{\sqrt{a}} + \dfrac{1}{\sqrt{b}}\right)$
27. $\dfrac{2}{1 + \sqrt{2}}$
28. $\dfrac{2 + \sqrt{3}}{2 - \sqrt{3}}$
29. $\left(\dfrac{\sqrt{2}}{\sqrt{3}} + \dfrac{1}{\sqrt{3}}\right)\left(\dfrac{\sqrt{2}}{\sqrt{3}} - \dfrac{1}{\sqrt{3}}\right)$
30. $\dfrac{2(5 - \sqrt{2})}{3(5 + \sqrt{2})}$
31. $\dfrac{\sqrt{3}}{\sqrt{5} - \sqrt{3}}$
32. $\dfrac{\sqrt{a}}{\sqrt{a} + \sqrt{b}}$
33. $\dfrac{\sqrt{x}}{\sqrt{x} - \sqrt{y}}$
34. $\dfrac{-1}{\sqrt{2} + 3}$
35. $\dfrac{-2}{4 - \sqrt{3}}$
36. $\dfrac{1 + \sqrt{2}}{\sqrt{2} - 1}$
37. $\dfrac{\sqrt{3} + \sqrt{2}}{1 + \sqrt{5}}$
38. $\dfrac{\sqrt{5} - \sqrt{2}}{\sqrt{5} + \sqrt{2}}$
39. $\dfrac{\sqrt{7} + \sqrt{2}}{\sqrt{5} + \sqrt{3}}$
40. $\dfrac{\sqrt{a} + \sqrt{b}}{\sqrt{a} - \sqrt{b}}$

5 Numbers as Exponents

41. Big Brother does not like radicals disturbing his smooth-running system. As you know, everyone in 1984 is given a number. To eradicate some rowdy radicals, Big Brother would like you to "rationalize" the following:

   a. $\dfrac{\sqrt{3}}{\sqrt{7}}$ b. $\dfrac{\sqrt{4}\sqrt{16}}{\sqrt{9}}$ c. $\dfrac{\sqrt{10}}{\sqrt{8}\sqrt{3}}$

42. It had to happen—another disease has been discovered! Some math teachers develop radicalphobia, a great fear of radicals in any form. Try to arrange the following as decimal approximations so as not to frighten your instructor.

   a. $\dfrac{10}{6+\sqrt{3}}$ b. $\dfrac{9}{\sqrt{7}+\sqrt{2}}$

CHAPTER 5 TEST

1. Your bank has just inaugurated a novel system that allows good clients to eradicate debts by a technique known as squaring. But the small print says that if the debt is over $10, cubing is introduced. Not knowing anything about math, you run up a debt of $12, thinking it will be eradicated. What is your new debt if debts are indicated by a negative number?

Compute the answers:

2. $(-2)^3$ 3. $(-2)^4$ 4. $(\tfrac{2}{3})^2$
5. $(-\tfrac{4}{5})^3$ 6. $(.1)^3$ 7. $(-.01)^2$
8. $(-6 \cdot \tfrac{1}{2})^2$ 9. $100 \times (.1)^3$

For each of the following, give a simpler equivalent expression:

10. $x^4 x^7$ 11. yy^5 12. $(x^3)^4$
13. $(-n^3)^3$ 14. $\left(\dfrac{m^3}{m}\right)^4$ 15. $\left(\dfrac{-5t^2}{25t}\right)^5$

By merely looking at each of the following, tell whether it names a positive or a negative number. Do not compute the answers.

16. $(-25)^{17}$ 17. $(-25)^{22}$ 18. $(-\tfrac{1}{2})^{46}$
19. $(-\tfrac{1}{2})^{51}$ 20. $(-1)^{200}$ 21. $(-1)^{301}$
22. $(-1)^3(-1)^{17}$ 23. $(-1)^{15}(-1)^{16}$

5.5 Square Root

24. A polar bear who wishes to traverse the Swiss Alps the easy way decides to slide up and down the mountains. His path is as follows:

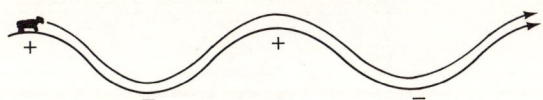

If a downward journey is indicated by a minus sign and an upward journey by a positive sign, and the maximum number of "slides" he can make on one push is given by $(-\frac{10}{3})^4$, will he stop at the top of a hill or at the bottom?

25. Give an argument why $(-x)^m(-x)^n = (-x)^{m+n}$ is true for all real number replacements of x and all whole number replacements of m and n.

26. Explain why $-(3)^2 \neq (-3)^2$, but $-(3)^3 = (-3)^3$.

Tell which of the following are true and which are false:

27. $-(6)^3 = (-6)^3$ **28.** $(\frac{1}{2})^2 < \frac{1}{2}$ **29.** $(1.1)^4 > 1.1$

30. $(-5)^3 > 0$ **31.** $(-5)^4 > 0$ **32.** $\dfrac{5^6}{5^2} = 5^3$

33. $(-2 \cdot 7)^3 = -2^3 \cdot 7^3$

Tell which of the following are true for all real number replacements of the variable:

34. $3x^2 = 9x^2$ **35.** $(3x)^2 = 9x^2$ **36.** $-5x^2 = 25x^2$

37. $(-5x)^3 = 125x^3$ **38.** $(x^3)^2 = x^5$ **39.** $-(x)^3 = (-x)^3$

Give a simpler equivalent expression for each of the following:

40. $\dfrac{x^{12}}{x^3}$ **41.** $\dfrac{x^3(x+y)^4}{x(x+y)}$ **42.** $\dfrac{-48\,a^2(a-b)^2}{24\,a(a-b)}$

Simplify each expression:

43. $\sqrt{128}$ **44.** $\dfrac{\sqrt{12}}{2}$ **45.** $\dfrac{2+\sqrt{5}}{2-\sqrt{5}}$

46. $\dfrac{\sqrt{2}}{1-\sqrt{2}}$ **47.** $(a+\sqrt{b})(a-\sqrt{b})$ **48.** $\dfrac{4(3-\sqrt{2})}{-2(3+\sqrt{2})}$

6

Real Numbers, Equations, and Inequalities

6.1 NUMBERS THAT AREN'T RATIONAL

REVIEW

- A rational number has a name of the form a/b, where a and b are integers and $b \neq 0$
- Equivalent equations have the same solution. Example: $x - 2 = 0$ and $x = 2$ are equivalent equations
- A right triangle has one right angle (90°)

OBJECTIVES

- Solve equations which have irrational number solutions
- Demonstrate the existence of points on the number line which have irrational numbers for their coordinates
- State and use the Pythagorean Theorem in finding the length of the third side of a right triangle knowing the lengths of the other two sides
- Locate, by construction, points on the number line corresponding to given irrational numbers

The most inclusive set of numbers we have encountered so far was the set of rational numbers. Rational numbers alone, however, would not take care of some rather fundamental problems. For example, there are

6.1 Numbers That Aren't Rational

some simple equations for which the set of rational numbers is not sufficient. The equation

$$x^2 - 2 = 0$$

has no solution in the set of rational numbers. This equation is equivalent to the equation $x^2 = 2$. Therefore, to be a solution of $x^2 = 2$, the square of the number has to be equal to 2. There are two such numbers: $\sqrt{2}$ and $-\sqrt{2}$.

$\sqrt{2}$ is the *positive* square root of two
$-\sqrt{2}$ is the *negative* square root of two

Each of $\sqrt{2}$ and $-\sqrt{2}$ is a solution of $x^2 = 2$, since $(\sqrt{2})^2 = 2$ and $(-\sqrt{2})^2 = 2$. Furthermore, $\sqrt{2}$ and $-\sqrt{2}$ are not rational numbers; they are *irrational numbers*. This will be proved later by showing that there is no name for $\sqrt{2}$, which is of the form a/b with a and b integers and $b \neq 0$.

It is instructive to look at the sets of rational and irrational numbers in terms of the points on a line. In attempting to assign to each point on a line a rational number, we will see that some points cannot be assigned any rational numbers. The argument proving this is contained in the following discussion.

Let us *assume* for now that $\sqrt{2}$ is not a rational number. It will be demonstrated that there is a point on a line which corresponds to $\sqrt{2}$, thus showing the existence of a point which does not have a rational number corresponding to it.

To accomplish this, we need to recall a relation which exists between the lengths of the sides in a right triangle. (See Figure 6.1.)

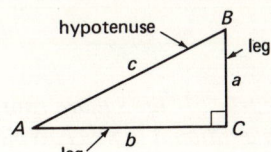

FIGURE 6.1

In a right triangle, the sum of the squares of the lengths of legs is equal to the square of the length of the hypotenuse.

For the triangle in Figure 6.1, this relation means that

$$c^2 = a^2 + b^2$$

The name of this relation is the *Pythagorean Theorem,* named after Pythagoras, the Greek philosopher who investigated it.

6 Real Numbers, Equations, and Inequalities

If the length of each leg in a right triangle is one unit, then the length of the hypotenuse is $\sqrt{2}$, as can be seen in Figure 6.2 and the computations below:

$$x^2 = 1^2 + 1^2$$
$$= 1 + 1$$
$$= 2$$
$$x = \sqrt{2}$$

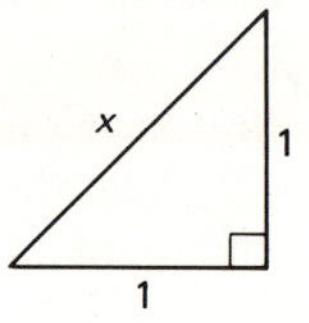

FIGURE 6.2

Only the positive square root of 2 is used, since it is used to describe a length of a segment.

In Figure 6.3 we demonstrate the existence of a point on a line corresponding to $\sqrt{2}$. In triangle ABC, each leg is one unit long; thus the hypotenuse is $\sqrt{2}$ units long. Place the hypotenuse on the number line so that point A corresponds to the number 0, making point B correspond to the number $\sqrt{2}$ on the number line. Thus there is a point on the number line which corresponds to an irrational number, in this case $\sqrt{2}$.

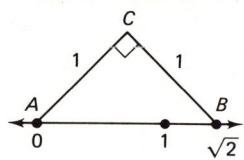

FIGURE 6.3

Exercises 6.1

Assume the following to be true: If the square root of a whole number is not a whole number, then it is an irrational number. Now tell which of the following are irrational numbers:

1. $\sqrt{49}$ 2. $\sqrt{30}$ 3. $\sqrt{14}$ 4. $\sqrt{121}$
5. $\sqrt{81}$ 6. $\sqrt{82}$ 7. $\sqrt{625}$ 8. $\sqrt{700}$
9. Big Brother makes sure everyone has a number related to the person's individual traits. A rational person is a well-behaved member of society. An irrational person is just the opposite. By examining the following ID numbers, see if the people are well-behaved citizens or troublemakers.

    a. $\sqrt{144}$ b. $\sqrt{603}$ c. $\sqrt{196}$ d. $\sqrt{224}$ e. $\sqrt{890}$

6.1 Numbers That Aren't Rational

For each of the following equations, give the two irrational number solutions:

10. $x^2 = 3$
11. $x^2 = 5$
12. $x^2 - 8 = 0$
13. $x^2 - 15 = 0$
14. $x^2 + 3 = 13$
15. $x^2 - 4 = 3$

Using the Pythagorean Theorem, compute the length of the side marked x of each right triangle pictured below, knowing the lengths of the two sides as indicated in each picture:

16.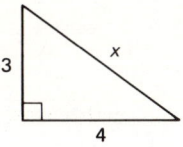

17.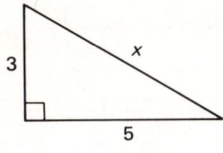

18.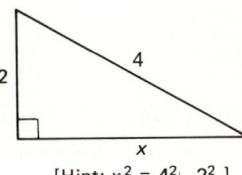

[Hint: $x^2 = 4^2 - 2^2$.]

19.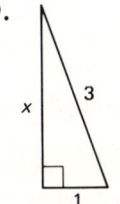

20.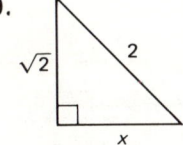

21.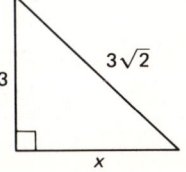

22. An ingenious mountaineer has entered a mountain-climbing contest in which he must reach the top of a mountain using only 60 ft of line and a trained carrier pigeon that flies at a steady rate of 25 ft per second. Midway up the mountain, the climber discovers that the path of the contest requires him to climb a 30 ft high cliff that is separated from him by a lake 40 ft wide. He intends to achieve this feat by making a loop in one end of his line, using 3 feet of it, and giving this loop to the pigeon, who is to fly to the top of the cliff and drop the loop over a secure rock. The other end the climber intends to secure to a tree, using another 3 feet of line. He will then scale the line to the top of the cliff. Before he begins, he must make certain he has enough line. Does he?

23. As a young man, a rancher bought a piece of property that formed a right triangle 4 miles long and 3 miles wide. As he became more prosperous, he purchased a perfect square of land on each side of the initial triangle. He wanted to divide the land among his three sons, leaving the eldest with the most land, the youngest with the least, and himself with a nice place to retire. He decided to keep the original triangle where his house was and give the three squares to his sons. How many square miles did each son get? Can you think of a graphic way to use this problem to prove or disprove the Pythagorean Theorem?

6 Real Numbers, Equations, and Inequalities

24. Using the triangle pictured, make a construction that will locate the point corresponding to $\sqrt{3}$ on the number line.

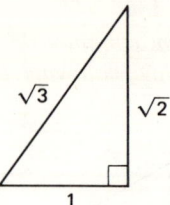

25. Using the triangle pictured, make a construction that will locate the point corresponding to $\sqrt{5}$ on the number line.

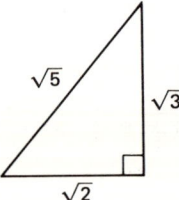

26. Using the triangle pictured, make a construction that will locate the point corresponding to $\sqrt{6}$ on the number line.

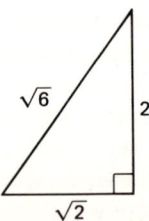

27. List as many practical applications of the Pythagorean Theorem as you can think of. Give some worked examples to illustrate why you consider the particular problems applications of the Pythagorean Theorem. (For example: You have been in a wreck. Your car, a small imported model, used to be 7 ft long. Now you observe that the tow truck has lifted the car 3 ft off the ground, and the distance from the end of your car along the street to the base of the tow truck is 4 ft. How long is your car now?)

6.2 PROVING A NUMBER IRRATIONAL

REVIEW

- There is a one-to-one correspondence between the points on a line and all real numbers
- The square of every even number is even
- The square of every odd number is odd

OBJECTIVES

- Relate the points on a number line to rational numbers, irrational numbers, and real numbers
- Prove that $\sqrt{2}$ is an irrational number

We know that there are points on the number line to which irrational numbers are assigned. In fact, there are lots of them. This means that if we could imagine a number line with all the rational numbers assigned to points, there would still be lots of "holes" left. But if we assign points to all of the real numbers, rational and irrational, the entire line is covered—no point is unaccounted for. Here are some of the rational numbers and some irrational numbers shown on the number line.

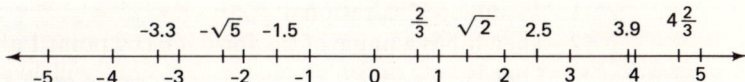

Just a little while ago, we assumed that $\sqrt{2}$ was an irrational number. Before actually proving this, let's establish four points.

1. Recall that a rational number has a name of the form a/b, where a and b are integers and $b \neq 0$. Furthermore, every rational number has a name a/b in which a and b are *relatively prime*; that is, the only common factor of a and b is 1 or -1. To see this, assume that a and b are not relatively prime. In that case, they have a common factor which is greater than 1. Just divide both numerator and denominator by this factor and keep dividing until no such common factors are left. The result is a/b, where a and b are relatively prime.

2. Look back to Theorem 3.1. There we proved that the square of an even number is even.

3. Look back to Theorem 3.2. We proved that the square of an odd number is odd.

4. Every integer is either even or odd—there is no other choice. It follows that if the square of some integer is even, then that integer is even, since the square of each odd integer is odd. That is,

If a^2 is even, then a is even

These four results will be used in the proof below.

Theorem 6.1 $\sqrt{2}$ is an irrational number.

Proof   We shall assume that $\sqrt{2}$ is a rational number. Of course, we are taking it for granted that it is a *real* number. Our plan is to show that the assumption that $\sqrt{2}$ is rational leads, by a logical argument, to a contradiction. Since our argument will be faultless (we hope!), the contradiction that we reach will imply that our assumption was false. Thus, we will establish that $\sqrt{2}$ is not a rational number, which means it is irrational. Since the proof will be somewhat lengthy, we shall number the statements for easy reference.

1. Assume $\sqrt{2}$ is rational.
2. Then it has a name of the form a/b (by point 1 above). That is, $\sqrt{2} = a/b$.
3. Squaring: $(a/b)^2 = (\sqrt{2})^2$ or $a^2/b^2 = 2$. Thus, $a^2 = 2b^2$.
4. Observe that $2b^2$ has a factor of 2. Therefore, it is an even number.
5. Since a^2 and $2b^2$ are the same number (by step 3), a^2 is even. Therefore, a is even (by point 4 above).
6. Since a is even, it can be shown as a product of 2 and some integer: $a = 2c$.
7. Squaring: $a^2 = (2c)^2$ or $a^2 = 4c^2$.
8. But $a^2 = 2b^2$ (by step 3). Therefore, $2b^2 = 4c^2$ or $b^2 = 2c^2$.
9. $2c^2$ is an even number, since it is a product of 2 and some integer.
10. Since $b^2 = 2c^2$, b^2 is even. Therefore, b is even (by point 4 above).
11. We have now established that a is even and b is even. This is a contradiction of our assumption in step 1, since this assumption implies that a and b are relatively prime.
12. It follows that $\sqrt{2}$ cannot be a rational number.
13. Therefore, $\sqrt{2}$ is an irrational number, which we set out to prove.

6.3 Irrational Numbers and Decimals

Exercises 6.2

If Q = the set of rational numbers, T = the set of irrational numbers, and R = the set of real numbers, tell which of the following are true and which are false:

1. $Q \subseteq T$
2. $T \subseteq Q$
3. $Q \subseteq R$
4. $T \subseteq R$
5. $R \subseteq T$
6. $R \subseteq Q$
7. $T \cap R = T$
8. $Q \cap R = Q$
9. $T \cap Q = R$
10. $T \cup Q = R$
11. Although mathematicians like to keep the number lines in good condition, invariably a few termites find their way into the lines. Trace the path of this particular termite by joining up the following points on a number line.
    a. $\sqrt{5} \rightarrow 2$ b. $2 \rightarrow 3.9$ c. $3.9 \rightarrow -3.3$ d. $-3.3 \rightarrow -\sqrt{5}$
    e. $-\sqrt{5} \rightarrow \sqrt{5}$
12. Describe the meaning of a one-to-one correspondence between the set of real numbers and the number line.
13. The set of irrational numbers is not closed under addition because the sums of some pairs of irrational numbers are not irrational numbers. Prove this by giving an example.
14. The irrational numbers think they are very exclusive and say to anyone who wishes to join, "Sorry, we're closed." But you immediately reply, "Oh no, you aren't—you're not closed under multiplication!" Give an example.
15. Prove that the set of irrational numbers is not closed under multiplication.

For each rational number below give another name of the form a/b, where a and b are relatively prime:

16. $\frac{8}{24}$
17. $\frac{7}{49}$
18. $\frac{11}{121}$
19. $\frac{21}{93}$
20. $\frac{16}{256}$
21. $\frac{9}{120}$
22. $\frac{6}{22}$
23. $\frac{8}{42}$
24. Using the method that proved $\sqrt{2}$ to be an irrational number, prove the following numbers to be rational or irrational.
    a. $\sqrt{5}$ b. $\sqrt{16}$ c. $\sqrt{3}$

6.3 IRRATIONAL NUMBERS AND DECIMALS

REVIEW

- Every terminating decimal and every repeating decimal names a rational number
- $\sqrt{2}$ is an irrational number

OBJECTIVES

- Create nonterminating and nonrepeating decimals
- Examine some approximations to $\sqrt{2}$

- The set of irrational numbers is not closed under addition, subtraction, multiplication, or division
- The sum of a rational number and an irrational number is an irrational number

- Add two irrational numbers and get a rational number for the sum
- Create irrational numbers

Before studying this section, quickly review Section 4.7 about decimals. In that section we were concerned with repeating decimals. In this section we are interested in nonrepeating decimals.

It is rather easy to play with decimals and establish some rules for forming nonrepeating decimals. For example, put down a decimal point, write 2 to the right of it, then 0, then 2 again, and now 2 zeros, then 2 again, and now 3 zeros, then 2, and so on—each time one more 0:

.2020020002 . . .

By writing one more 0 each time, we make sure that we never get a repeating block of digits. So, we created a new decimal—nonterminating and nonrepeating.

We know that no irrational number has a name of the form a/b. When computing with irrational numbers, we just simply use some decimal approximations. The more digits in the decimal, the more accurate is the approximation. To get a little feeling for how this business of approximations works, let's find some approximations to $\sqrt{2}$. We will use the symbol $\doteq$ to mean *is approximately equal to*.

Here are three approximations to $\sqrt{2}$ and their squares, with the difference of the number 2 and the square of the approximation computed. Keep in mind that $(\sqrt{2})^2 = 2$.

$$1.4 \doteq \sqrt{2} \qquad (1.4)^2 = 1.96 \qquad 2 - 1.96 = .04$$
$$1.41 \doteq \sqrt{2} \qquad (1.41)^2 = 1.9881 \qquad 2 - 1.9881 = .0119$$
$$1.414 \doteq \sqrt{2} \qquad (1.414)^2 = 1.999396 \qquad 2 - 1.999396 = .000604$$

A seven-decimal place approximation to $\sqrt{2}$ is 1.4142135, and $(1.4142135)^2$ = 1.99999982358225. The approximation 1.4142135 is quite close to $\sqrt{2}$,

6.3 Irrational Numbers and Decimals

since

$$2 - 1.99999982358225 = .00000017641775$$

Thus, using more and more decimal places gives a better and better approximation to $\sqrt{2}$, but no decimal is exactly equal to $\sqrt{2}$.

In exercise 13, Section 6.2, you showed that the set of irrational numbers is not closed under addition by displaying an example of two irrational numbers whose sum is a rational number. The same can now be done in terms of nonrepeating decimal numerals. Here is one such example:

$$\begin{array}{r} .3737737773\ldots \\ +\ .5151151115\ldots \\ \hline .8888888888\ldots \end{array}$$

Notice that the two numbers which are added are irrational numbers, since their names are nonterminating and nonrepeating. Their sum, however, is a repeating decimal. Since every repeating decimal numeral names a rational number, the sum of the two irrational numbers is a rational number.

The set of irrational numbers is infinite. As a matter of fact, it is a very plentiful set. To get a feeling for this, imagine the sum of each natural number and $\sqrt{2}$:

$$1 + \sqrt{2}$$
$$2 + \sqrt{2}$$
$$3 + \sqrt{2}$$
$$\cdot$$
$$\cdot$$
$$\cdot$$

Each number displayed above is irrational. Since there are infinitely many natural numbers, there are infinitely many numbers of the form $n + \sqrt{2}$, where n is a natural number. The same can be done for every other irrational number, that is, every number of the form

$$n + t$$

where n is a natural number and t is an irrational number.

Furthermore, every number of the form

$$q + t$$

where q is a *rational* number and t is an irrational number is an irrational number. Thus, the irrational numbers are plentiful indeed!

Exercises 6.3

1. Give two examples of irrational numbers, one in the form $\sqrt{i}$ and the other in the form of a nonterminating decimal.
2. Below are three approximations to $\sqrt{3}$:

    $1.7 \doteq \sqrt{3}$
    $1.73 \doteq \sqrt{3}$
    $1.732 \doteq \sqrt{3}$

    Square each approximation, and subtract it from 3. The square of which approximation is closest to 3?
3. In a recent math talent contest, the contestants were asked to guess the value of $\sqrt{11}$. Below are the answers they gave. As a judge, it is your job to see which is closest.
    a. 3.2   b. 3.3   c. 3.32   d. 3.318
    e. 3.3175   f. 3.317   g. 3.3168   h. 3.3166
4. Taking any 5 pairs of numbers (choose some square roots) that you have not already investigated, determine the following.
    a. Are the numbers irrational?
    b. If they are irrational, what are their sum and difference?
    c. Of what decimal forms are the sum and difference?
5. Give an example of two irrational numbers, each in the form of a nonterminating nonrepeating decimal, whose sum is a repeating decimal.
6. Give an example of two irrational numbers, each in the form of a nonterminating nonrepeating decimal, whose difference is in the form of a repeating decimal.
7. a. Give an example of two irrational numbers, each in the form of a nonterminating nonrepeating decimal, whose sum is also such a decimal.
    b. Is the sum in this case an irrational number?
8. a. Give an example of two irrational numbers, each in the form of a nonterminating nonrepeating decimal, whose difference is also such a decimal.
    b. Is the difference in this case an irrational number?
9. Consider the following sequence of rational numbers:

    $1, \frac{1}{2}, \frac{1}{3}, \ldots$

    and the irrational number $\sqrt{5}$. Construct the sequence of irrational numbers of the form $q + \sqrt{5}$, where q is a rational number taken from the given sequence in the order they are listed.

6.4 TWO IMPORTANT RELATIONS: IS EQUAL TO (=), IS LESS THAN (<)

REVIEW

- Two lines in the same plane are parallel if they do not intersect
- Two triangles are congruent if they have three pairs of corresponding sides and angles congruent

OBJECTIVES

- Define the relation *is equal to*
- State the reflexive, symmetric and transitive properties of a relation
- Define an equivalence relation
- Define substitution
- State four properties for equations
- Define the relation *is less than*
- Show that < is not an equivalence relation
- Define the relation *is greater than*
- Examine nonmathematical relations for equivalence

A large part of mathematics consists of the study of equations and inequalities. In order to have a basis on which to build, certain properties of the relations involved in equations and inequalities are needed. These relations are *is equal to* (=) and *is less than* (<).

> *a is equal to b*, written $a = b$, means that a and b are names for the same thing.

The relation *is equal to* has three very obvious properties.

1. $x = x$ for every number x; this is called the *reflexive property*.
2. If $x = y$, then $y = x$; this is called the *symmetric property*.
3. If $x = y$ and $y = z$, then $x = z$; this is called the *transitive property*.

Any relation which has these three properties is called an *equivalence relation*.

When solving equations, we also make *substitutions*. What this amounts to is simply replacing one name for something with another name for it.

6 Real Numbers, Equations, and Inequalities

There are four properties which are very essential in solving equations. You used them many times before. Let's state them for our future reference.

ADDITION PROPERTY FOR EQUATIONS
If $a = b$, then $a + c = b + c$

MULTIPLICATION PROPERTY FOR EQUATIONS
If $a = b$, then $ac = bc$

CANCELLATION PROPERTY FOR ADDITION
If $a + c = b + c$, then $a = b$

CANCELLATION PROPERTY FOR MULTIPLICATION
If $ac = bc$ and $c \neq 0$, then $a = b$

We also want to examine the relation *is less than* ($<$). This relation is used to compare such things as weights of two individuals, scores of two teams, distances between towns, and so on. It is helpful to look at this relation in terms of equality.

x is less than y, $x < y$, means that there is a positive number p such that $x + p = y$.

Here are some specific examples.

Example 1 $-3 < 1$ means that there is a positive real number p such that $-3 + p = 1$. This number is 4, since $-3 + 4 = 1$.

Example 2 $-7 < 0$ means that there is a positive real number p such that $-7 + p = 0$. This number is 7, since $-7 + 7 = 0$.

Example 3 $2 < 17$ means that there is a positive number p such that $2 + p = 17$. This number is 15, since $2 + 15 = 17$.

If you wish, you can also use the relation *is greater than* ($>$). As a matter of fact, you used it to compare rational numbers.

$x > y$ simply means that $y < x$.

So, $x > y$ can be also translated into equality.

Is the relation *is less than* an equivalence relation? To be an equivalence relation, it must have the three properties stated for equality. If it fails to have one property, it is not an equivalence relation:

$5 < 5$ is false

This is sufficient to prove that $<$ is not an equivalence relation.

6.4 Two Important Relations: Is Equal to (=), Is Less Than (<)

It is interesting to see that < does not have the symmetric property either. For example, $7 < 9$ is true, but $9 < 7$ is false. But it does have the transitive property for numbers:

If $x < y$ and $y < z$, then $x < z$
is true for all numbers x, y, and z.

From the three relations we studied, three other relations can be derived:

$\neq$ is not equal to $(5 \neq 7)$
$\not<$ is not less than $(3 \not< 1)$
$\not>$ is not greater than $(5 \not> 8)$

We can also examine nonmathematical relations in terms of the three properties. Let's see whether *is a sister of* is an equivalence relation.

Reflexive property: no (Mary is a sister of Mary is false.)
Symmetric property: no (If Mary is a sister of John, then John is a sister of Mary is false.)
Transitive property: no (If Mary is a sister of Susan and Susan is a sister of Mary, then Mary is a sister of Mary is false.)

Thus, we see that *is a sister of* is not an equivalence relation.

Exercises 6.4

For each of the following relations, determine whether it is (1) reflexive, (2) symmetric, (3) transitive. Then conclude whether or not it is an equivalence relation:

1. is a cousin of
2. is a brother of
3. is the father of
4. is married to
5. $\neq$ for real numbers
6. $\not<$ for real numbers
7. *is parallel to* for lines in a plane
8. *is congruent to* for triangles
9. *has the same area as* for squares
10. Check the validity of the following statements.
    a. Earth > Sun (in size)
    b. 6" > 2"
    c. 10 years old < 5 years old
    d. $100 $\not>$ $101

6.5 THE SYSTEM OF REAL NUMBERS

REVIEW

- Real numbers consist of rational numbers and irrational numbers
- $\sqrt{x} \cdot \sqrt{x} = x$

OBJECTIVES

- State the Trichotomy Property of real numbers
- Prove that if $x < 0$, then $\sqrt{x}$ is not a real number
- State the eleven properties of a field
- Interpret the properties of a field

The study of square roots leads to some significant observations about numbers. The square root of a whole number can be a whole number. For example, $\sqrt{16} = 4$. But if the square root of a whole number is not a whole number, then it can be proved that the square root is irrational. For example, $\sqrt{5}$ is irrational.

It is appropriate to ask whether the square root of a real number is always a real number. In other words, we are asking whether the set of real numbers is closed under the operation of taking the square root. The answer to this question is *no*. Square roots of some real numbers are not real numbers. We shall reason this out in a very logical way. But first, let's observe that, given a real number x, only one of the following is true of x (this is called the *Trichotomy Property* of real numbers):

$$x < 0, \quad x = 0, \quad x > 0$$

All we are saying is that a real number is either negative, or 0, or positive. These are the only possibilities. Seems reasonable enough.

This observation will serve as our basis in the argument that the square root of a negative number is not a real number. The argument is kind of tricky, so think hard as you move through it.

Consider $\sqrt{x}$ and assume that x is negative, that is $x < 0$. We know that $\sqrt{x} \cdot \sqrt{x} = x$, since that's the meaning of the square root. For example, $\sqrt{5} \cdot \sqrt{5} = 5$. If $\sqrt{x}$ is a real number, then only one of the following is true:

$$\sqrt{x} < 0, \quad \sqrt{x} = 0, \quad \sqrt{x} > 0$$

We will show that none of the above is true, if $x < 0$. This will show that $\sqrt{x}$ is not a real number for $x < 0$.

6.5 The System of Real Numbers

1. Suppose $\sqrt{x} < 0$. We know that the product of two negative numbers is positive, so $\sqrt{x} \cdot \sqrt{x} > 0$. But we assumed that x is negative. Since $\sqrt{x} \cdot \sqrt{x} = x$, we have reached a contradiction. Therefore, $\sqrt{x}$ cannot be negative.
2. Suppose $\sqrt{x} = 0$. Then $\sqrt{x} \cdot \sqrt{x} = 0$. This contradicts the fact that $x < 0$.
3. Suppose $\sqrt{x} > 0$. We know that the product of two positive numbers is positive. So, $\sqrt{x} \cdot \sqrt{x} > 0$. This also contradicts the fact that $x < 0$. So, $\sqrt{x}$ cannot be positive.

Thus, we have a number, $\sqrt{x}$, where x is negative, such that it cannot be negative, it cannot be 0, and it cannot be positive. Obviously, it is not a real number. It happens to be a *complex number*. We won't study complex numbers any further in this book, but it is useful to know that these numbers exist.

For ease of reference, the basic properties of the set of real numbers, with which you are already familiar, are summarized below.

CLOSURE PROPERTY UNDER ADDITION
    For all real numbers x and y, $x + y$ is a real number.

CLOSURE PROPERTY UNDER MULTIPLICATION
    For all real numbers x and y, $x \cdot y$ is a real number.

COMMUTATIVE PROPERTY OF ADDITION
$$x + y = y + x$$

COMMUTATIVE PROPERTY OF MULTIPLICATION
$$x \cdot y = y \cdot x$$

ASSOCIATIVE PROPERTY OF ADDITION
$$(x + y) + z = x + (y + z)$$

ASSOCIATIVE PROPERTY OF MULTIPLICATION
$$(x \cdot y) \cdot z = x \cdot (y \cdot z)$$

DISTRIBUTIVE PROPERTY OF MULTIPLICATION OVER ADDITION
$$x \cdot (y + z) = (x \cdot y) + (x \cdot z)$$

PROPERTY OF ZERO FOR ADDITION
    There exists a unique real number 0 such that $x + 0 = 0 + x = x$. 0 is called the *additive identity*.

PROPERTY OF ONE FOR MULTIPLICATION
    There exists a unique real number 1 such that $x \cdot 1 = 1 \cdot x = x$. 1 is called the *multiplicative identity*.

6 Real Numbers, Equations, and Inequalities

PROPERTY OF ADDITIVE INVERSE

For every real number x, there exists a unique real number $-x$ such that $x + (-x) = 0$. We call $-x$ the *additive inverse* or *opposite of x*.

PROPERTY OF MULTIPLICATIVE INVERSE

For every real number $x \neq 0$, there exists a unique real number $\frac{1}{x}$ such that $x \cdot \frac{1}{x} = 1$. We call $\frac{1}{x}$ the *multiplicative inverse* or *reciprocal of x*.

Any set of elements which has two operations with these eleven properties is called a *field*. Generally, the elements do not have to be numbers and the two operations do not have to be addition and multiplication.

Exercises 6.5

1. If $x = -4$, then what can be said about $\sqrt{x}$?
2. What does the property of zero for addition tell about $x + 0$ for every real number x?
3. What does the property of one for multiplication tell about $x \cdot 1$ for every real number x?
4. What does the property of additive inverse tell about x for every real number x?

Give the additive inverse of each of the following real numbers:

5. 3 6. $\sqrt{7}$ 7. 0
8. $-\frac{1}{2}$ 9. $-\sqrt{10}$ 10. $-.37$

11. What does the property of multiplicative inverse tell about every real number $x \neq 0$?

Give the multiplicative inverse of each of the following real numbers:

12. $\frac{3}{4}$ 13. $-\frac{4}{7}$ 14. $\sqrt{3}$
15. $\frac{1}{\sqrt{5}}$ 16. 7 17. -2.5

18. a. Examine the properties of a field listed in this section. Which two are lacking for addition and multiplication in the set of whole numbers?
    b. Why is the set of whole numbers not a field?
19. That well-known author Lewis Carrott has written a new fairy tale entitled "Alice through the Inverse Glass." It is the story of a young girl who is transported to a world of inverses. She is an avid mathematician. Upon seeing these numbers in the inverse world, she decides to find out what they are in the real world. Show what she found for:
    a. 2 b. $\frac{8}{9}$ c. $\frac{1}{1000}$ d. $\frac{10}{3}$ e. $\frac{15}{2}$

6.6 INEQUALITIES

REVIEW

- $x < y$ means that there is a positive number p such that $x + p = y$
- Addition property for equations: if $x = y$, then $x + z = y + z$
- Commutative properties:
  $x + y = y + x$
  $xy = yx$
- Associative properties:
  $(x + y) + z = x + (y + z)$
  $(xy)z = x(yz)$
- Multiplication property for equations: if $x = y$, then $xz = yz$
- Distributive property of multiplication over addition:
  $(x + y)z = xz + yz$
- Closure of positive numbers under multiplication: the product of any two positive numbers is a positive number
- Property of additive inverse:
  $x + (-x) = 0$
- Property of zero for addition:
  $x + 0 = x$
- Symmetric property of $=$:
  if $x = y$, then $y = x$
- The product of a positive number and a negative number is negative
- Transitive property of $=$:
  if $x = y$ and $y = z$, then $x = z$

OBJECTIVES

- Prove that the same number can be added to each side of an inequality
- Prove that each side of an inequality can be multiplied by the same positive number
- Prove that multiplying each side of an inequality by the same negative number reverses the sense of the inequality
- Prove the transitive property of $<$
- Abbreviate $x < y$ or $x = y$
- Abbreviate $x < y$ and $y < z$
- Solve inequalities using the theorems proved in this section

6   Real Numbers, Equations, and Inequalities

We have formulated some properties of the relations *is equal to* ($=$), *is less than* ($<$), and *is greater than* ($>$). Now we can prove some theorems about inequalities which we can use in solving inequalities.

The first theorem about the *is less than* relation is like the addition property for equations. According to this property, you can add the same number to each side of a given equation. For example:

$$\text{if } 6 = 1 + 5, \text{ then } 6 + 4 = (1 + 5) + 4$$

For the *is less than* relation, we state that you can add the same number to each side of an inequality.

Theorem 6.2   If $x < y$, then $x + z < y + z$

Before starting to prove this theorem, recall that $x + z < y + z$ means that there is some positive number which when added to $x + z$ will give $y + z$. We need to prove the existence of such a number. To help you understand the proof, we state the reason for each statement to its right.

Proof

$x < y$	GIVEN
$x + w = y \ [w > 0]$	Meaning of $<$
$(x + w) + z = y + z$	Addition property for equations
$x + (w + z) = y + z$	Associative property of addition
$x + (z + w) = y + z$	Commutative property of addition
$(x + z) + w = y + z$	Associative property of addition
$x + z < y + z$	Meaning of $<$

So, starting with $x < y$, the conclusion that $x + z < y + z$ was reached, thus proving the theorem.

In attempting to decide whether inequalities have something that is analogous to the multiplication property for equations, let us examine some examples.

Example 1 $2 < 5$ and $2 \cdot 6 < 5 \cdot 6$ because $12 < 30$.
Example 2 $3 < 10$ and $3 \cdot 5 < 10 \cdot 5$ because $15 < 50$.
Example 3 $4 < 6$, but $4 \cdot (-2) > 6 \cdot (-2)$ because $-8 > -12$.
Example 4 $-2 < 3$, but $-2 \cdot (-1) > 3 \cdot (-1)$ because $2 > -3$.

These examples suggest that the multiplication in inequalities is somewhat more complicated than in equations. It is necessary to distinguish between two cases:

6.6 Inequalities

CASE 1 Multiplication by a positive number
CASE 2 Multiplication by a negative number

The two cases are presented in the next two theorems.

Theorem 6.3 For each *positive* real number z, if $x < y$, then $xz < yz$.

Proof

$x < y$	GIVEN
$x + w = y [w > 0]$	Meaning of $>$
$(x + w)z = yz [z > 0]$	Multiplication property for equations
$xz + wz = yz$	Distributive property of multiplication over addition
$xz < yz$	Closure of positive numbers under multiplication; meaning of $<$

Thus, it is proved that if $x < y$, then $xz < yz [z > 0]$.

Observe that it is essential to know for step 4 of the proof that wz is positive. This is so, because both w and z are positive, as is stated in steps 2 and 3 of the proof. Now the case of multiplication by a negative number is considered.

Theorem 6.4 For each *negative* real number z, if $x < y$, then $xz > yz$.

Proof

$x < y$	GIVEN
$x + w = y [w > 0]$	Meaning of $<$
$(x + w)z = yz [z < 0]$	Multiplication property for equations
$xz + wz = yz$	Distributive property of multiplication over addition
$(xz + wz) + [-(wz)]$ $= yz + [-(wz)]$	Addition property for equations
$xz = yz + [-(wz)]$	Associative property for addition; property of additive inverse; property of zero for addition
$yz + [-(wz)] = xz$	Symmetric property of $=$
$-(wz) > 0$	Since $w > 0$ and $z < 0$, $wz < 0$; therefore $-(wz) > 0$
$yz < xz$	Meaning of $<$
$xz > yz$	Meaning of $>$

Thus, it is proved that if $x < y$ and $z < 0$, then $xz > yz$.

6 Real Numbers, Equations, and Inequalities

A theorem that is analogous to the transitive property of equality will now be proved.

Theorem 6.5   If $x < y$ and $y < z$, then $x < z$.

Proof

$x < y$ and $y < z$	GIVEN
$x + s = y, y + t = z$ $[s > 0$ and $t > 0]$	Meaning of $<$
$(x + s) + t = z$	Substitution property. ($x + s$ is substituted for y in $y + t = z$)
$x + (s + t) = z$	Associative property of addition
$s + t > 0$	The sum of two positive numbers is positive
$x < z$	Meaning of $<$

Thus, it is proved that if $x < y$ and $y < z$, then $x < z$; that is, the transitivity of the relation *is less than* was proved. Note that in the third statement of proof, $x + s$ was substituted for y, since $x + s = y$, according to step 2 of the proof.

The results proved above will be used when solving inequalities. In that context, sentences which consist of two inequalities will occur. To simplify writing, some abbreviations will be used:

$x < y$ or $x = y$

will be abbreviated as

$x \leq y$ Read: x is less than or equal to y

$x < y$ and $y < z$

will be abbreviated as

$x < y < z$ Read: x is less than y and y is less than z

Now look at the following example to see how we can use what we have just proved in solving an inequality.

Example 1   Solve $4x + 7 < x - 10$.

6.6 Inequalities

Solution $4x + 7 + (-x) < x - 10 + (-x)$ Theorem 6.2
$\qquad\qquad 3x + 7 < -10$ Adding
$\qquad\qquad 3x + 7 + (-7) < -10 + (-7)$ Theorem 6.2
$\qquad\qquad 3x < -17$ Adding
$\qquad\qquad x < -\frac{17}{3}$ Theorem 6.3 (multiplying by $\frac{1}{3}$)

Thus, the solution set of $4x + 7 < x - 10$ is the set of all numbers which are less than $-\frac{17}{3}$; that is, $\{x | x < -\frac{17}{3}\}$.

Exercises 6.6

For each inequality below, a number or an expression is given. Use this number or expression as a replacement for z in Theorem 6.2 to obtain a new inequality. Simplify the new inequality.

$\qquad$ Example $6x - 1 > 3x + 7;\quad -3x$
$\qquad\qquad\qquad 6x - 1 + (-3x) > 3x + 7 + (-3x)$
$\qquad\qquad\qquad 3x - 1 > 7$

1. $5x + 9 < 4x - 2;\quad -4x$ $\qquad$ 2. $3x + 5 < 9;\quad -9$
3. $1 - 2x > 3 - 5x;\quad 5x$ $\qquad$ 4. $9 + 7x > -8x - 2;\quad 8x$

For each inequality below a number is given. Use this number as a replacement for z in Theorem 6.3 or 6.4 to obtain a new inequality.

$\qquad$ Example $8(x + 1) > 4(x - 2);\quad \frac{1}{4}$
$\qquad\qquad\qquad 8(x + 1) \cdot \frac{1}{4} > 4(x - 2) \cdot \frac{1}{4}$
$\qquad\qquad\qquad 2(x + 1) > x - 2$

$\qquad$ Example $9(2 - x) < 3(2x + 1);\quad -\frac{1}{3}$
$\qquad\qquad\qquad 9(2 - x)(-\frac{1}{3}) > 3(2x + 1)(-\frac{1}{3})$
$\qquad\qquad\qquad -3(2 - x) > -(2x + 1)$

5. $12(x + 3) > 6(x - 4);\quad \frac{1}{6}$ $\qquad$ 6. $\frac{1}{3}(2 - x) < \frac{1}{4}(4 + 3x);\quad 12$
7. $-4(2x + 5) > -2(x - 6);\quad -\frac{1}{2}$
8. $-\frac{1}{5}(2x + 3) < -\frac{1}{3}(3x + 5);\quad -15$
9. Determine the solution set of each inequality in exercises 1–8.

Write each of the following in an abbreviated form:

10. $5 > 2$ or $5 = 2$ $\qquad$ 11. $9 < 12$ and $12 < 16$
12. $0 < 1$ and $1 \leq 5$ $\qquad$ 13. $0 \leq 1$ and $1 \leq 5$
14. $r < s$ or $r = s$ $\qquad$ 15. $r < s$ and $s < t$
16. $r \leq s$ and $s < t$ $\qquad$ 17. $r \leq s$ and $s \leq t$

★18. Prove: If $x + z < y + z$, then $x < y$. (*Hint:* Use the meaning of $<$ and the cancellation property for addition.)

★19. Prove: If $xz < yz [z > 0]$, then $x < y$. (*Hint:* First prove that $xz < yz$ means that there is a positive real number wz such that $xz + wz = yz$.)

6.7 SOLVING EQUATIONS AND INEQUALITIES

REVIEW

- If $x = y$, then $x + z = y + z$
- If $x = y$, then $xz = yz$
- Distributive property:
  $$x(y + z) = xy + xz$$
- If $x < y$, then $x + z < y + z$
- For each *positive* real number z,

  if $x < y$, then $xz < yz$
- For each *negative* real number z,

  if $x < y$, then $xz > yz$
- If $x = y$, then $x^2 = y^2$

OBJECTIVES

- Summarize the number systems studied so far
- Give examples of equations that have solutions in some systems but not in others
- Solve equations
- Solve inequalities

Thus far, we have considered several number systems:

Natural numbers:
   $N = \{1, 2, 3, \ldots\}$

Whole numbers:
   $W = \{0, 1, 2, 3, \ldots\}$

Integers:
   $I = \{\ldots, -2, -1, 0, 1, 2, 3, \ldots\}$

Rational numbers:
   All numbers that have a name of the form a/b, where a and b are integers and $b \neq 0$. We shall use Q for the set of rational numbers.

Real numbers:
   All rational and irrational numbers. We shall use R for the set of real numbers.

There are equations which have solutions in one number system but not in another. Let us illustrate this.

6.7 Solving Equations and Inequalities

Example 1 $x + 7 = 5$ has a solution in I but *not* in W. The solution is -2.

Example 2 $2x = 9$ has a solution in Q but *not* in I. This solution is $\frac{9}{2}$.

Example 3 $x^2 = 3$ has solutions in R but *not* in Q. These solutions are $\sqrt{3}$ and $-\sqrt{3}$.

The equations in Examples 1 and 2 are called *linear* (first degree) because the highest degree of the variable x is 1. (x means x^1.) The equation in Example 3 is *quadratic* because the highest degree of the variable x is 2. It is true that every linear equation has a solution in the set of real numbers.

Study the examples below to see how various properties are used in solving equations.

Example 1 Find the solution set of $3x + 7 = 6$.

$3x + 7 = 6$	GIVEN
$3x + 7 + (-7) = 6 + (-7)$	Adding -7 to each side
$3x = -1$	Adding
$\frac{1}{3} \cdot 3x = \frac{1}{3} \cdot (-1)$	Multiplying each side by $\frac{1}{3}$
$x = -\frac{1}{3}$	Multiplying

The solution seems to be $-\frac{1}{3}$.

CHECK To check, replace x by $-\frac{1}{3}$, and see whether each side of the equation will yield the same number.

$3x + 7$	6
$3(-\frac{1}{3}) + 7$	6
$-1 + 7$	
6	

Thus, the solution set of $3x + 7 = 6$ is $\{-\frac{1}{3}\}$.

Example 2 Find the solution set of $3(2x - 1) + 2 = -2(3 - x) - 5$.

$3(2x - 1) + 2 = -2(3 - x) - 5$	GIVEN
$6x - 3 + 2 = -6 + 2x - 5$	Distributive property of multiplication over addition
$6x - 1 = 2x - 11$	Simplifying
$6x - 1 + (-2x) = 2x - 11 + (-2x)$	Adding $-2x$ to each side
$4x - 1 = -11$	Simplifying
$4x - 1 + 1 = -11 + 1$	Adding 1 to each side
$4x = -10$	Simplifying
$x = -\frac{10}{4}$ or $-\frac{5}{2}$	Multiplying each side by $\frac{1}{4}$

The solution seems to be $-\frac{5}{2}$.

6 Real Numbers, Equations, and Inequalities

CHECK

$3(2x - 1) + 2$	$-2(3 - x) - 5$
$3[2 \cdot (-\frac{5}{2}) - 1] + 2$	$-2[3 - (-\frac{5}{2})] - 5$
$3(-5 - 1) + 2$	$-2(3 + \frac{5}{2}) - 5$
$3 \cdot (-6) + 2$	$-2 \cdot \frac{11}{2} - 5$
$-18 + 2$	$-11 - 5$
-16	-16

Thus, the solution set of the given equation is $\{-\frac{5}{2}\}$.

Example 3 Find the solution set of $5x - 1 < 2x + 3$.

$5x - 1 < 2x + 3$	GIVEN
$5x - 1 + (-2x) < 2x + 3 + (-2x)$	Adding $-2x$ to each side
$3x - 1 < 3$	Simplifying
$3x - 1 + 1 < 3 + 1$	Adding 1 to each side
$3x < 4$	Simplifying
$\frac{1}{3} \cdot 3x < \frac{1}{3} \cdot 4$	Multiplying each side by $\frac{1}{3}$
$x < \frac{4}{3}$	Simplifiying

Thus, the solution set seems to be $\{x | x < \frac{4}{3}\}$. This is read: the set of all x such that $x < \frac{4}{3}$.

Obviously, this is an infinite set. Therefore, checking inequalities is a little more complicated than checking equations. We could not possibly verify an infinite number of solutions. But there is a way to check. We first replace the variable with $\frac{4}{3}$. This should result in the same number on each side of the inequality. Then we try a number less than $\frac{4}{3}$. This should give us a true statement: the number resulting from $5x - 1$ should be less than the number resulting from $2x + 3$. At that point we would feel pretty confident that we have the solution set.

First, try $\frac{4}{3}$ for x:

$5x - 1$	$2x + 3$
$5 \cdot \frac{4}{3} - 1$	$2 \cdot \frac{4}{3} + 3$
$\frac{20}{3} - 1$	$\frac{8}{3} + 3$
$\frac{17}{3}$	$\frac{17}{3}$

Now try a number less than $\frac{4}{3}$ for x, say 1.

$5x - 1$	$2x + 3$
$5 \cdot 1 - 1$	$2 \cdot 1 + 3$
$5 - 1$	$2 + 3$
4	5

And indeed we have a true statement: $4 < 5$. We conclude that $\{x | x < \frac{4}{3}\}$ is the solution set.

6.7 Solving Equations and Inequalities

Example 4 Find the solution set of $\dfrac{5x-3}{-4} > -5$.

$$\dfrac{5x-3}{-4} > -5 \qquad \text{GIVEN}$$

$$-4 \cdot \dfrac{5x-3}{-4} < -4 \cdot (-5) \qquad \text{Multiplying each side by } -4; \text{ reverse the inequality symbol}$$

$$5x - 3 < 20 \qquad \text{Simplifying}$$

$$5x - 3 + 3 < 20 + 3 \qquad \text{Adding 3 to each side}$$

$$5x < 23 \qquad \text{Simplifying}$$

$$\dfrac{1}{5} \cdot 5x < \dfrac{1}{5} \cdot 23 \qquad \text{Multiplying each side by } \dfrac{1}{5}$$

$$x < \dfrac{23}{5} \qquad \text{Simplifying}$$

The solution set seems to be $\{x \mid x < \frac{23}{5}\}$.

CHECK First replace x by $\frac{23}{5}$, hoping to obtain the same number in each member of the inequality.

$\dfrac{5x-3}{-4}$	-5
$\dfrac{5 \cdot \frac{23}{5} - 3}{-4}$	-5
$\dfrac{23 - 3}{-4}$	
$\dfrac{20}{-4}$	
-5	

Now choose a number which is less than $\frac{23}{5}$, say 4. The left-hand member of the inequality should yield a number which is greater than -5.

$\dfrac{5x-3}{-4}$	-5
$\dfrac{5 \cdot 4 - 3}{-4}$	-5
$\dfrac{20-3}{-4}$	
$-\dfrac{17}{4}$	

Since $-\frac{17}{4} > -5$ is true, we are confident that $\{x \mid x < \frac{23}{5}\}$ is the solution set.

6 Real Numbers, Equations, and Inequalities

In solving *radical* equations, that is equations which contain radicals (square roots, for example), some special techniques have to be used. They are based on the principle that for each natural number n:

if $x = y$, then $x^n = y^n$

The examples below illustrate the techniques to be used.

Example 1 Solve:
$\sqrt{x} - 8 = 17$
$\sqrt{x} = 25$ Adding 8 to each side
$(\sqrt{x})^2 = 25^2$ Squaring each side
$x = 625$ Simplifying

The solution set *seems* to be $\{625\}$.

CHECK

$\sqrt{x} - 8$	17
$\sqrt{625} - 8$	17
$25 - 8$	
17	

Thus, $\{625\}$ *is* the solution set.

Example 2 Solve:
$\sqrt{x + 29} = \sqrt{-3x + 13}$
$(\sqrt{x + 29})^2 = (\sqrt{-3x + 13})^2$ Squaring each side
$x + 29 = -3x + 13$ Simplifying
$4x = -16$ Adding -29 to each side
$x = -4$ Simplifying

The solution set *seems* to be $\{-4\}$.

CHECK

$\sqrt{x + 29}$	$\sqrt{-3x + 13}$
$\sqrt{-4 + 29}$	$\sqrt{-3(-4) + 13}$
$\sqrt{25}$	$\sqrt{25}$
5	5

Thus, $\{-4\}$ *is* the solution set.

Example 3 Solve:
$2\sqrt{x} + 3 = 1$
$2\sqrt{x} = -2$ Adding -3 to each side
$\sqrt{x} = -1$ Multiplying each side by $\frac{1}{2}$
$(\sqrt{x})^2 = (-1)^2$ Squaring each side
$x = 1$ Simplifying

The solution set seems to be $\{1\}$.

6.7 Solving Equations and Inequalities

CHECK

$$\begin{array}{c|c} 2\sqrt{x}+3 & 1 \\ \hline 2\sqrt{1}+3 & 1 \\ 2+3 & \\ 5 & \end{array}$$

The last statement, $5 = 1$, is false. Therefore $\{1\}$ is not the solution set. We conclude that the solution set of $2\sqrt{x} + 3 = 1$ in the set of real numbers is the empty set, $\emptyset$. It is easy to see why this is so in the third step of the solving procedure above: $\sqrt{x} = -1$. It is clear that no real number will satisfy this equation since there is no real number whose principal square root is -1.

The technique used in solving radical equations leads to new equations which may or may not be equivalent to the equation with which we started. In Examples 1 and 2 equivalent equations were obtained, but in Example 3 the final equation $x = 1$ was not equivalent to the original equation, $2\sqrt{x} + 3 = 1$.

Exercises 6.7

1. The minimum deposit a bank will accept to open an account is given by the inequality $6x - 1 > 3x + 7$, in which x is the minimum deposit. Your aunt has sent you $3 with instructions that you must *save* the money. Can you open a savings account?
2. Your 5-year-old brother, who is an expert in rocketry, has heard that the maximum height a rocket can attain is given by the formula $s = 600t + 20$, where t is the time after launch in seconds and s is in feet. Because he hasn't yet learned to add, he wants you to do some calculations for him. How long does it take you to determine the following?
   a. How high is the rocket after 9 seconds?
   b. How long after launch will it take to reach a height of 10,000 ft?

Determine the solution set of each of the following equations:

3. $3x + 4 = 13$
4. $2x - 5 = -10$
5. $3(x - 1) = -9$
6. $2(3 - x) = 5$
7. $-2(4x - 5) = 8$
8. $3(2x + 1) - 3 = -2(x - 2) + 3$
9. $-4(1 - 2x) + 1 = 5(2 - x) - 3$
10. $\dfrac{2x + 3}{2} = \dfrac{2 - 5x}{3}$

Determine the solution set of each of the following inequalities:

11. $2x > 12$
12. $3x < -15$
13. $x + 4 > 16$
14. $x - 3 < 17$

15. $2x + 4 > 0$
16. $3x - 2 < -1$
17. $2 - 4x > 5$
18. $2x + 7 > x + 6$
19. $5x + 3 < 3x - 13$
20. $2(2x + 3) > 3(x - 1)$
21. $5(3 - 5x) < -3(2 + 7x)$
22. $\dfrac{2(x - 1)}{3} > \dfrac{-3(2 + 3x)}{4}$

Determine the real number solution sets of the following equations:

23. $\sqrt{x} - 9 = -8$
24. $2\sqrt{x} + 11 = 61$
25. $3 - 2\sqrt{x} = -5$
26. $\sqrt{x + 3} = \sqrt{22}$
27. $\dfrac{\sqrt{2x - 1}}{3} = 1$
28. $\dfrac{\sqrt{x + 2}}{3} = 0$
29. $\dfrac{\sqrt{x + 4}}{2} = \dfrac{\sqrt{7x + 1}}{4}$
30. $\sqrt{3x + 4} = \sqrt{3x + 1}$
31. $\sqrt{x + 1} = \sqrt{3x + 3}$
32. $\sqrt{\dfrac{4 + x}{3}} = \sqrt{\dfrac{7x + 23}{6}}$
33. $\sqrt{\dfrac{1}{9 - x}} = \sqrt{\dfrac{2}{17x + 18}}$

34. If I take the square root of the number of dollars in my wallet, multiply by 3, and subtract 2 from the product obtaining 4, how much did I originally have in my wallet?

35. A long-distance runner who had never taken any math wanted to know how far he had run. The judge told him that he had run $\sqrt{x} - 9 = 3$ miles. Not wishing to demonstrate his ignorance, he took a quick course in math and found out how many miles he had run. Can you find out too?

36. A number is multiplied by 6, and 5 is subtracted from the product. The square root of this difference is 5. What is the number?

37. The square root of the sum of twice a number and 3 is equal to 5. What is the number?

38. The square root of the sum of a number and 1 is equal to the square root of the difference of the number and 15. What is the number?

CHAPTER 6 TEST

For each of the following equations, give the two irrational number solutions:

1. $x^2 = 8$
2. $x^2 = 11$
3. $x^2 - 15 = 0$
4. $x^2 - 2 = 17$

5. In a recent map-reading contest, each contestant was given two equations to solve. The answers were then to be plotted upon a specially designed map. Given the two equations, determine the position of the points (do not draw; just calculate).
   a. $x^2 - 6 = 31$ b. $x^2 + 20 = 22$

6.7 Solving Equations and Inequalities

Using the Pythagorean Theorem, compute the length of the side marked x of each right triangle pictured below, knowing the length of the two sides as is indicated in each picture.

6.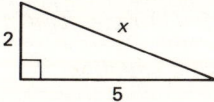

7.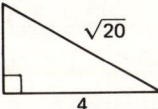

8.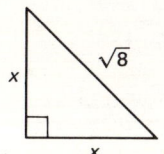

9.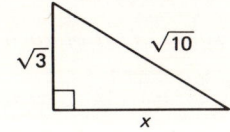

10. A large oil tanker got stuck underneath a drawbridge, distorting the roadway. A surveyor prepared the following diagram:

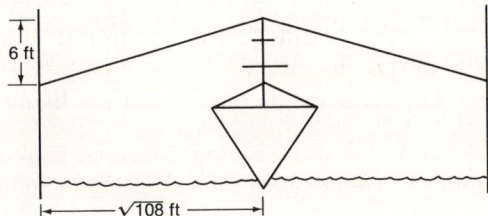

Find how far you would have to drive to traverse the distorted bridge.

11. By construction, locate on the number line the point corresponding to $\sqrt{3}$.
12. Give examples of two irrational numbers whose product is a rational number.

For each of the following rational numbers give another name of the form a/b, where a and b are relatively prime:

13. $\frac{9}{120}$ 14. $\frac{16}{84}$ 15. $\frac{12}{153}$ 16. $\frac{45}{105}$

17. Give an example of two irrational numbers, each in the form of a non-terminating nonrepeating decimal numeral, whose sum is in the form of a repeating decimal.
18. Consider the set of all whole numbers $W = \{0, 1, 2, 3, \ldots\}$. State all properties which W lacks in order to be a field.

Give the additive inverse of each of the following numbers:

19. -3 20. $\sqrt{2}$ 21. 0 22. -9.8

6 Real Numbers, Equations, and Inequalities

Give the multiplicative inverse (reciprocal) of each of the following numbers:

23. $\frac{5}{6}$
24. -5
25. 3
26. $\sqrt{3}$
27. 1.3

For each of the following relations, tell whether it is (1) reflexive (2) symmetric (3) transitive, (4) an equivalence relation.

28. $\not<$ for real numbers
29. $\neq$ for real numbers
30. *is similar to* for triangles

Determine the solution set of each of the following equations:

31. $x + 9 = 4$
32. $2x + 3 = 2$
33. $4(2x - 3) = -28$
34. $3(1 - x) + 1 = 2(2x + 3) - 4$

Determine the solution set of each of the following inequalities.

35. $3x < 36$
36. $x + 7 > -2$
37. $4x - 3 < 21$
38. $2(4x - 1) > 34$
39. $5(4 - 2x) < -21$
40. $2(2x + 1) > -7(3 - x)$
41. The minimum safe cruising altitude for an aircraft in which you are traveling is given by the inequality $48x - 7000 > x - 1000$ (all dimensions in feet). Solve the inequality. If you are flying at 200 ft, are you safe?

Determine the real number solution sets of the following equations:

42. $\sqrt{x - 2} = 3$
43. $\sqrt{x + 1} = 5$
44. $\sqrt{x - 1} = \sqrt{2x + 3}$
45. $\sqrt{2x - 20} = \sqrt{x - 5}$
46. $\sqrt{\dfrac{1}{2 - x}} = \sqrt{\dfrac{2}{3 + 2x}}$

7

Geometry

7.1 A VARIETY OF GEOMETRIES

REVIEW

None

OBJECTIVES

- Name five geometries
- Briefly describe the basic characterizations of the five geometries

In mathematical studies, several geometries are considered. *Topology* is the most general of them. It deals with geometric concepts of a great variety. For example, one concern in topology is with the transformations of one set of points into another. Figure 7.1 portrays a situation where one set of points is transformed into two triangles.

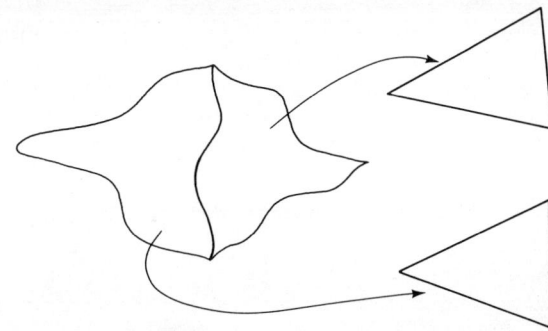

FIGURE 7.1

This kind of transformation is given the name *homeomorphism*. Today topology is a very important well-developed branch of mathematics. It is studied by all students majoring in mathematics.

Another geometry, studied in elementary school, is *nonmetric geometry*. It is related, in some sense, to topology. Each geometric figure, in this geometry, is considered to be a *set of points*. The word nonmetric means that we are not concerned with measurement when studying nonmetric geometry. Some typical ideas studied in this geometry are given in the next section.

The geometry in which we measure and are concerned with the size of objects is *metric geometry*. This is also studied in elementary school. Since the United States is moving into the general use of the metric system, metric units of measure need to be studied. They are presented in Section 7.4.

Another way to approach the study of geometric ideas is through logic. This is called *deductive geometry*, and it is presented in Section 7.6.

Motion geometry is a dynamic way of looking at geometric things. Of particular interest are motions that displace sets of points without distorting them, that is, without changing their shape or size.

Exercises 7.1

1. Name the five geometries discussed in this section.
2. Briefly characterize each of the five geometries.

7.2 NONMETRIC GEOMETRY: FIGURES IN A PLANE

REVIEW

- Intersection of sets
- Union of sets

OBJECTIVES

- Recognize and correctly name the following geometric figures: point, segment, line, half-line, ray, angle, triangle
- Identify and write statements for various unions and intersections of these sets of points
- Distinguish between a triangle and a triangular region

7.2 Nonmetric Geometry: Figures in a Plane

The development of plane geometry dates back to a Greek mathematician by the name of Euclid (about 300 B.C.). He developed geometry as a *deductive system*—a system in which many results are logically deduced from some accepted assumptions.

In this section we will view geometric figures as sets of points. For now we will limit ourselves to sets of points which are subsets of a plane. For example, Figure 7.2 shows a point A located in the plane p. A *point* is the simplest geometric concept.

The next concept, still rather simple, is a *segment*. Figure 7.3 shows a segment, which we denote by $\overline{AB}$, located in plane p.

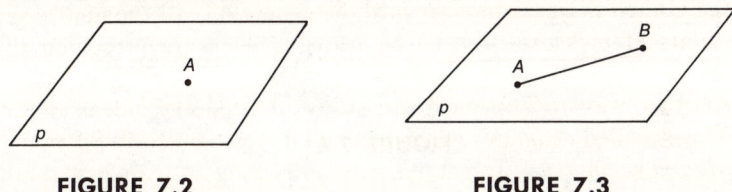

FIGURE 7.2   **FIGURE 7.3**

Since geometric figures are sets of points, the terminology and symbolism of sets will be used in dealing with them. You may wish to refer to Chapter 2 to refresh your memory on some of that symbolism. To write that two segments, shown in Figure 7.4, intersect in one point, we use set notation:

$$\overline{AB} \cap \overline{CD} = \{E\}$$

It is important to recall that the intersection of two sets is a set. For this reason, E is included within braces. This indicates that the intersection is a *set* consisting of point E.

It is possible for two segments not to intersect. In this case, the intersection is the empty set. See Figure 7.5, showing that $\overline{MR} \cap \overline{ST} = \emptyset$.

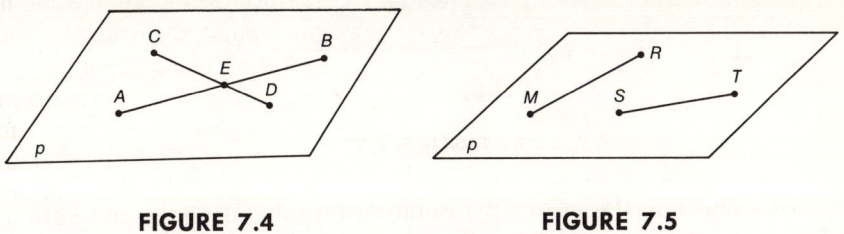

FIGURE 7.4   **FIGURE 7.5**

Some other geometric figures (sets of points) are important in studying geometry:

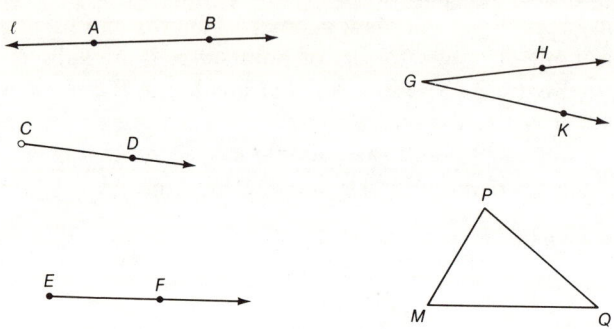

FIGURE 7.6

A *line*, denoted by $\overleftrightarrow{AB}$ or ℓ: extends indefinitely in two directions.
A *half-line*, denoted by $\overset{\circ}{\overrightarrow{CD}}$: all points to one side of a chosen point on a line.
A *ray*, denoted by $\overrightarrow{EF}$: a half-line $\overset{\circ}{\overrightarrow{EF}}$ with point E included.
An *angle*, denoted by $\angle G$, or $\angle HGK$, or $\angle KGH$: a union of two rays, $\overrightarrow{GH}$ and $\overrightarrow{GK}$, with a common endpoint, G.
A *triangle*, denoted by $\triangle MPQ$: a union of three segments, each pair having a common endpoint.

It is important to note that a triangle consists of segments only. If we take a union of a triangle and its interior, $\triangle MPQ \cup I_{\triangle MPQ}$ (see Figure 7.7), we obtain a *triangular region*, $R_{\triangle MPQ}$.

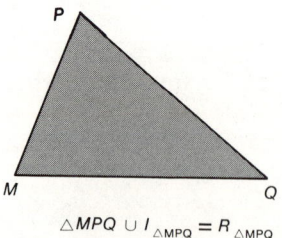

$\triangle MPQ \cup I_{\triangle MPQ} = R_{\triangle MPQ}$

FIGURE 7.7

To illustrate the use of the notation introduced above and various relationships between sets of points, Figure 7.8 is given. Study the state-

7.2 Nonmetric Geometry: Figures in a Plane

ments next to the figure, and relate them to the appropriate parts of the picture.

1. $\overrightarrow{AC} \cup \overrightarrow{AE} = \angle A$
2. $\overline{DC} \cap \overline{CE} = \{C\}$
3. $\overline{BC} \cup \overline{CD} \cup \overline{DB} = \triangle BCD$
4. $\overline{BD} \cap \overset{\circ}{DF} = \emptyset$
5. $\overrightarrow{FE} \cap \overset{\circ}{FA} = \emptyset$
6. $\overrightarrow{FE} \cap \overrightarrow{FA} = \{F\}$
7. $\overrightarrow{FA} \cup \overrightarrow{FE} = \overleftrightarrow{AE}$
8. $\overline{AD} \cup \overline{DC} = \overline{AC}$
9. $\overline{EF} \cup \overrightarrow{FA} = \overrightarrow{EA}$
10. $\overline{EF} \cap \overline{CD} = \emptyset$

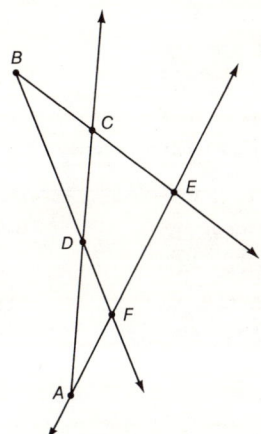

FIGURE 7.8

Exercises 7.2

1. What is the simplest geometric figure?
2. Describe a segment.
3. Identify eight different geometric figures in this picture.

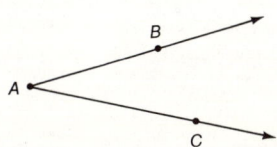

4. Complete the following statement in relation to the picture:

$$R_{\triangle ABC} \cap R_{\triangle DEF} = \underline{\qquad}$$

7 Geometry

5. Complete each of the following statements in relation to the given picture.
   a. $\overline{BC} \cup \overline{CD} = $ _____
   b. $\overrightarrow{AF} \cup \overrightarrow{AB} = $ _____
   c. $\overset{\circ}{DE} \cap \overset{\circ}{DC} = $ _____
   d. $\overrightarrow{DE} \cap \overrightarrow{DC} = $ _____
   e. $\overline{AE} \cap \overrightarrow{DC} = $ _____
   f. $\overrightarrow{AC} \cup \overrightarrow{CG} = $ _____
   g. $\overrightarrow{AF} \cup \overrightarrow{AB} = $ _____
   h. $\overline{AB} \cup \overline{BC} \cup \overline{CA} = $ _____
   i. $\triangle ADC \cup I_{\triangle ADC} = $ _____
   j. $\overline{CA} \cap \overrightarrow{CG} = $ _____

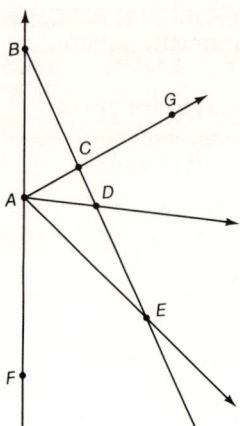

★ 6. Below is a picture of a quadrilateral (4 sides), $ABCD$. It has two diagonals, $\overline{AC}$ and $\overline{BD}$.

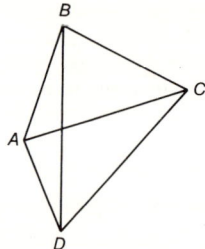

This is a picture of a pentagon (5 sides). It has five diagonals.

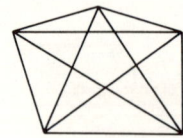

Draw a picture of a hexagon (6 sides) and find out how many diagonals it has. Try to discover a pattern for deciding the number of diagonals any polygon has. How many diagonals does a 10-sided polygon have? 15-sided polygon? 20-sided polygon? n-sided polygon?

7.3 SETS OF POINTS IN SPACE

REVIEW

- Meaning of segment, square, polygon
- Meaning of radius of a circle
- Meaning of congruent

OBJECTIVES

- Define space
- Form a cube in stages, starting with a point
- Know the five regular polyhedra
- Build a cylinder from its component parts
- Build a cone from its component parts
- Define a sphere

Thinking of objects in space is very natural for us. That space is our physical world. In mathematical geometry, we think of *space* as the *set of all points*.

We will extend geometric ideas from the simple to the more complicated, starting with a single point.

Choose a point A.

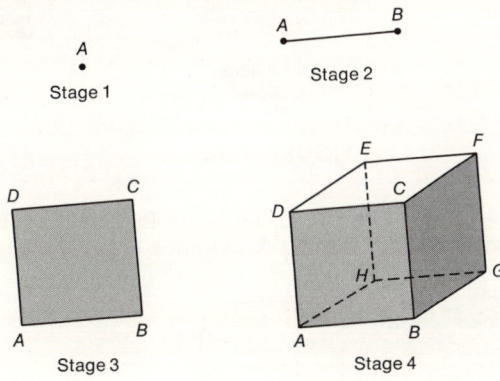

FIGURE 7.9

Move the point a certain distance; a segment is formed, $\overline{AB}$.

Move the segment $\overline{AB}$ parallel to itself the distance equal to the length of the segment; a square is formed. This square determines a plane.

Move the square $ABCD$ parallel to itself the distance equal to the length of the side of the square; a cube is formed. The cube is a space figure.

A cube belongs to a general class of *polyhedra* (singular, *polyhedron*). Special polyhedra are of interest to us, the *regular polyhedra*. They have congruent regular regions for their faces, and the same number of edges meet at each vertex. In Figure 7.10 five regular polyhedra are shown.

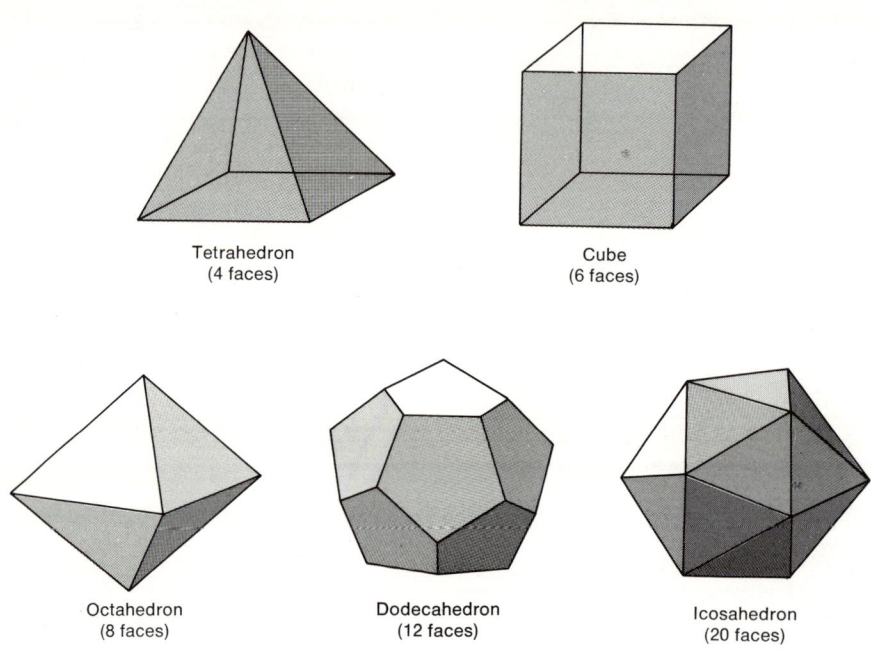

Tetrahedron
(4 faces)

Cube
(6 faces)

Octahedron
(8 faces)

Dodecahedron
(12 faces)

Icosahedron
(20 faces)

FIGURE 7.10

The faces of all the regular polyhedra are polygons. Another kind of space figure involves circular shapes. A cylinder is such a space figure.

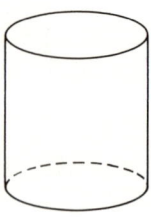

FIGURE 7.11

The *cylinder* is a subset of a more general set of points in space that are a chosen distance from a chosen line. See Figure 7.12.

7.3 Sets of Points in Space

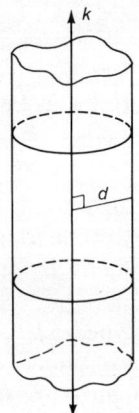

FIGURE 7.12

Line k is the *axis*. The *cylindrical surface*, which is the set of all points the chosen distance from the axis, extends infinitely up and down. If we imagine planes perpendicular to the axis cutting the cylindrical surface, then they intersect the surface in circles.

Another space figure with many uses is a *cone*. You can cut two pieces out of paper so that, put together, they form a cone. What would these pieces be?

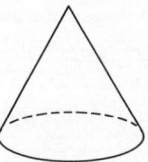

FIGURE 7.13

The shape with the widest reflection in everyday life is the *sphere*. In mathematical geometry the sphere is the set of all points that are a chosen distance from a chosen point. The chosen point is the *center* of the sphere, and the chosen distance is the length of its *radius*.

FIGURE 7.14

7 Geometry

Exercises 7.3

1. What is space in mathematical geometry?
2. a. Imagine a point in space. It is moved along a straight line a distance of 10 cm. What figure is formed? How long is it?
   b. A segment 10 cm long is moved parallel to itself a distance of 10 cm. What figure is formed? How long is each of its sides?
   c. A square with each side 10 cm long is moved parallel to itself a distance of 10 cm. What figure is formed? How long is each of its edges?
3. What are the five regular polyhedra?
4. Cut three parts out of a sheet of paper so that, put together, they form a cylinder. What geometric figures are the three parts?
5. Cut two parts out of a sheet of paper so that, put together, they form a cone. What geometric figures are the two parts?
6. Define a sphere.
★ 7. The picture shows a *dihedral angle*. What geometric figures does this angle consist of? Comparing it with a plane angle, which of its parts correspond to a vertex? Which part corresponds to the sides of a plane angle?

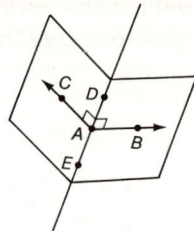

7.4 METRIC GEOMETRY AND THE METRIC SYSTEM

REVIEW

- Positive and negative powers of 10

OBJECTIVES

- Know the relationships between the metric units of linear measure
- Express the relationships between the metric units of length using exponents

7.4 Metric Geometry and the Metric System

- Know the approximate relationships between the English and the metric units of length
- Compute the perimeters of polygons using metric units
- Measure the lengths of objects using metric units
- Compute the circumferences of circles, given the lengths of the radii

One of the most practical kinds of geometry is metric geometry, in which we are concerned with measurement. The three most basic things we measure are length, area, and volume.

For measuring length, the United States is now in the process of changing to the metric system, which is based on the number 10. The regularity of this system is apparent from the units listed in Table 7.1.

Table 7.1 The Metric System

Unit		Abbreviation
	millimeters	mm
10 millimeters	= 1 centimeter	cm
10 centimeters	= 1 decimeter	dm
10 decimeters	= 1 meter	m
10 meters	= 1 decameter	dkm
10 decameters	= 1 hectometer	hm
10 hectometers	= 1 kilometer	km

The relationships between these units can be shown in a step arrangement. The smallest unit is the millimeter, and all the other units are built from it. In each case we multiply by 10. Thus, the relationships between the metric units can be shown as powers of 10. For example,

$1 \text{ m} = 10^2 \text{ cm}$
$1 \text{ mm} = 10^{-3} \text{ m}$

During the transitional period, it is necessary to have some idea how the units in the English system compare with the metric units. This comparison is based on the established exact relationship between the yard and the meter:

1 yd = 0.9144 m (exactly)
1 m = 1.093613 yd

For everyday purposes approximate relationships are sufficient. Using ≐ to denote "is approximately equal to," we have the following:

1 in ≐ 2.5 cm	1 mm ≐ 0.04 in
1 ft ≐ 30 cm	1 cm ≐ 0.4 in
1 yd ≐ 0.9 m	1 m ≐ 3.3 ft
1 mi ≐ 1.6 km	1 m ≐ 1.1 yd
	1 km ≐ 0.6 mi

Once the transition to the metric system is fully accomplished, we can begin to *think metric* and abandon the English system.

The nature of measurement is such that measures are always approximate. They are accurate to the nearest unit used for measuring. For example, using a centimeter ruler, the length of the pencil pictured in Figure 7.15 would be given as 12 cm with accuracy to 1 cm. It is generally agreed to round the measures to the nearest unit. Measures falling half-way between two numbers are rounded *up*. Thus, the length of a pencil falling at 12.5 cm would be given as 13 cm to the nearest centimeter.

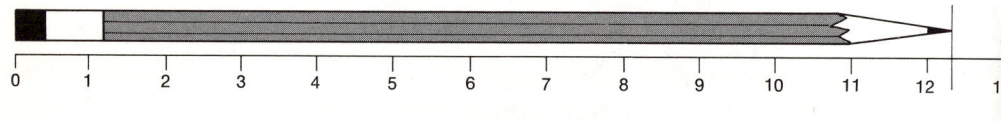

FIGURE 7.15

In geometry, we use linear measures to tell length of segments, perimeters of polygons, and circumferences of circles. Study these examples. The perimeter of this triangle is 2.5 cm + 3.4 cm + 3.8 cm = 9.7 cm

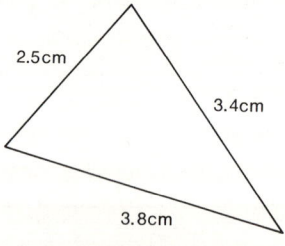

FIGURE 7.16

7.4 Metric Geometry and the Metric System

The perimeter of this rectangle is 2 × 51 mm + 2 × 16 mm = 134 mm

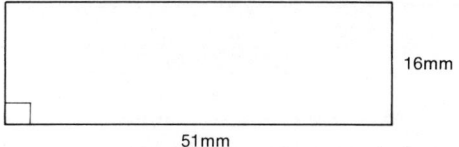

FIGURE 7.17

The length of the radius of this circle is 20 mm. Its circumference is approximately

$$2\pi \cdot 20 \text{ m} = 40\pi \text{ mm}$$
$$\doteq 40 \times 3.14 \text{ mm}$$
$$= 125.6 \text{ mm}$$

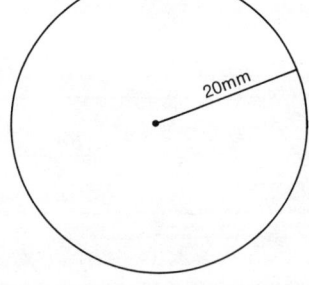

FIGURE 7.18

Pi (π) is an irrational number approximately equal to 3.14 (to two decimal places).

Exercises 7.4

Complete:

1. 1 dm = _____ mm
2. 1 m = _____ cm
3. 1 m = _____ mm
4. 1 km = _____ m
5. 1 km = _____ cm
6. 1 km = _____ mm
7. 1 mm = _____ cm
8. 1 mm = _____ m
9. 1 m = _____ km
10. 1 cm = _____ km
11. 1 mm = _____ km
12. 1 m = _____ mm

Express as powers of 10:

13. 1 mm = _____ cm
14. 1 km = _____ m
15. 1 m = _____ mm
16. 1 km = _____ cm
17. Measure your foot in centimeters.
18. Measure your height in centimeters.

7 Geometry

Measure the sides of the following polygons with accuracy to 1 mm and compute their perimeters:

19.

20.

21.

22.

23.

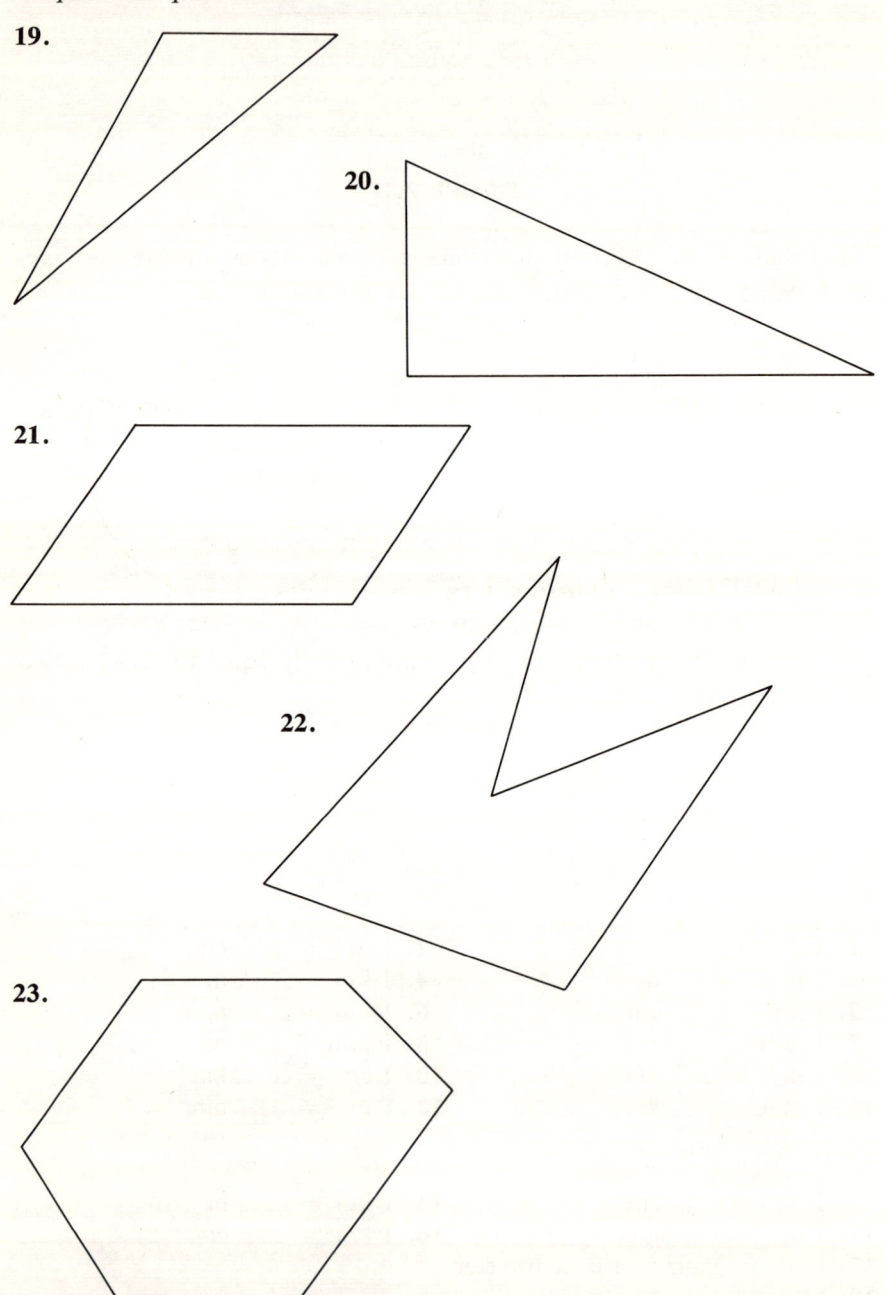

7.5 Areas and Volumes

24. Measure the diameter of this circle with accuracy to 1 mm and compute the circumference, using $\pi \doteq 3.1$.

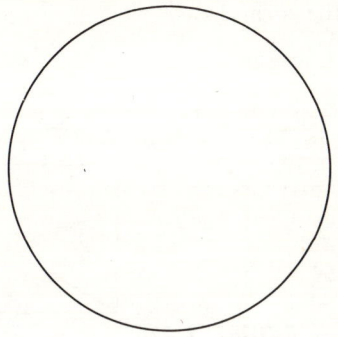

7.5 AREAS AND VOLUMES

REVIEW

- Know about how long a centimeter (cm) is
- Meaning of rectangle, triangle, and parallelogram

OBJECTIVES

- Know the meaning of a square unit
- Compute the areas of rectangles, triangles, parallelograms, trapezoids, and circles
- Compute the volume of a rectangular solid

For measuring areas, we use square units. Figure 7.19 shows the *square centimeter*. When reporting the area of a region, we give the number of square units that would fit into the given region. For example, in Figure 7.20, the area of the rectangular region is 15 sq cm. Generally, the area of a rectangular region is found by multiplying the length and the width. We shall, for simplicity, say *area of a rectangle* and mean the area of the rectangular region.

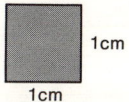

FIGURE 7.19

FIGURE 7.20

We can easily observe that a diagonal of a rectangle divides the rectangle into two right triangles of the same area. Therefore, since the area of a rectangle is given by the formula

$$A = \ell w$$

the area of a right triangle is

$$A = \tfrac{1}{2}\ell w$$

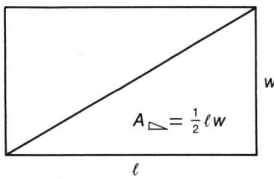

FIGURE 7.21

In considering the area of a parallelogram, we rearrange some parts to make the parallelogram into a rectangle having the same area. In Figure 7.22, we have parallelogram $ABCD$. Its base is b units long and its height is h. Observe that triangles AED and BFC have the same area. Therefore, the area of parallelogram $ABCD$ is the same as the area of rectangle $EFCD$, which is bh. Thus,

$$A_{\square} = bh$$

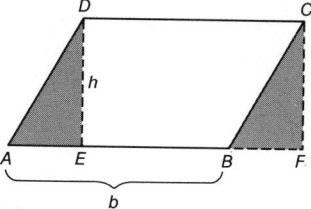

FIGURE 7.22

That is, the area of a parallelogram is the product of its base and height. Since a diagonal of a parallelogram divides it into two triangles of the same area, the area of any triangle is

$$A_{\triangle} = \tfrac{1}{2}bh$$

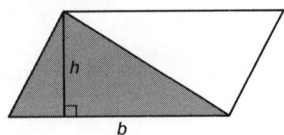

FIGURE 7.23

This formula can be used when computing the area of a triangle, given the length of one of its sides and the altitude upon this side. There is

7.5 Areas and Volumes

another interesting formula, which can be used to compute the area of a triangle given the lengths of its sides. It is called Hero's formula:

$$A = \sqrt{s(s-a)(s-b)(s-c)}$$

where $s = (a+b+c)/2$, that is, s is one-half of the perimeter of the triangle.

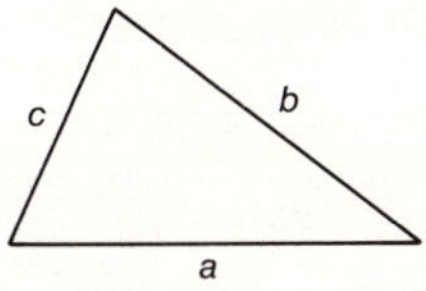

FIGURE 7.24

There is another quadrilateral that is of interest in geometry. It is a quadrilateral with at least one pair of parallel sides, called a *trapezoid*. To find the area of a trapezoid, we take one-half of the product of the height and the sum of the two bases.

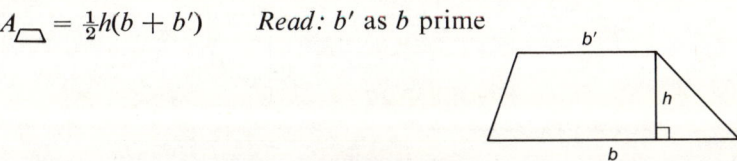

$A_\square = \frac{1}{2}h(b + b')$ Read: b' as b prime

FIGURE 7.25

Since it is rather difficult to reason out the formula for the area of a circle, we will just state it:

$$A_\odot = \pi r^2$$

where r is the length of the radius.

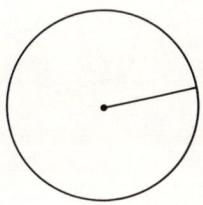

FIGURE 7.26

A cube whose edge has the length of a given unit is a *cubic unit*. The cubic unit is used to measure volume. Figure 7.27 shows a *cubic centimeter*.

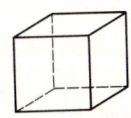

FIGURE 7.27

7 Geometry

When giving the volume of a rectangular solid, we use a number which tells the number of cubic units that fit into the solid. For example, the volume of the rectangular solid in Figure 7.28 is $4 \times 3 \times 2$ cu cm, or 24 cu cm. In general, the volume of a rectangular solid is found by multiplying its length, width, and height.

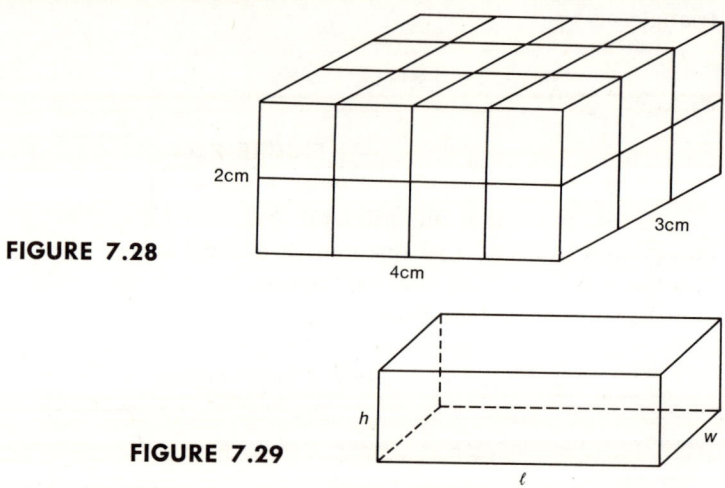

FIGURE 7.28

FIGURE 7.29

Exercises 7.5

1. What is the area of a square with the length of the side 1 inch?
2. Using 1 in $\doteq$ 2.5 cm, approximately how many square centimeters are there in a square inch?

Compute the area of each of the following:

3.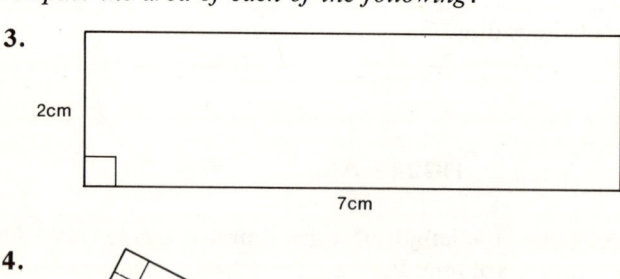

4.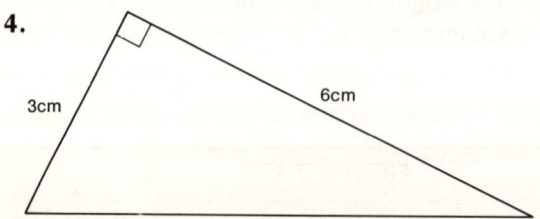

7.5 Areas and Volumes

5.

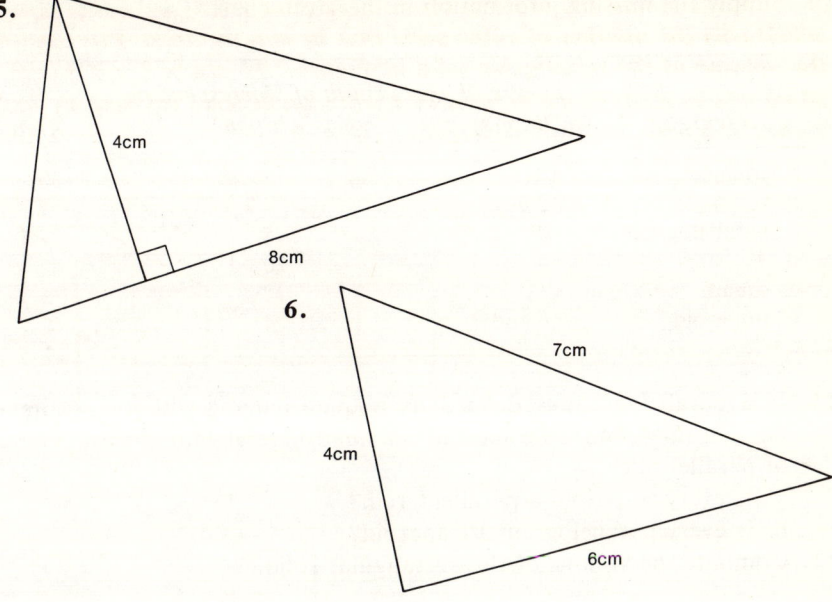

6.

7.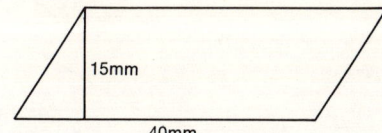

8.

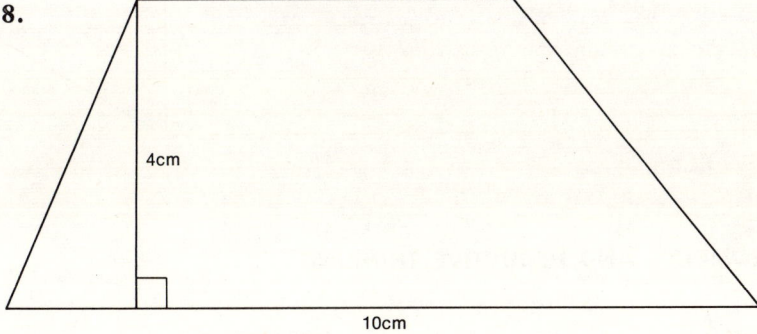

9.

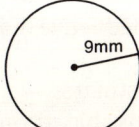

7 Geometry

10. Supply the missing information in the given chart.

Figure	Length of Base(s)	Length of Corresponding Altitude	Area
a. triangle	10	?	30
b. rectangle	12.6	4.2	?
c. parallelogram	?	5	37.5
d. ?	4	20	40
e. square	$\sqrt{3}$	?	?
f. trapezoid	10 and ?	4	60

11. If a parallelogram is defined as a quadrilateral with two pairs of parallel sides, and if a trapezoid is a quadrilateral with at least one pair of parallel sides,
   a. Is every trapezoid a parallelogram?
   b. Is every parallelogram a trapezoid?

12. Compute the volume of this rectangular solid.

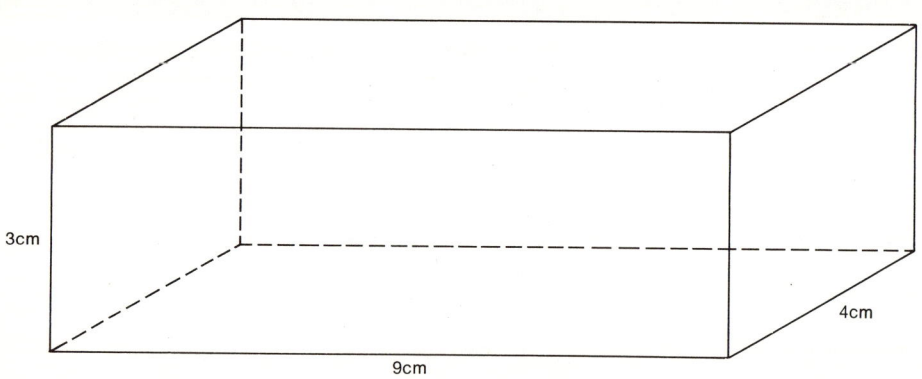

7.6 GEOMETRY AND DEDUCTIVE THINKING

REVIEW

• Intersection of sets

OBJECTIVES

• Explore the deductive nature of geometry
• State eight postulates
• Prove theorems, based on the accepted postulates

7.6 Geometry and Deductive Thinking

Geometry, like other parts of mathematics, is often viewed as a *deductive system*, that is, as a system in which we study the logical consequences of accepted assumptions. The statements we prove are called *theorems*, and the statements we accept as true without proof are called *postulates*.

To illustrate this kind of geometry, we shall state some postulates and show how they are used in proving a theorem. You should have no trouble seeing that these postulates agree with your intuitive observations.

POSTULATES
1. Any three noncollinear points are contained in a unique plane.
2. Any two parallel lines are contained in a unique plane.
3. A line contains at least two points.
4. A plane contains at least three noncollinear points.
5. Space contains at least four noncoplanar points.
6. Any two points are contained in exactly one line.
7. If a line contains two points of a plane, then the line is a subset of the plane.

A few key terms used above need some clarification.

Noncollinear points are points which cannot belong to the same line.
Parallel lines are coplanar (contained in the same plane) and they do not intersect.
Noncoplanar points are points which cannot belong to the same plane.

We shall now prove a theorem. For each statement in our proof, we state the reason which supports the statement.

STATEMENT	REASON
1. k_1 and k_2 are distinct intersecting lines ($k_1 \neq k_2$)	1. Given
2. Let k_1 and k_2 intersect in two points, P_1 and P_2	2. We are making this assumption (hoping that it will lead to a contradiction)

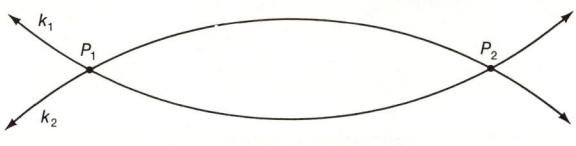

FIGURE 7.30

3. P_1 and P_2 are contained in exactly one line ($k_1 = k_2$)	3. Postulate 6

4. Statements 1 and 3 contradict each other	4. $k_1 \neq k_2$ (statement 1) and $k_1 = k_2$ (statement 3)
5. Assumption in statement 2 is false	5. This assumption leads to contradictory statements
6. Therefore, k_1 and k_2 intersect in exactly one point	

Note that we proved that two lines cannot intersect in *two* points. This means that they cannot intersect in more than two points either, since if two lines had more than two points in common, they would have two points in common as well. However, they must intersect, according to the given part of the theorem. Thus, they must intersect in one point.

Exercises 7.6

1. We proved that if two lines intersect, then their intersection is a point. Consider the possible intersections of two planes. Can two planes intersect in
    a. exactly one point?
    b. exactly two points?
    c. exactly three noncollinear points?
    d. exactly one line?
2. How many different lines are determined by three noncollinear points?
3. How many different planes can contain three collinear points?
4. In how many different ways can a line and a plane intersect?
5. Let's add another postulate to the seven stated in this section:

    POSTULATE 8. If two planes intersect, then their intersection is exactly one line.

    Now prove the following theorem: If a line intersects a plane not containing the line, then the intersection is exactly one point.
6. Prove: There is exactly one plane containing a line and a point not contained in the line.
7. Prove: If two lines intersect, then there is exactly one plane containing them.

7.7 MOTION GEOMETRY

REVIEW

None

OBJECTIVES

- Know the meaning of a rigid motion
- Know the meaning of the three motions (transformations): slide (translation), flip (reflection), and turn (rotation)
- Recognize which of the three transformations is represented by a given picture

The last kind of geometry we consider is one in which geometric figures are thought of in terms of motion. We are particularly interested in those motions that leave the size and the shape of a figure unchanged. These motions are called *rigid motions*. Technically we say that a rigid motion preserves congruence. A more technical term for a motion is *transformation*.

One transformation is a *slide*, more technically called a *translation*. This is a displacement of a figure along the straight line, as illustrated in Figure 7.31. It is easy to see that "sliding" a figure does not change its shape or size.

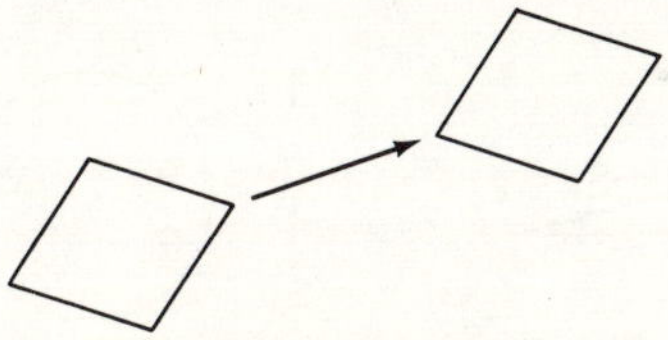

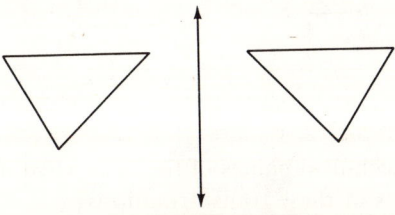

FIGURE 7.31

Another transformation is a *flip* or, more technically, a *reflection*. A flip is just like a mirror reflection about a line, called the *axis of reflection*, as illustrated in Figure 7.32.

FIGURE 7.32

7 Geometry

The third transformation is a *turn*, given the technical name of *rotation*. As both terms suggest, this is a turn of a figure about a point of rotation. Turn also leaves a figure unchanged, as suggested in Figure 7.33.

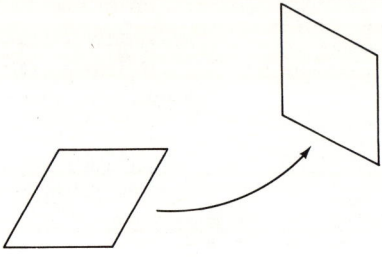

FIGURE 7.33

Exercises 7.7

1. Tell whether each picture is a translation, reflection, or rotation.

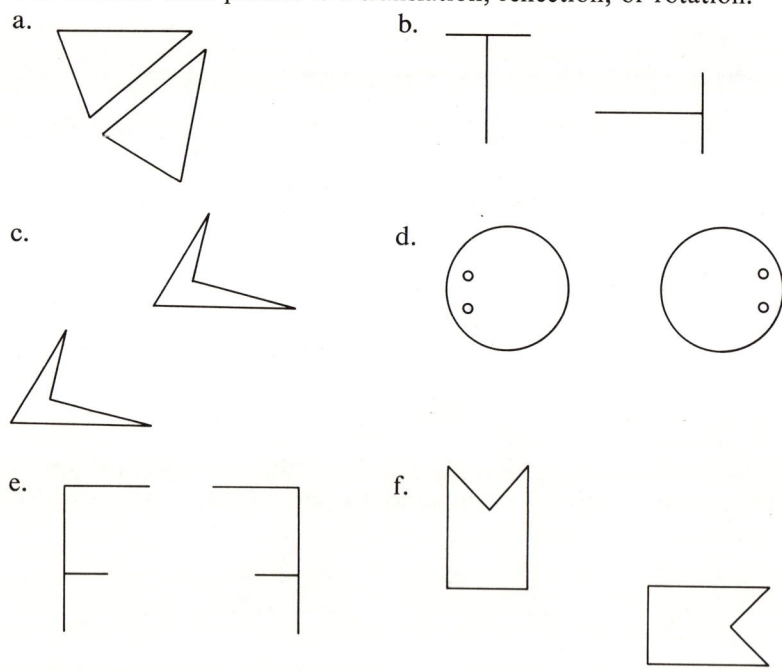

2. What are the nontechnical names of the three rigid motions? What are the technical names of these transformations?
3. Why are these transformations called rigid motions?

CHAPTER 7 TEST

1. What are the five geometries discussed in this chapter?
2. Identify all possible sets of points in this figure.

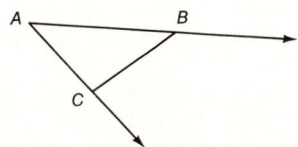

3. Complete each of the following statements in relation to the given picture.
   a. $\overline{CD} \cup \overline{DE} =$ ____
   b. $\overline{AC} \cup \overrightarrow{CE} =$ ____
   c. $\overline{CB} \cap \overline{DB} =$ ____
   d. $\triangle BCD \cap \triangle BDE =$ ____
   e. $R_{\triangle BCD} \cap R_{\triangle CBE} =$ ____
   f. $\overrightarrow{AB} \cap \overrightarrow{DB} =$ ____
   g. $\overrightarrow{AC} \cup \overrightarrow{AD} =$ ____
   h. $\overrightarrow{DC} \cup \overrightarrow{DE} =$ ____
   i. $\overline{CD} \cap \overset{\circ}{\overrightarrow{DE}} =$ ____
   j. $\overset{\circ}{\overrightarrow{DC}} \cap \overset{\circ}{\overrightarrow{DE}} =$ ____

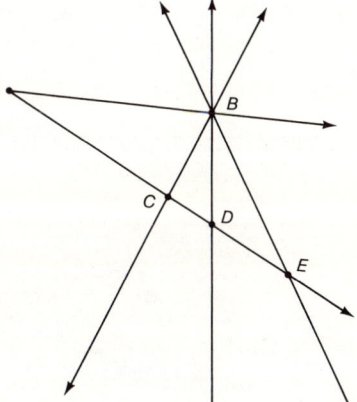

4. Sketch a picture of each of the following.
   a. a cube
   b. a cylinder
   c. a cone
   d. a sphere
   e. a tetrahedron
   f. a rectangular solid
5. Complete.
   a. 1 m = ____ cm
   b. 1 km = ____ m
   c. 1 dm = ____ cm
   d. 1 mm = ____ cm
   e. 1 cm = ____ m
   f. 1 m = ____ km
6. What is the area of a rectangle with length 13 cm and width 5 cm?
7. What is the area of a trapezoid with two bases 12 cm and 26 cm long and altitude 7 cm?
8. What is the area of a circle with radius of length 10 cm? Compute the answer using $\pi \doteq 3.14$.
9. What is the intersection of two planes, if they do intersect?

10. For each picture, tell whether it is a translation, reflection, or rotation.

a.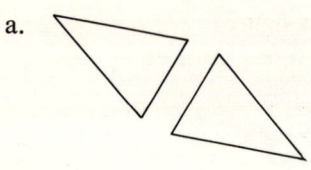

b.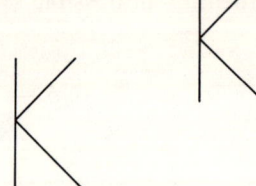

c.

d.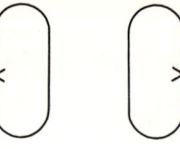

8

Finite Systems and Matrices

8.1 ADDITION IN A THREE-NUMBER SYSTEM

REVIEW

- Closure property of an operation in set S: the result of operating on any members in S is in S
- Commutative property of addition:

  $x + y = y + x$

- Associative property of addition:

  $(x + y) + z = x + (y + z)$

- Additive identity: there is a unique member e in S such that for every x in S,

  $x + e = x$

- Additive inverse: for every element x in S there is a unique element y such that

  $x + y = e$

  where e is the additive identity

OBJECTIVES

- Define a three-member system, S_3
- Define addition in S_3
- Build an addition table for modulo 3 arithmetic
- State and verify the following properties of addition modulo 3:
  closure property
  commutative property
  associative property
  property of additive identity
  property of additive inverse

8 Finite Systems and Matrices

So far we have been working with infinite sets of numbers. It is not necessary to have infinite sets, however, in order to see how a set can be a *system*, that is, a set of objects with one or more operations.

To demonstrate this, let's choose a set with three numbers in it. We'll call it S_3 *(read: S sub-three)*.

$$S_3 = \{0, 1, 2\}$$

To perform operations in this system, it will be helpful to have a pictorial model of it. A three-hour clock is a convenient model. (See Figure 8.1.) Think of addition as motions in the clockwise direction, as indicated in Figure 8.2.

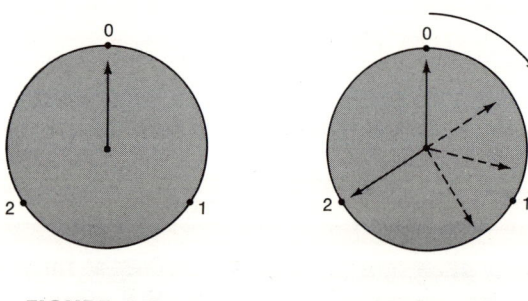

FIGURE 8.1   **FIGURE 8.2**

Here is a simple interpretation of addition in terms of our model:

$0 + 2$ means start at 0 and move 2 notches in the clockwise direction

$0 + 2 = 2$

Another example:

$2 + 2$ means start at 2 and move 2 notches in the clockwise direction

$2 + 2 = 1$

We sometimes call this system *arithmetic modulo 3*. In ordinary arithmetic, $2 + 2$ would be 4. If you subtract 3, you have the answer in S_3. We use a new notation to indicate this:

$4 = 1 \pmod{3}$ *Read:* four is equal to one *modulo* three

8.1 Addition in a Three-Number System

We can perform all possible additions in S_3 and obtain the addition table:

+	0	1	2
0	0	1	2
1	1	2	0
2	2	0	1

Addition table for
$S_3 = \{0, 1, 2\}$

There are advantages in working with a small finite system. The various properties of operations can be proved by simply verifying all cases. For example, in the real number system we had to assume commutativity of addition—there is no way to verify an infinite number of cases. Let's state the essential properties of addition in S_3. You should work out some examples of your own to see that these properties are indeed true.

1. Notice that every possible addition combination of members of S_3 has an answer in S_3. That is,

    for every $x \in S_3$ and every $y \in S_3$, $(x + y) \in S_3$

    Thus, S_3 has the *closure property of addition*.

2. It is easy to verify that for every $x \in S_3$ and every $y \in S_3$,

    $x + y = y + x$

    Thus, S_3 has the *commutative property of addition*.

3. The following computations suggest another property of addition.

    $(2 + 1) + 2 = 0 + 2 = 2$
    $2 + (1 + 2) = 2 + 0 = 2$

    $(1 + 0) + 2 = 1 + 2 = 0$
    $1 + (0 + 2) = 1 + 2 = 0$

    These two examples illustrate the *associative property of addition*. It is easy to verify that

    for every $x \in S_3$ and every $y \in S_3$ and every $z \in S_3$, $(x + y) + z = x + (y + z)$

4. There is one number in S_3 which has the following property: the sum of *any* number in S_3 and that number is equal to the former number. This number is 0. This property can be stated as follows:

for every $x \in S_3$, $x + 0 = 0 + x = x$

Thus, 0 is the *additive identity* in S_3.

5. For each number in S_3, there is a number such that the sum of the two is equal to 0:

$$0 + 0 = 0$$
$$1 + 2 = 0$$
$$2 + 1 = 0$$

Thus,

for every $x \in S_3$, there exists a unique *additive inverse* $y \in S_3$ such that $x + y = 0$

These five properties will be discussed from time to time in connection with other mathematical systems.

Exercises 8.1

1. State the closure property for the addition of whole numbers.
2. Does the addition of whole numbers have the associative property? Give two examples to illustrate it, using the numbers 5, 16, and 184 for the first example and 0, 275, and 25 for the second example.
3. How many hours does the hand of a normal clock sweep in representing $0 + 2$ in the S_3 system?
4. You come from the planet Tripos, where only S_3 arithmetic is used. Upon visiting Earth, you are amazed at our cumbersome base-ten system. Quickly show us poor Earthlings how easy your system is by computing the following sums in S_3.
   a. $1 + 2$ b. $0 + 1$ c. $2 + 2$ d. $1 + 1$ e. $0 + 2$
5. After delivering a paper on behalf of the planet Tripos, you get into a heated debate with one of the senior mathematicians at the convention. He argues that the additive inverses of the following numbers are as shown.
   a. 2, −2   b. 4, −4   c. 1, −1
   What do you say the additive inverses are?

8.2 Other Operations and Their Properties in S_3

Give answers to the following in S_3:

6. $1 + 2$
7. $0 + 1$
8. $2 + 2$
9. $1 + 1$
10. $0 + 2$
11. What is the additive inverse of 2 in S_3?
12. What does it mean to say that 0 is the additive identity in S_3?
13. Is 0 the additive identity in the set of whole numbers?
14. Does every whole number have an additive inverse which is also a whole number?
15. How many whole numbers have additive inverses in the set of whole numbers?
16. In what infinite set of numbers does every number have a unique additive inverse?

8.2 OTHER OPERATIONS AND THEIR PROPERTIES IN S_3

REVIEW

- Multiplicative identity: there is a unique member e in a set S such that for every x in S,

  $xe = x$

- Multiplicative inverse: for every element x in S there is a unique element y such that

  $xy = e$

  where e is the multiplicative identity.

OBJECTIVES

- State all cases of subtraction in S_3 by relating subtraction to addition
- Define multiplication in S_3 in terms of addition
- State all cases of multiplication in S_3
- State all cases of division in S_3 by relating it to multiplication
- State all properties of addition and multiplication in S_3
- Show that S_3 is a field under addition and multiplication

The operation of subtraction is related to the operation of addition in the manner shown by the following example:

If $1 + 2 = 3$, then $3 - 2 = 1$

In general,

If $a + b = c$, then $c - b = a$

8 Finite Systems and Matrices

On the basis of this relationship, the following are all combinations of subtraction in S_3.

$$
\begin{array}{ll}
0 + 0 = 0 \rightarrow & 0 - 0 = 0 \\
1 + 0 = 1 \rightarrow & 1 - 0 = 1 \\
2 + 0 = 2 \rightarrow & 2 - 0 = 2 \\
0 + 1 = 1 \rightarrow & 1 - 1 = 0 \\
1 + 1 = 2 \rightarrow & 2 - 1 = 1 \\
2 + 1 = 0 \rightarrow & 0 - 1 = 2 \\
0 + 2 = 2 \rightarrow & 2 - 2 = 0 \\
1 + 2 = 0 \rightarrow & 0 - 2 = 1 \\
2 + 2 = 1 \rightarrow & 1 - 2 = 2
\end{array}
$$

It is clear that subtraction has the property of closure in S_3, since each difference of two numbers in S_3 belongs to S_3.

For multiplication in S_3, the following are assumed to be true:

For every number x in S_3, $x \cdot 0 = 0$
For every number x in S_3, $x \cdot 1 = x$

The second statement reveals that 1 is the *multiplicative identity* in S_3. The table contains all multiplication facts in S_3.

×	0	1	2
0	0	0	0
1	0	1	2
2	0	2	1

Multiplication table for S_3

The entry $2 \times 2 = 1$ in the table above can be thought of in terms of addition: 2×2 means two 2s or $2 + 2$, which is equal to 1. Observe in the multiplication table that each number in S_3, except 0, has a *multiplicative inverse*; that is, there is one number for 1, which when multiplied by 1, gives 1 for the product. The same is true for the number 2: $2 \times 2 = 1$. It can be stated that

For every $x \neq 0$ in S_3, there exists a unique y such that $xy = 1$

Just as addition and subtraction are related (they are inverses of each other), so are multiplication and division, excluding, of course, division by

8.2 Other Operations and Their Properties in S_3

0. The following display shows this relationship, which gives a method of determining quotients in S_3.

$$1 \times 0 = 0 \rightarrow \quad 0 \div 1 = 0$$
$$1 \times 1 = 1 \rightarrow \quad 1 \div 1 = 1$$
$$1 \times 2 = 2 \rightarrow \quad 2 \div 2 = 1 \text{ and } 2 \div 1 = 2$$
$$2 \times 0 = 0 \rightarrow \quad 0 \div 2 = 0$$
$$2 \times 2 = 1 \rightarrow \quad 1 \div 2 = 2$$

In general,

If $ab = c$, then $c \div a = b (a \neq 0)$ and $c \div b = a (b \neq 0)$

The following is a summary of all of the essential properties of addition and multiplication in S_3.

1. **CLOSURE PROPERTY OF ADDITION**

    For all numbers x and y in S_3, $(x + y) \in S_3$

2. **CLOSURE PROPERTY OF MULTIPLICATION**

    For all numbers x and y in S_3, $(xy) \in S_3$

3. **COMMUTATIVE PROPERTY OF ADDITION**

    For all numbers x and y in S_3, $x + y = y + x$

4. **COMMUTATIVE PROPERTY OF MULTIPLICATION**

    For all numbers x and y in S_3, $xy = yx$

5. **ASSOCIATIVE PROPERTY OF ADDITION**

    For all numbers x, y, and z in S_3, $(x + y) + z = x + (y + z)$

6. **ASSOCIATIVE PROPERTY OF MULTIPLICATION**

    For all numbers x, y, and z in S_3, $(xy)z = x(yz)$

7. **DISTRIBUTIVE PROPERTY OF MULTIPLICATION OVER ADDITION**

    For all numbers x, y, and z in S_3, $x(y + z) = (xy) + (xz)$

8. **PROPERTY OF ADDITIVE IDENTITY**

    There is an additive identity in S_3, namely 0, such that for every number x in S_3, $x + 0 = 0 + x = x$

8 Finite Systems and Matrices

9. PROPERTY OF MULTIPLICATIVE IDENTITY

There is a multiplicative identity in S_3, namely 1, such that for every number x in S_3, $x \cdot 1 = 1 \cdot x = x$

10. PROPERTY OF ADDITIVE INVERSE

For every x in S_3, there is an additive inverse of x, say y, such that $x + y = 0$

11. PROPERTY OF MULTIPLICATIVE INVERSE

For every number $x \neq 0$ in S_3, there is a multiplicative inverse of x, say z, such that $xz = 1$

Checking the properties listed above with those stated for the set of real numbers, we can conclude that S_3 is a *field* under the operations of addition and multiplication. Thus, a set does not have to have lots of members in order to be a field.

Exercises 8.2

1. Use the three-hour clock and define subtraction in S_3 in terms of this clock.
2. Using the three-hour clock, show that $1 - 2 = 2$.
3. Define multiplication in S_3 in terms of addition, and then show by addition that $2 \times 2 = 1$.
4. List the additive inverse for each member of S_3.
5. List the multiplicative inverse for each member of S_3.
6. Verify the distributive property of multiplication over addition for the following case: $x = 1$, $y = 0$, $z = 2$.
7. The planet Tripos wishes to invade the planet Earth and impose its will upon the gentle people who live here. Before they can invade, the Tripans must calculate the following areas in terms of S_3, their math system. As chief Tripan battle planner, it is your job to do this task. Proceed.
   a. 2×1 b. 2×3 c. 4×2
   (All are in Earth notation. Change to S_3.)

8.3 ANOTHER SYSTEM—S_5

REVIEW

- The distributive property of multiplication over addition:

  $x(y + z) = (xy) + (xz)$

OBJECTIVES

- Add in S_5 in terms of a five-hour clock
- Complete an addition table in S_5
- Define subtraction in S_5 in terms of addition
- Define multiplication in S_5 in terms of addition
- Verify the associative properties of addition and multiplication in S_5
- Verify the distributive property of multiplication over addition in S_5
- Verify the distributive property of multiplication over subtraction in S_5
- Observe properties of abstract operations in an abstract system

For reasons which will become apparent a little later, we will consider S_5 before looking at S_4. S_5 has five elements:

$S_5 = \{0, 1, 2, 3, 4\}$

Just as in the case of S_3, we can use a clock as a model for addition in S_5. Figure 8.3 shows the five-hour clock.

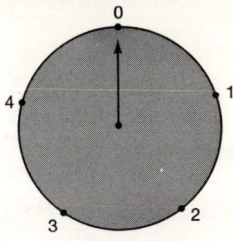

FIGURE 8.3

8 Finite Systems and Matrices

Using the same approach as in S_3, you should be able to complete the addition table below.

+	0	1	2	3	4
0				3	
1					0
2					
3				1	
4	4			2	

Addition table for S_5
(partially completed)

From the completed table, it should be obvious that 0 is the additive identity in S_5.

The statement $0 + 0 = 0$ means that 0 is the additive inverse of 0.

$1 + 4 = 0$ means that 1 is the additive inverse of 4 and 4 is the additive inverse of 1.

$2 + 3 = 0$ means that 2 is the additive inverse of 3 and 3 is the additive inverse of 2.

Thus, every member of S_5 has a unique additive inverse.

Compute the answers to the following to verify that addition in S_5 has the associative property. The first example is completed.

Example 1 $(0 + 4) + 1, \quad 0 + (4 + 1)$

Solution $(0 + 4) + 1 = 4 + 1 = 0$
$0 + (4 + 1) = 0 + 0 = 0$

Example 2 $(2 + 3) + 3, \quad 2 + (3 + 3)$
Example 3 $(4 + 1) + 2, \quad 4 + (1 + 2)$
Example 4 $(1 + 1) + 4, \quad 1 + (1 + 4)$
Example 5 $(4 + 3) + 3, \quad 4 + (3 + 3)$
Example 6 $(2 + 3) + 4, \quad 2 + (3 + 4)$

The addition table above can be used to perform subtraction in S_5. From the relation

If $a + b = c$, then $c - a = b$ and $c - b = a$

it follows that

If $4 + 3 = 2$, then $2 - 4 = 3$ and $2 - 3 = 4$

To construct a multiplication table for S_5, think of multiplication in terms of addition. For example, 4×2 means four 2s or $2 + 2 + 2 + 2$.

8.3 Another System—S_5

Now, $2 + 2 + 2 + 2 = 4 + 2 + 2 = 1 + 2 = 3$. Complete the multiplication table.

×	0	1	2	3	4
0			0		
1					4
2				1	
3				4	
4			3		

Multiplication table for S_5
(partially completed)

The completed multiplication table should reveal that 1 is the multiplicative identity in S_5.

The statement $1 \times 1 = 1$ means that 1 is the multiplicative inverse of 1.

$2 \times 3 = 1$ means that 2 is the multiplicative inverse of 3 and 3 is the multiplicative inverse of 2.

$4 \times 4 = 1$ means that 4 is the multiplicative inverse of 4.

Thus, every number of S_5, except 0, has a unique multiplicative inverse in S_5.

Compute the answers to the following to verify that multiplication in S_5 has the associative property. The first example is completed.

Example 1 $(2 \times 2) \times 3$, $2 \times (2 \times 3)$

Solution $(2 \times 2) \times 3 = 4 \times 3 = 2$
$2 \times (2 \times 3) = 2 \times 1 = 2$

Example 2 $(0 \times 4) \times 2$, $0 \times (4 \times 2)$
Example 3 $(1 \times 3) \times 3$, $1 \times (3 \times 3)$
Example 4 $(3 \times 1) \times 4$, $3 \times (1 \times 4)$
Example 5 $(2 \times 3) \times 4$, $2 \times (3 \times 4)$
Example 6 $(3 \times 4) \times 4$, $3 \times (4 \times 4)$

The multiplication table above can be used to perform division in S_5. From the relation

If $a \times b = c$, then $c \div b = a$ and $c \div a = b$

it follows that

If $4 \times 2 = 3$, then $3 \div 4 = 2$ and $3 \div 2 = 4$

8 Finite Systems and Matrices

The distributive property of multiplication over addition can also be verified by working out specific cases. Complete the cases below (the first example is worked out).

Example 1 $2 \times (1 + 0)$, $(2 \times 1) + (2 \times 0)$

Solution $2 \times (1 + 0) = 2 \times 1 = 2$
$(2 \times 1) + (2 \times 0) = 2 + 0 = 2$

Example 2 $0 \times (3 + 4)$, $(0 \times 3) + (0 \times 4)$
Example 3 $2 \times (4 + 1)$, $(2 \times 4) + (2 \times 1)$
Example 4 $3 \times (0 + 1)$, $(3 \times 0) + (3 \times 1)$
Example 5 $1 \times (4 + 2)$, $(1 \times 4) + (1 \times 2)$
Example 6 $4 \times (4 + 0)$, $(4 \times 4) + (4 \times 0)$

Working with finite systems reveals an important difference between infinite and finite systems. To establish that something is true for all members of an infinite system, we must either assume it or prove logically that it is true. In a finite system, it is possible to verify all cases. And, of course, verifying all cases is easy if the number of elements is small.

Exercises 8.3

1. Verify the eleven properties in Section 8.2 to show that S_5 is a field under addition and multiplication.
2. Using a clock face for the S_5 system, calculate the number of hours between the following. (*Note:* Use a normal 12-hour clock superimposed on an S_5 clock.)
   a. $4 + 2$ b. $1 + 3$ c. $3 + 4$

Use the addition and the multiplication tables in this section to determine the answers to the following:

3. $2 \times (3 - 1)$, $(2 \times 3) - (2 \times 1)$
4. $4 \times (1 - 2)$, $(4 \times 1) - (4 \times 2)$
5. $1 \times (1 - 4)$, $(1 \times 1) - (1 \times 4)$
6. $2 \times (4 - 2)$, $(2 \times 4) - (2 \times 2)$
7. $3 \times (3 - 4)$, $(3 \times 3) - (3 \times 4)$
8. $4 \times (0 - 4)$, $(4 \times 0) - (4 \times 4)$
9. $2 \times (2 - 4)$, $(2 \times 2) - (2 \times 4)$
10. We have already met the warlike people of Tripos. An even more warlike people are the Pentans from Pentos. They have also decided to conquer the planet Earth. They have to convert some of our earthly measures to the S_5 system in order to be able to deploy their forces

successfully. This time you are battle commander for Pentos. Work out the following areas in S_5.
a. $(2 \times (3 - 1)) \times (2 - 1)$ b. 3×3
c. $(3 + 1) \times (3 + 3)$ d. $(4 + (3 - 1)) \times (4 - 2)$
11. What property is suggested by exercises 3–9?

Suppose you are given a system consisting of the elements a, b, c, d, and e, and two operations α and β. The operations are defined by the tables below.

α	a	b	c	d	e
a	a	b	c	d	e
b	b	c	d	e	a
c	c	d	e	a	b
d	d	e	a	b	c
e	e	a	b	c	d

β	a	b	c	d	e
a	a	a	a	a	a
b	a	b	c	d	e
c	a	c	e	b	d
d	a	d	b	e	c
e	a	e	d	c	b

12. Is there an identity for α in this system? What is it?
13. Is there an identity for β in this system? What is it?
14. Does the system have closure under α? Under β?
15. Do you think β is distributive over α? Try a few examples.
16. Do you think α is distributive over β? Try a few examples.

8.4 THE SYSTEM S_4

REVIEW

- S_n is a system of numbers with n elements: $0, 1, 2, \ldots, n - 1$

OBJECTIVES

- Define addition in S_4
- Identify the additive identity and additive inverses in S_4
- Define multiplication in S_4
- Explore properties of addition and multiplication in S_4
- Observe that S_4 is not a field under addition and multiplication
- Explore addition and multiplication in S_6
- Explore factorizations in S_6
- Explore square roots in S_6
- Explore a three-letter system under an abstract operation

8 Finite Systems and Matrices

We found that S_3 and S_5 are both fields under the operations of addition and multiplication. Could it be that S_n, where n is any natural number greater than 2, is a field? Before leaping to any conclusions, let's explore $S_4 = \{0, 1, 2, 3\}$.

Addition in S_4 can be done with the help of an appropriate clock—a four-hour clock in this case. Figure 8.4 is a picture of such a clock.

You should be able to complete the addition table below. Use the clock if necessary.

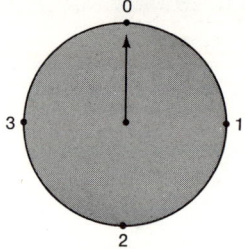

+	0	1	2	3
0				
1			2	
2			0	
3				2

Addition table for S_4
(partially completed)

FIGURE 8.4

The table should show that 0 is the additive identity in S_4: for every x in S_4, $x + 0 = x$. Does every member of S_4 have a unique additive inverse?

Compute the answers to the following to verify that addition in S_4 is associative. The first example is completed.

Example 1 $(1 + 2) + 3$, $1 + (2 + 3)$

Solution $(1 + 2) + 3 = 3 + 3 = 2$
$1 + (2 + 3) = 1 + 1 = 2$

Example 2 $(3 + 3) + 2$, $3 + (3 + 2)$
Example 3 $(0 + 3) + 1$, $0 + (3 + 1)$

Now on to the multiplication in S_4. The table for this has been completed.

×	0	1	2	3
0	0	0	0	0
1	0	1	2	3
2	0	2	0	2
3	0	3	2	1

Multiplication table for S_4

8.4 The System S_4

Compute the answers to the following to see that multiplication in S_4 has the associative property. The first example is completed.

Example 1 $(1 \times 2) \times 3, \quad 1 \times (2 \times 3)$

Solution $(1 \times 2) \times 3 = 2 \times 3 = 2$
$1 \times (2 \times 3) = 1 \times 2 = 2$

Example 2 $(2 \times 0) \times 3, \quad 2 \times (0 \times 3)$
Example 3 $(2 \times 3) \times 1, \quad 2 \times (3 \times 1)$
Example 4 $(3 \times 3) \times 2, \quad 3 \times (3 \times 2)$

Now compute the following to verify that multiplication is distributive over addition. The first example is completed.

Example 1 $2 \times (1 + 3), \quad (2 \times 1) + (2 \times 3)$

Solution $2 \times (1 + 3) = 2 \times 0 = 0$
$(2 \times 1) + (2 \times 3) = 2 + 2 = 0$

Example 2 $3 \times (0 + 2), \quad (3 \times 0) + (3 \times 2)$
Example 3 $2 \times (2 + 2), \quad (2 \times 2) + (2 \times 2)$
Example 4 $3 \times (2 + 2), \quad (3 \times 2) + (3 \times 2)$
Example 5 $2 \times (3 + 2), \quad (2 \times 3) + (2 \times 2)$
Example 6 $1 \times (2 + 3), \quad (1 \times 2) + (1 \times 3)$
Example 7 $0 \times (1 + 2), \quad (0 \times 1) + (0 \times 2)$

Is it true that for every number x in S_4, $x \cdot 0 = 0$? Is it true that for every number x in S_4, $x \cdot 1 = x$?

From $1 \cdot 1 = 1$, it follows that the multiplicative inverse of 1 is 1.

But now observe something strange! There is no number to be multiplied by 2 to obtain 1 for the product. This means that 2 in S_4 does *not* have a multiplicative inverse. Recall that for a system to be a field, every nonzero element in it must have a multiplicative inverse. Thus, if follows that

S_4 is *not* a field

Another strange thing can be observed in the multiplication table. To see it, consider the following situation: In the arithmetic of whole numbers, if $3 \cdot x = 12$ and $3 \cdot y = 12$, then $x = y = 4$, since $3 \cdot 4 = 12$. Now observe the following from the multiplication table for S_4:

$2 \cdot 0 = 0$ and $2 \cdot 2 = 0$

Does it follow that $0 = 2$? Certainly not, since 0 and 2 are different members of S_4.

8 Finite Systems and Matrices

This situation leads to another observation. In the arithmetic of whole numbers, if $xy = 0$, then $x = 0$ or $y = 0$ or both x and y are 0. That is, to obtain 0 for the product of two numbers, at least one of the numbers must be 0. This is not the case in S_4, since

$$2 \cdot 2 = 0$$

The product of two numbers is equal to 0, but neither of the numbers is 0! Thus, S_4 is a system which has some unusual properties not encountered so far in other number systems.

Exercises 8.4

Using the addition table for S_4, compute the following differences.

1. $0 - 0$
2. $1 - 0$
3. $0 - 1$
4. $1 - 1$
5. $2 - 1$
6. $1 - 2$
7. $3 - 1$
8. $1 - 3$
9. $3 - 2$
10. $2 - 3$
11. Earth has recently become the subject of many proposed attacks by the rowdy planets Tripos and Pentos. Coming from the planet Quartos, it is your aim to show Earthlings that, if they adopt the Quartan math system, S_4, they will have nothing to fear from the two invaders joining forces to conquer them. Why? (*Hint:* The trick involves S_4 and the number 2.)

Using the multiplication table for S_4, compute the following quotients. (There may be more than one answer.)

12. $2 \div 2$
13. $3 \div 3$
14. $2 \div 3$
15. $1 \div 1$
16. $0 \div 3$
17. $1 \div 3$
18. Prisoners during World War II used S_6 as a code. You are the camp commander, and it is your job to crack this S_6 code. Prepare an addition table for S_6 so you can decipher the prisoners' code.
19. Make an addition table for $S_6 = \{0, 1, 2, 3, 4, 5\}$.

Compute answers to the following in S_6.

20. $(2 + 3) + 1, \quad 2 + (3 + 1)$
21. $(0 + 5) + 1, \quad 0 + (5 + 1)$
22. $(4 + 2) + 5, \quad 4 + (2 + 5)$
23. $(1 + 4) + 4, \quad 1 + (4 + 4)$
24. $(5 + 4) + 3, \quad 5 + (4 + 3)$
25. Do you think that addition in S_6 has the associative property?
26. What is the additive identity in S_6?

8.4 The System S_4

Solve the following equations, and answer the questions. The statements refer to members of S_6.

27. $0 + x = 0$; what is the additive inverse of 0?
28. $1 + x = 0$; what is the additive inverse of 1?
29. $2 + x = 0$; what is the additive inverse of 2?
30. $3 + x = 0$; what is the additive inverse of 3?
31. $4 + x = 0$; what is the additive inverse of 4?
32. $5 + x = 0$; what is the additive inverse of 5?
33. Due to the POW camp commander's efforts to decipher the S_6 addition code, we are now in a position to communicate with the newly discovered planet Sixatos. All we know about Sixatans is that they use S_6. Prepare a multiplication table in S_6 so we can send them an example of our math.
34. Does each member of S_6 have a unique additive inverse?
35. Make a multiplication table for $S_6 = \{0, 1, 2, 3, 4, 5\}$.
36. Is it true that for every number x in S_6, $x \cdot 0 = 0$?
37. What is the multiplicative identity in S_6?

In each of the following tell whether there is a replacement for x which will result in a true statement. If there is such a replacement, tell what it is. All the statements are in S_6. Use the multiplication table you made for exercise 35.

38. $1 \times x = 1$
39. $2 \times x = 1$
40. $3 \times x = 1$
41. $4 \times x = 1$
42. $5 \times x = 1$
43. Examine your answers in exercises 38–42. Now answer the question: What elements of S_6 have no multiplicative inverses?
44. After communicating with Sixatos, Earth diplomats learn that an alliance is possible, but that Sixatans are offended by the number 4. The example of Earth math sent (by slow interplanetary freighter) to Sixatos contained many 4s. Quick! Jump into your photon-powered spaceship, intercept the mail freighter, and change all the 4s to acceptable S_6 expressions.
45. Give all factorizations of 4 in S_6. (*Hint:* one factorization of 4 is 5×2 because $5 \times 2 = 4$.)
46. Give all factorizations of 2 in S_6.

Compute answers to each pair of the following statements in S_6.

47. $3 \times (1 + 3)$, $(3 \times 1) + (3 \times 3)$
48. $1 \times (5 + 5)$, $(1 \times 5) + (1 \times 5)$
49. $0 \times (3 + 4)$, $(0 \times 3) + (0 \times 4)$
50. $4 \times (1 + 2)$, $(4 \times 1) + (4 \times 2)$
51. $5 \times (3 + 0)$, $(5 \times 3) + (5 \times 0)$
52. $2 \times (5 + 4)$, $(2 \times 5) + (2 \times 4)$
53. $5 \times (5 + 5)$, $(5 \times 5) + (5 \times 5)$

8 Finite Systems and Matrices

54. In S_6, is multiplication distributive over addition?

55. Review the eleven properties of a field in Section 8.2 and tell what property or properties S_6 lacks to be a field.

In ordinary arithmetic we say that the square root of 16 is 4 because $4 \times 4 = 16$. We write $\sqrt{16} = 4$. Tell the square root of each of the following numbers in ordinary arithmetic:

56. $\sqrt{100}$ **57.** $\sqrt{49}$ **58.** $\sqrt{81}$ **59.** $\sqrt{225}$
60. $\sqrt{900}$

Look at the multiplication table for S_6 which you completed in exercise 35, and answer the following:

61. How many numbers are equal to a square root of 1? (*Hint:* $5 \times 5 = 1$ and also $1 \times 1 = 1$.)

62. How many numbers are equal to a square root of 4?

63. Is $\sqrt{0}$ unique? What is it?

64. Is $\sqrt{3}$ unique? What is it?

65. Which numbers do not have a square root in S_6?

66. Is S_6 closed under the operation of *taking square root*? Why?

*Below is a table for the system consisting of a, b, and c, and the operation *.*

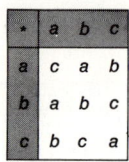

67. What is the identity element for the operation $*$?

68. What is the inverse of a for the operation $*$? Of b? Of c?

69. Find the answer to $(a * b) * c$.

8.5 SOLVING EQUATIONS IN S_5

REVIEW

- Terms used in connection with equations: for the equation $x + 1 = 4$

  3 is the *solution*
  {3} is the *solution set*

- Review the addition and multiplication tables for S_5 (Section 8.3)

OBJECTIVES

- Solve linear and quadratic equations in S_5

8.5 Solving Equations in S_5

To solve an equation means to find all replacements for a variable that will result in true statements. For example, the solution of the equation

$$2x + 1 = 7$$

is 3, since 3 in place of x yields

$$2 \cdot 3 + 1 = 7$$

which is a true statement. We say that $\{3\}$ is the *solution set* of the equation $2x + 1 = 7$. In this case, 3 is the only number for which $2x + 1 = 7$ becomes a true statement.

Now suppose that the only available set is $S_5 = \{0, 1, 2, 3, 4\}$. In making up equations, only these five numbers can be used and their solutions must also come from S_5.

To solve equations in S_5, the addition and the multiplication tables developed in Section 8.3 are helpful. Simple equations can be solved by means of informal techniques using common sense. For example, to solve

$$2x + 3 = 0$$

observe that $2x$ is to be replaced by a number such that the sum of this number and 3 is equal to 0. This number is found in the addition table for S_5 to be 2, since $2 + 3 = 0$. Thus, we have

$$2x = 2$$

It is easy to see that 1 is the replacement for x which yields a true statement from $2x = 2$. Thus, $\{1\}$ should be the solution set of $2x + 3 = 0$.

CHECK

$2x + 3$	0
$2 \cdot 1 + 3$	0
$2 + 3$	
0	

To solve

$$x^2 - 1 = 2$$

in S_5, proceed in a similar manner. Since $3 - 1 = 2$, $x^2 = 3$. Searching the multiplication table, we find no number in S_5 which when squared yields 3. So, the solution set of $x^2 - 1 = 2$ in S_5 is the empty set, $\emptyset$.

8 Finite Systems and Matrices

Exercises 8.5

Solve each of the following equations in S_5. In some cases there may be more than one solution.

1. $x + 2 = 4$
2. $3 + x = 2$
3. $x + 4 = 0$
4. $2x + 1 = 0$
5. $4x + 4 = 1$
6. $x - 1 = 3$
7. $x - 4 = 2$
8. $3 - 2x = 3$
9. $4x - 1 = 0$
10. $x^2 = 1$
11. $x^2 = 4$
12. $2x^2 = 3$
13. $x^2 + 2 = 1$
14. $2x^2 + 3 = 0$
15. $3x^2 - 1 = 4$
16. $4x^2 + 2 = 2$

17. The golfer Hot Shot Harry uses a number 5 iron. He wishes to determine how far his ball will go when he hits it. Because he is using a number 5 iron, the equation must be solved in S_5. Equation: $2x + 4 - 3 = 3$, where x is the distance in multiples of 100 yd.

18. Hot Shot Harry realizes that 100 yards is not far enough, so he changes to a number 7 iron. The equation is now $2x + 1 = 5$. Will Hot Shot now get better mileage? (x is still the distance in multiples of 100 yd.)

19. Make an addition table and a multiplication table for $S_7 = \{0, 1, 2, 3, 4, 5, 6\}$.

Using the addition table you made in exercise 19 and the relation

If $a + b = c$, then $c - b = a$

give the answers to the following in S_7:

20. $5 - 3$
21. $6 - 2$
22. $2 - 5$
23. $1 - 6$
24. $3 - 4$
25. $5 - 1$
26. $3 - 5$
27. $1 - 3$
28. $1 - 4$

Using the multiplication table you made in exercise 19 and the relation

If $a \times b = c$, then $c \div b = a$

give the answers to the following in S_7:

29. $4 \div 2$
30. $6 \div 3$
31. $1 \div 2$
32. $6 \div 4$
33. $2 \div 6$
34. $2 \div 3$
35. $2 \div 5$
36. $1 \div 6$
37. $3 \div 5$
38. $4 \div 6$
39. $5 \div 3$
40. $5 \div 2$

Using the tables you made in exercise 19 solve the following equations in S_7:

41. $x + 6 = 2$
42. $x = 3 = 0$
43. $2x + 1 = 5$
44. $2 + 3x = 0$
45. $2x - 5 = 1$
46. $3 - 5x = 6$
47. $1 - 6x = 0$
48. $x \cdot 3 = 4$
49. $x \cdot 5 = 6$
50. $2 \div x = 4$
51. $1 \div x = 6$
52. $3 \div x = 2$

53. It has been said that we developed base ten because we have 10 fingers. Consider Arnold, the spider. He lost one leg to a praying mantis and now has only 7 legs. What would his numeration system probably be if he spins his web according to the equation $6x - 1 = 3$, where x is the number of sides to the web? Since this number must be an integer, how many sides does his web have?
54. You have now dealt extensively with many numeration systems. Discuss the merits of several. Draw some conclusions regarding the system you would like to see used. Give reasons and examples for your choice.

8.6 TWO-BY-TWO MATRICES

REVIEW

- 23_{five} = 2 fives + 3 ones
  = 13_{ten}
- $.11_{\text{three}} = 3^{-1} + 3^{-2}$
  $= \frac{1}{3} + \frac{1}{9}$
  $= \frac{4}{9}$
- 111_{two} = 1 four + 1 two + 1 one
  = 7
- $.11_{\text{two}} = 2^{-1} + 2^{-2}$
  $= \frac{1}{2} + \frac{1}{4}$
  $= \frac{3}{4}$

OBJECTIVES

- Know some historical background for the development of matrices
- Define equality of matrices

The study of the history of mathematics reveals that new developments in mathematics were usually motivated by two considerations. In some cases mathematical concepts were introduced in order to solve some practical problem. In other cases, mathematical concepts were the result of the creative drive of an individual, without any regard for practical needs.

In 1857 a British mathematician named Arthur Cayley invented a new algebra, known today as *matrix algebra*. In part he was motivated by the desire to invent something in mathematics that would have no chance to ever become useful.

In 1925 Werner Heisenberg recognized that the algebra of matrices was exactly the tool he needed for his pioneering work in quantum mechanics. Since then, matrix algebra has proved to be an indispensable tool for work in nuclear physics. Its applications in today's business are also quite numerous.

Thus, the invention of matrix algebra is only one of many examples of the creation of a new mathematical concept of no apparent usefulness, but which is later discovered to be of great importance for the scientist and the businessman.

One example of a matrix is a *2-by-2 matrix*. It consists of four real numbers, arranged in two *rows* and two *columns* as in the example below:

$$\begin{array}{c} \text{First} \quad \text{Second} \\ \text{column} \quad \text{column} \\ \downarrow \qquad \downarrow \end{array}$$

First row $\rightarrow$
Second row $\rightarrow$
$\begin{pmatrix} -5 & 0 \\ -\sqrt{2} & \frac{2}{3} \end{pmatrix}$

The number in the first row and the first column is -5; 0 is in the first row and the second column; $-\sqrt{2}$ is in the second row and the first column; $\frac{2}{3}$ is in the second row and the second column.

We shall study matrices made up of real numbers, although sometimes the entries may be limited to whole numbers or rational numbers. The entries can also be complex numbers or even other matrices.

When dealing with real numbers, we say that $a = b$ means that a and b are two names for the same number. For example, $\frac{1}{2} = \frac{2}{4}$ is true because $\frac{1}{2}$ and $\frac{2}{4}$ name the same number. To work with matrices, it is also necessary to decide when two matrices are equal. The following is an example of equal 2-by-2 matrices.

$$\begin{pmatrix} \frac{1}{\sqrt{2}} & \frac{1}{2} \\ 0 & -.75 \end{pmatrix} \quad \text{and} \quad \begin{pmatrix} \frac{\sqrt{2}}{2} & \frac{2}{4} \\ \frac{0}{5} & -\frac{3}{4} \end{pmatrix}$$

These two matrices are equal because they have equal real numbers for their corresponding entries:

$\frac{1}{\sqrt{2}} = \frac{\sqrt{2}}{2}$ (first row, first column)

$\frac{1}{2} = \frac{2}{4}$ (first row, second column)

$0 = \frac{0}{5}$ (second row, first column)

$-.75 = -\frac{3}{4}$ (second row, second column)

8.6 Two-by-Two Matrices

In general,

$$\begin{pmatrix} a & b \\ c & d \end{pmatrix} = \begin{pmatrix} e & f \\ g & h \end{pmatrix}$$

if each of the following is true:

$$a = e \quad b = f$$
$$c = g \quad d = h$$

In other words, for two matrices to be equal, each pair of corresponding entries must be equal.

It is somewhat cumbersome to refer to the particular entries in a matrix by saying "the entry in the first row and the second column," and so on. To make such references simple, the following notation will be used.

$$\begin{pmatrix} a_{11} & a_{12} \\ a_{21} & a_{22} \end{pmatrix}$$

Thus, in a_{11}, the subscript means that the entry is in the first row and the first column. In a_{12}, the subscript means that the entry is in the first row and the second column. In general, the first subscript number tells the row and the second subscript number tells the column.

Exercises 8.6

In

$$\begin{pmatrix} \sqrt{3} & 5 \\ -6 & 0 \end{pmatrix}$$

what entry is

1. In the second row and the first column?
2. In the second row and the second column?
3. In the first row and the second column?
4. In the first row and the first column?
5. At the recent Hot Rod Speed Machine Exhibition, four exhibits were set up as below:

Ⓐ Ⓑ
Ⓒ Ⓓ

Each exhibit was given a number so it would be easier to find:

$$\begin{matrix} 3 & \frac{2}{3} \\ \sqrt{2} & 1 \end{matrix}$$

You visit the exhibition and see the exhibits in the following order:

$$a_{22},\ a_{12},\ a_{11},\ a_{21}$$

In what order did you visit them (list by the "easier to find" numbers)?

6. Decipher the calculations below from the following matrices:

$$\begin{pmatrix} 3, & 4 \\ 5, & 6 \end{pmatrix} \quad \begin{pmatrix} 2, & 5 \\ \tfrac{1}{2}, & 3 \end{pmatrix} \quad \begin{pmatrix} \tfrac{3}{4}, & \tfrac{3}{8} \\ 6, & 9 \end{pmatrix}$$

$$a_{22} \quad + \quad a_{21} \quad - \quad a_{11}$$

Tell which of the following are true and which are false:

7. $\begin{pmatrix} -\tfrac{3}{10} & 5 \\ 1 & -4 \end{pmatrix} = \begin{pmatrix} -.3 & \sqrt{25} \\ \sqrt{2} & -\sqrt{16} \end{pmatrix}$

8. $\begin{pmatrix} 23_{\text{five}} & 11_{\text{five}} \\ 14_{\text{five}} & 31_{\text{five}} \end{pmatrix} = \begin{pmatrix} 13 & 6 \\ 9 & 16 \end{pmatrix}$

9. $\begin{pmatrix} .1_{\text{three}} & .2_{\text{three}} \\ .01_{\text{three}} & .11_{\text{three}} \end{pmatrix} = \begin{pmatrix} \tfrac{1}{3} & \tfrac{2}{3} \\ \tfrac{1}{9} & \tfrac{4}{9} \end{pmatrix}$

10. $\begin{pmatrix} 10_{\text{two}} & 101_{\text{two}} \\ 11_{\text{two}} & 111_{\text{two}} \end{pmatrix} = \begin{pmatrix} 4 & 5 \\ 2 & 10 \end{pmatrix}$

11. $\begin{pmatrix} .1_{\text{two}} & .01_{\text{two}} \\ .11_{\text{two}} & .001_{\text{two}} \end{pmatrix} = \begin{pmatrix} \tfrac{1}{2} & \tfrac{1}{4} \\ \tfrac{3}{4} & \tfrac{1}{8} \end{pmatrix}$

12. $\begin{pmatrix} -\sqrt{256} & \sqrt{\tfrac{1}{4}} \\ \sqrt{\tfrac{1}{9}} & \sqrt{\tfrac{1}{25}} \end{pmatrix} = \begin{pmatrix} -16 & \tfrac{1}{2} \\ \tfrac{1}{3} & \tfrac{1}{5} \end{pmatrix}$

13. By substituting letters for the following numbers (number the letters in the alphabet 1–26), decipher this highly secret message sent out by one of the math department's spies:

$$\begin{pmatrix} 20 & 13 \\ 1 & 8 \end{pmatrix} \quad \begin{pmatrix} 9 & 3 \\ 2 & 19 \end{pmatrix} \quad \begin{pmatrix} 6 & 14 \\ 21 & 26 \end{pmatrix}$$

If the message is read in the following order, what does it say?

$$a_{12} \quad a_{21} \quad a_{11} \quad a_{22} \quad a_{11} \quad a_{22} \quad a_{11} \quad a_{21} \quad a_{12}$$

Tell the replacements for the variables for which the following will be true:

14. $\begin{pmatrix} x+1 & y-2 \\ 2w & 2z-1 \end{pmatrix} = \begin{pmatrix} -3 & 5 \\ 0 & 1 \end{pmatrix}$

15. $\begin{pmatrix} \tfrac{x}{2} & \tfrac{y}{3} \\ \tfrac{w}{4} & \tfrac{z}{5} \end{pmatrix} = \begin{pmatrix} 1 & 2 \\ 3 & 4 \end{pmatrix}$

8.7 Addition and Subtraction of 2-by-2 Matrices

16. $\begin{pmatrix} \dfrac{x+1}{2} & \dfrac{y-1}{3} \\ \dfrac{2w+1}{2} & \dfrac{2z-1}{2} \end{pmatrix} = \begin{pmatrix} -1 & -2 \\ 0 & 0 \end{pmatrix}$

17. $\begin{pmatrix} \sqrt{x} & \sqrt[3]{y} \\ \sqrt{w+1} & \sqrt{z-1} \end{pmatrix} = \begin{pmatrix} 2 & -2 \\ 2 & 5 \end{pmatrix}$

18. $\begin{pmatrix} \dfrac{\sqrt{x}}{2} & \dfrac{\sqrt[3]{y}}{2} \\ \dfrac{\sqrt{w+1}}{3} & \dfrac{\sqrt{z-1}}{3} \end{pmatrix} = \begin{pmatrix} 1 & -1 \\ 1 & 1 \end{pmatrix}$

19. $\begin{pmatrix} 2(x+1) & -3(y-1) \\ 3(2w+1) & 4(z-1) \end{pmatrix} = \begin{pmatrix} -8 & 9 \\ 12 & -4 \end{pmatrix}$

8.7 ADDITION AND SUBTRACTION OF 2-BY-2 MATRICES

REVIEW

- $32_{\text{four}} = 3 \text{ fours} + 2 \text{ ones}$
  $= 12_{\text{ten}} + 2$
  $= 14_{\text{ten}}$

- $62_{\text{seven}} = 6 \text{ sevens} + 2 \text{ ones}$
  $= 42_{\text{ten}} + 2$
  $= 44_{\text{ten}}$

- The additive inverse of x is $-x$ for every number x

OBJECTIVES

- Add 2-by-2 matrices
- Determine the additive identity in the set of 2-by-2 matrices
- Determine additive inverses of 2-by-2 matrices

The following example suggests the procedure for adding 2-by-2 matrices:

$$\begin{pmatrix} 3 & -2 \\ 0 & 5 \end{pmatrix} + \begin{pmatrix} -2 & 5 \\ 6 & 3 \end{pmatrix} = \begin{pmatrix} 3+(-2) & -2+5 \\ 0+6 & 5+3 \end{pmatrix} = \begin{pmatrix} 1 & 3 \\ 6 & 8 \end{pmatrix}$$

In general,

$$\begin{pmatrix} a_{11} & a_{12} \\ a_{21} & a_{22} \end{pmatrix} + \begin{pmatrix} b_{11} & b_{12} \\ b_{21} & b_{22} \end{pmatrix} = \begin{pmatrix} a_{11}+b_{11} & a_{12}+b_{12} \\ a_{21}+b_{21} & a_{22}+b_{22} \end{pmatrix}$$

Thus, in the sum of two matrices, the entries are the sums of the corresponding entries in the addends. Remember that the subscript 21 in a_{21} means that the entry is in the second row and the first column. The first digit of the subscript tells the row and the second digit the column.

It should be easy to conclude that subtraction would be defined in a manner analogous to the definition of addition. This is shown in the following example:

$$\begin{pmatrix} 9 & 5 \\ 7 & 4 \end{pmatrix} - \begin{pmatrix} 6 & 5 \\ 1 & 1 \end{pmatrix} = \begin{pmatrix} 9-6 & 5-5 \\ 7-1 & 4-1 \end{pmatrix} = \begin{pmatrix} 3 & 0 \\ 6 & 3 \end{pmatrix}$$

In general,

$$\begin{pmatrix} a_{11} & b_{12} \\ a_{21} & a_{22} \end{pmatrix} - \begin{pmatrix} b_{11} & b_{12} \\ b_{21} & b_{22} \end{pmatrix} = \begin{pmatrix} a_{11} - b_{11} & a_{12} - b_{12} \\ a_{21} - b_{21} & a_{22} - b_{22} \end{pmatrix}$$

In the set of real numbers, the number 0 is called the additive identity since for every real number x, $x + 0 = 0 + x = x$. It is easy to see that

$$\begin{pmatrix} 0 & 0 \\ 0 & 0 \end{pmatrix}$$

is the additive identity in the set of 2-by-2 matrices, since

$$\begin{pmatrix} a & b \\ c & d \end{pmatrix} + \begin{pmatrix} 0 & 0 \\ 0 & 0 \end{pmatrix} = \begin{pmatrix} a & b \\ c & d \end{pmatrix}$$

for all numbers a, b, c, and d.

In the set of real numbers, each real number has a unique additive inverse. That is, for each real number x, there exists a unique real number $-x$, such that $x + (-x) = 0$. Is there a 2-by-2 matrix with such a property? That is, given a 2-by-2 matrix

$$\begin{pmatrix} a_{11} & a_{12} \\ a_{21} & a_{22} \end{pmatrix}$$

is there a unique matrix

$$\begin{pmatrix} x_{11} & x_{12} \\ x_{21} & x_{22} \end{pmatrix}$$

such that

$$\begin{pmatrix} a_{11} & a_{12} \\ a_{21} & a_{22} \end{pmatrix} + \begin{pmatrix} x_{11} & x_{12} \\ x_{21} & x_{22} \end{pmatrix} = \begin{pmatrix} 0 & 0 \\ 0 & 0 \end{pmatrix}$$

8.7 Addition and Subtraction of 2-by-2 Matrices

It is not difficult to see that the additive inverse of

$$\begin{pmatrix} a_{11} & a_{12} \\ a_{21} & a_{22} \end{pmatrix} \quad \text{is} \quad \begin{pmatrix} -a_{11} & -a_{12} \\ -a_{21} & -a_{22} \end{pmatrix}$$

since

$$\begin{pmatrix} a_{11} & a_{12} \\ a_{21} & a_{22} \end{pmatrix} + \begin{pmatrix} -a_{11} & -a_{12} \\ -a_{21} & -a_{22} \end{pmatrix} = \begin{pmatrix} 0 & 0 \\ 0 & 0 \end{pmatrix}$$

Exercises 8.7

Compute the sum of each pair of matrices:

1. $\begin{pmatrix} -1 & 2 \\ 0 & -\frac{1}{2} \end{pmatrix} + \begin{pmatrix} -3 & -5 \\ -4 & \frac{1}{4} \end{pmatrix}$

2. $\begin{pmatrix} \sqrt{2} & \sqrt{2} \\ -\sqrt{2} & -\sqrt{2} \end{pmatrix} + \begin{pmatrix} 2\sqrt{2} & -2\sqrt{3} \\ 4\sqrt{2} & -3\sqrt{3} \end{pmatrix}$

3. $\begin{pmatrix} \frac{1}{3} & -\frac{2}{3} \\ \frac{1}{4} & \frac{3}{4} \end{pmatrix} + \begin{pmatrix} -\frac{5}{6} & \frac{4}{9} \\ -\frac{1}{12} & -\frac{5}{8} \end{pmatrix}$

4. $\begin{pmatrix} \frac{1}{2} & -\frac{1}{3} \\ \frac{1}{4} & -\frac{1}{5} \end{pmatrix} + \begin{pmatrix} -\frac{2}{3} & \frac{1}{4} \\ -\frac{2}{5} & \frac{3}{4} \end{pmatrix}$

5. $\begin{pmatrix} .25 & -.16 \\ -.34 & .09 \end{pmatrix} + \begin{pmatrix} -.36 & .52 \\ .75 & -.21 \end{pmatrix}$

6. $\begin{pmatrix} 32_{\text{four}} & 11_{\text{four}} \\ 13_{\text{four}} & 23_{\text{four}} \end{pmatrix} + \begin{pmatrix} 13_{\text{four}} & 33_{\text{four}} \\ 22_{\text{four}} & 32_{\text{four}} \end{pmatrix}$

7. $\begin{pmatrix} 62_{\text{seven}} & 15_{\text{seven}} \\ 44_{\text{seven}} & 43_{\text{seven}} \end{pmatrix} + \begin{pmatrix} 15_{\text{seven}} & 62_{\text{seven}} \\ 23_{\text{seven}} & 66_{\text{seven}} \end{pmatrix}$

8. You are flying over a football field in your helicopter, and you realize that eight of the players are standing as in the following diagram:

$$\begin{pmatrix} 72 & 3 \\ 10 & 8 \end{pmatrix}$$
$$\begin{pmatrix} 84 & 80 \\ 70 & 74 \end{pmatrix}$$

If after the play you see

$$\begin{pmatrix} 156 & 83 \\ 80 & 82 \end{pmatrix}$$

what terrible accident happened?

8 Finite Systems and Matrices

Add and simplify:

9. $\begin{pmatrix} 2(x-1) & 3(x-2) \\ -(3-x) & -(2x+5) \end{pmatrix} + \begin{pmatrix} 3(x+2) & -2(2-x) \\ -3(1-x) & 5(x-1) \end{pmatrix}$

10. $\begin{pmatrix} \dfrac{x-2}{2} & \dfrac{3-x}{3} \\ \dfrac{2x+1}{4} & \dfrac{2x-3}{2} \end{pmatrix} + \begin{pmatrix} \dfrac{x+2}{2} & \dfrac{3x-1}{3} \\ \dfrac{x+3}{4} & \dfrac{-2x+5}{2} \end{pmatrix}$

11. $\begin{pmatrix} \dfrac{x-1}{4} & \dfrac{x-2}{3} \\ \dfrac{x-3}{2} & \dfrac{x-4}{4} \end{pmatrix} + \begin{pmatrix} \dfrac{x+1}{3} & \dfrac{x-1}{2} \\ \dfrac{x-1}{3} & \dfrac{x+3}{3} \end{pmatrix}$

12. $\begin{pmatrix} \dfrac{1}{2}(x+3) & -\dfrac{1}{3}(6x-2) \\ \dfrac{1}{4}(12x-5) & -\dfrac{1}{2}(2x+3) \end{pmatrix} + \begin{pmatrix} \dfrac{1}{6}(6x+4) & \dfrac{2}{9}(x+1) \\ -\dfrac{7}{12}(12x-6) & \dfrac{5}{4}(x-1) \end{pmatrix}$

Subtract and simplify wherever possible:

13. $\begin{pmatrix} 5 & -2 \\ 6 & -1 \end{pmatrix} - \begin{pmatrix} 3 & -4 \\ -5 & 2 \end{pmatrix}$

14. $\begin{pmatrix} 0 & -1 \\ 1 & 0 \end{pmatrix} - \begin{pmatrix} -1 & 0 \\ 0 & -2 \end{pmatrix}$

15. $\begin{pmatrix} \frac{3}{2} & -\frac{1}{3} \\ \frac{1}{4} & -\frac{1}{2} \end{pmatrix} - \begin{pmatrix} \frac{5}{2} & -\frac{4}{3} \\ \frac{5}{4} & -\frac{1}{2} \end{pmatrix}$

16. $\begin{pmatrix} 2+\sqrt{5} & 1-\sqrt{3} \\ \sqrt{2}-1 & \sqrt{3}+1 \end{pmatrix} - \begin{pmatrix} 1-\sqrt{5} & 2+\sqrt{3} \\ 1+\sqrt{2} & 1+\sqrt{3} \end{pmatrix}$

17. $\begin{pmatrix} \frac{2}{7} & -\frac{1}{3} \\ \frac{2}{5} & -\frac{3}{2} \end{pmatrix} - \begin{pmatrix} \frac{1}{2} & \frac{1}{4} \\ \frac{1}{2} & \frac{2}{3} \end{pmatrix}$

18. $\begin{pmatrix} 101_{\text{two}} & 100_{\text{two}} \\ 111_{\text{two}} & 110_{\text{two}} \end{pmatrix} - \begin{pmatrix} 11_{\text{two}} & 10_{\text{two}} \\ 11_{\text{two}} & 11_{\text{two}} \end{pmatrix}$

19. $\begin{pmatrix} 2x+3 & x-2 \\ 3x-1 & 2x-1 \end{pmatrix} - \begin{pmatrix} x+5 & 2x-3 \\ x+1 & x+3 \end{pmatrix}$

20. $\begin{pmatrix} 3(1-x) & 2(2x+3) \\ 4(2-3x) & -3(x+4) \end{pmatrix} - \begin{pmatrix} 4(2x-1) & -(x+2) \\ -2(1-x) & -2(2-x) \end{pmatrix}$

Determine the replacements for the variables for which the sum of the matrices in each case will be

$\begin{pmatrix} 0 & 0 \\ 0 & 0 \end{pmatrix}$

8.7 Addition and Subtraction of 2-by-2 Matrices

21. $\begin{pmatrix} x+1 & 2y-3 \\ 3w-1 & 4z+5 \end{pmatrix} + \begin{pmatrix} 3 & -5 \\ 2 & 1 \end{pmatrix}$

22. $\begin{pmatrix} 3(x+3) & -2(y-1) \\ -2(w-1) & -(2-z) \end{pmatrix} + \begin{pmatrix} -6 & 4 \\ 12 & -1 \end{pmatrix}$

23. $\begin{pmatrix} x-5 & y-1 \\ 2w+3 & 3z+4 \end{pmatrix} + \begin{pmatrix} 2x+4 & -3y+2 \\ -3w-1 & -4z-5 \end{pmatrix}$

24. $\begin{pmatrix} 2(x-3) & -2(2y+1) \\ -3(2-3w) & -5(1-z) \end{pmatrix} + \begin{pmatrix} -(3-x) & -(1-2y) \\ -(3w-2) & -(z-1) \end{pmatrix}$

25. You have come across a map of an old island in the Pacific. The natives who constructed the map knew a great deal about math. They hid all their treasures where no one (except a mathematician) could find them. The sum of the map coordinates is 0; therefore, you know that the numbers you are after are the additive inverses of the given numbers. From this partially filled-in reference system, find the coordinates of the treasure.

$$\begin{pmatrix} 2(y-1) & 2 \\ 3(z+1) & 18 \end{pmatrix} + \begin{pmatrix} 10 & 2 \\ -11 & 4(q+3) \end{pmatrix}$$

Give the additive inverse of each of the following matrices:

26. $\begin{pmatrix} -2 & 0 \\ 6 & -5 \end{pmatrix}$

27. $\begin{pmatrix} \frac{1}{2} & -\frac{1}{3} \\ \frac{2}{5} & -\frac{4}{7} \end{pmatrix}$

28. $\begin{pmatrix} 1.7 & -12.6 \\ -1.9 & 13.4 \end{pmatrix}$

29. $\begin{pmatrix} \sqrt{2} & 3+\sqrt{3} \\ 1-\sqrt{2} & 1+\sqrt{5} \end{pmatrix}$

30. $\begin{pmatrix} a & b \\ c & d \end{pmatrix}$

31. $\begin{pmatrix} x-1 & 2y+3 \\ z-1 & 3-w \end{pmatrix}$

32. $\begin{pmatrix} 1 & \frac{2+y}{-3} \\ \frac{1+x}{a+1} & \frac{b}{b+2} \end{pmatrix}$

33. $\begin{pmatrix} x+y & -(x+y) \\ 2x-y & x-2y \end{pmatrix}$

★34. Prove that addition of 2-by-2 matrices has the
   a. commutative property   b. associative property

8.8 MULTIPLICATION

REVIEW

None

OBJECTIVES

- Define multiplication of a real number and a matrix
- Define multiplication of two matrices
- Prove various properties of multiplication
- Determine the multiplicative identity in the set of 2-by-2 matrices

There are two kinds of multiplication to be considered in the set of matrices: multiplication of a real number and a matrix and multiplication of two matrices.

Here is an example of multiplication of the first kind.

$$3 \times \begin{pmatrix} 2 & 0 \\ -3 & 1 \end{pmatrix} = \begin{pmatrix} 3 \times 2 & 3 \times 0 \\ 3 \times (-3) & 3 \times 1 \end{pmatrix} = \begin{pmatrix} 6 & 0 \\ -9 & 3 \end{pmatrix}$$

Generally,

$$b \times \begin{pmatrix} a_{11} & a_{12} \\ a_{21} & a_{22} \end{pmatrix} = \begin{pmatrix} ba_{11} & ba_{12} \\ ba_{21} & ba_{22} \end{pmatrix}$$

Here is an example of multiplication of two matrices.

$$\begin{pmatrix} 2 & 3 \\ 5 & 4 \end{pmatrix} \times \begin{pmatrix} 6 & 0 \\ 1 & 7 \end{pmatrix} = \begin{pmatrix} 2 \cdot 6 + 3 \cdot 1 & 2 \cdot 0 + 3 \cdot 7 \\ 5 \cdot 6 + 4 \cdot 1 & 5 \cdot 0 + 4 \cdot 7 \end{pmatrix}$$
$$= \begin{pmatrix} 15 & 21 \\ 34 & 28 \end{pmatrix}$$

Generally,

$$\begin{pmatrix} a_{11} & a_{12} \\ a_{21} & a_{22} \end{pmatrix} \times \begin{pmatrix} b_{11} & b_{12} \\ b_{21} & b_{22} \end{pmatrix} = \begin{pmatrix} a_{11}b_{11} + a_{12}b_{21} & a_{11}b_{21} + a_{12}b_{22} \\ a_{21}b_{11} + a_{22}b_{21} & a_{21}b_{12} + a_{22}b_{22} \end{pmatrix}$$

Observe that, in multiplying matrices, the first row of the matrix on the left and the first column of the matrix on the right determine the entry in the first row and the first column of the product matrix. Similar statements about the remaining three entries in the product matrix can be made.

8.8 Multiplication

We are accustomed to accepting the idea that multiplication has the commutative property $ab = ba$. A check is made to see what the situation is with multiplication of matrices.

In the example above, we found that:

$$\begin{pmatrix} 2 & 3 \\ 5 & 4 \end{pmatrix} \times \begin{pmatrix} 6 & 0 \\ 1 & 7 \end{pmatrix} = \begin{pmatrix} 15 & 21 \\ 34 & 28 \end{pmatrix}$$
$$A \quad \times \quad B \quad = \quad C$$

Now we compute the product $B \times A$:

$$\begin{pmatrix} 6 & 0 \\ 1 & 7 \end{pmatrix} \times \begin{pmatrix} 2 & 3 \\ 5 & 4 \end{pmatrix} = \begin{pmatrix} 6 \cdot 2 + 0 \cdot 5 & 6 \cdot 3 + 0 \cdot 4 \\ 1 \cdot 2 + 7 \cdot 5 & 1 \cdot 3 + 7 \cdot 4 \end{pmatrix}$$
$$B \quad \times \quad A \quad = \begin{pmatrix} 12 & 18 \\ 37 & 31 \end{pmatrix}$$

It is clear that

$$\begin{pmatrix} 15 & 21 \\ 34 & 28 \end{pmatrix} \neq \begin{pmatrix} 12 & 18 \\ 37 & 31 \end{pmatrix}$$

Thus, we see that

$$A \times B \neq B \times A$$

One example is sufficient to prove that multiplication of 2-by-2 matrices does not have the commutative property.

Exercises 8.8

Compute the products:

1. $-2 \times \begin{pmatrix} 0 & -\frac{1}{2} \\ 2 & -4 \end{pmatrix}$

2. $\begin{pmatrix} 0 & -\frac{1}{2} \\ 2 & -4 \end{pmatrix} \times (-2)$

3. $5 \times \begin{pmatrix} -2 & 1 \\ 0 & 3 \end{pmatrix}$

4. $\begin{pmatrix} -2 & 1 \\ 0 & 3 \end{pmatrix} \times 5$

5. $1 \times \begin{pmatrix} -1 & 0 \\ \frac{1}{2} & -\frac{2}{3} \end{pmatrix}$

6. $\begin{pmatrix} -1 & 0 \\ \frac{1}{2} & -\frac{2}{3} \end{pmatrix} \times 1$

7. $-1 \times \begin{pmatrix} 6 & 2 \\ -1 & -5 \end{pmatrix}$

8. $\begin{pmatrix} 6 & 2 \\ -1 & -5 \end{pmatrix} \times (-1)$

9. $0 \times \begin{pmatrix} -5 & 6 \\ 7 & 8 \end{pmatrix}$

10. $\begin{pmatrix} -5 & 6 \\ 7 & 8 \end{pmatrix} \times 0$

11. $6 \times \begin{pmatrix} 0 & 0 \\ 0 & 0 \end{pmatrix}$

12. $\begin{pmatrix} 0 & 0 \\ 0 & 0 \end{pmatrix} \times 6$

13. You may have heard of the famous mathematician Mat Rix. If you have had a course in biology, you know that the genetic code for a person is stored in chromosomes. Biologists assign numbers to these chromosomes. Mat's code looks like this:

$$\begin{pmatrix} -3 & 4 \\ 2 & 0 \end{pmatrix}$$

If Mat has 5 children, what will be the *total* chromosome number for all the little Mat Rices? (*Hint:* Multiply the matrix by 5.)

14. Let $A = \begin{pmatrix} a_{11} & a_{12} \\ a_{21} & a_{22} \end{pmatrix}$

Prove that for each real number c and each 2-by-2 matrix A, $c \times A = A \times c$.

15. Prove that for all real numbers b and c and each 2-by-2 matrix A,

$$(bc) \times A = b(c \times A)$$

16. Prove that for each real number c and all matrices A and B,

$$c \times (A + B) = (A + B) \times c$$

17. Prove that, if Mat Rix had only one child, his chromosome code would not have been altered. (That is, prove that the multiplication can be performed in any way.)

18. Prove that for each 2-by-2 matrix A,

$$1 \times A = A \times 1 = A$$

19. Prove that for each 2-by-2 matrix A,

$$0 \times A = A \times 0 = \begin{pmatrix} 0 & 0 \\ 0 & 0 \end{pmatrix}$$

20. Prove that for each real number a,

$$a \times \begin{pmatrix} 0 & 0 \\ 0 & 0 \end{pmatrix} = \begin{pmatrix} 0 & 0 \\ 0 & 0 \end{pmatrix} \times a = \begin{pmatrix} 0 & 0 \\ 0 & 0 \end{pmatrix}$$

21. Prove that for each 2-by-2 matrix A,

$$-1 \times A = A \times (-1) = -A$$

8.8 Multiplication

22. Prove that for each 2-by-2 matrix A,

$$-1 \times A = 1 \times (-A)$$

Compute the products:

23. $\begin{pmatrix} -1 & 0 \\ 2 & 4 \end{pmatrix} \times \begin{pmatrix} 0 & 1 \\ 2 & -1 \end{pmatrix}$
24. $\begin{pmatrix} 2 & 1 \\ 0 & -1 \end{pmatrix} \times \begin{pmatrix} 1 & 1 \\ 1 & 1 \end{pmatrix}$

25. $\begin{pmatrix} 2 & 3 \\ 1 & -1 \end{pmatrix} \times \begin{pmatrix} 1 & 1 \\ 0 & 0 \end{pmatrix}$
26. $\begin{pmatrix} 1 & 2 \\ -1 & 1 \end{pmatrix} \times \begin{pmatrix} 0 & 0 \\ 1 & 1 \end{pmatrix}$

27. $\begin{pmatrix} 2 & 3 \\ 1 & 5 \end{pmatrix} \times \begin{pmatrix} 1 & 0 \\ 1 & 0 \end{pmatrix}$
28. $\begin{pmatrix} -1 & 2 \\ 3 & 0 \end{pmatrix} \times \begin{pmatrix} 0 & 1 \\ 0 & 1 \end{pmatrix}$

29. $\begin{pmatrix} 2 & -2 \\ 1 & 3 \end{pmatrix} \times \begin{pmatrix} 1 & 0 \\ 0 & 1 \end{pmatrix}$
30. $\begin{pmatrix} 2 & 3 \\ -2 & 1 \end{pmatrix} \times \begin{pmatrix} 0 & 1 \\ 1 & 0 \end{pmatrix}$

31. $\begin{pmatrix} 4 & 3 \\ -1 & 1 \end{pmatrix} \times \begin{pmatrix} 1 & 0 \\ 0 & 0 \end{pmatrix}$
32. $\begin{pmatrix} 2 & 2 \\ -1 & 3 \end{pmatrix} \times \begin{pmatrix} 0 & 1 \\ 0 & 0 \end{pmatrix}$

33. $\begin{pmatrix} -2 & -3 \\ 0 & 1 \end{pmatrix} \times \begin{pmatrix} 0 & 0 \\ 1 & 0 \end{pmatrix}$
34. $\begin{pmatrix} 4 & 1 \\ 0 & -2 \end{pmatrix} \times \begin{pmatrix} 0 & 0 \\ 0 & 1 \end{pmatrix}$

35. $\begin{pmatrix} 2 & -3 \\ -1 & -2 \end{pmatrix} \times \begin{pmatrix} 0 & 0 \\ 0 & 0 \end{pmatrix}$
36. $\begin{pmatrix} 3 & -1 \\ -2 & -3 \end{pmatrix} \times \begin{pmatrix} -1 & -1 \\ -1 & -1 \end{pmatrix}$

37. $\begin{pmatrix} 4 & -1 \\ 3 & -2 \end{pmatrix} \times \begin{pmatrix} -1 & -1 \\ 0 & 0 \end{pmatrix}$
38. $\begin{pmatrix} -2 & -1 \\ -1 & -2 \end{pmatrix} \times \begin{pmatrix} 0 & 0 \\ -1 & -1 \end{pmatrix}$

39. $\begin{pmatrix} 2 & 1 \\ 1 & 2 \end{pmatrix} \times \begin{pmatrix} -1 & 0 \\ -1 & 0 \end{pmatrix}$
40. $\begin{pmatrix} 3 & 2 \\ 1 & 1 \end{pmatrix} \times \begin{pmatrix} 0 & -1 \\ 0 & -1 \end{pmatrix}$

41. $\begin{pmatrix} 1 & 2 \\ 3 & 4 \end{pmatrix} \times \begin{pmatrix} -1 & 0 \\ 0 & -1 \end{pmatrix}$
42. $\begin{pmatrix} 0 & 1 \\ 2 & 1 \end{pmatrix} \times \begin{pmatrix} 0 & -1 \\ -1 & 0 \end{pmatrix}$

43. $\begin{pmatrix} 4 & 2 \\ 1 & 0 \end{pmatrix} \times \begin{pmatrix} -1 & 0 \\ 0 & 0 \end{pmatrix}$
44. $\begin{pmatrix} 2 & 3 \\ 1 & 2 \end{pmatrix} \times \begin{pmatrix} 0 & -1 \\ 0 & 0 \end{pmatrix}$

45. $\begin{pmatrix} 2 & 2 \\ 1 & 5 \end{pmatrix} \times \begin{pmatrix} 0 & 0 \\ -1 & 0 \end{pmatrix}$
46. $\begin{pmatrix} 5 & 6 \\ 1 & 1 \end{pmatrix} \times \begin{pmatrix} 0 & 0 \\ 0 & -1 \end{pmatrix}$

47. You may have heard the saying "they breed like rabbits." The following matrices wish to multiply. What will they produce?

    a. $\begin{pmatrix} 2 & 1 \\ 1 & 2 \end{pmatrix} \times \begin{pmatrix} -1 & 0 \\ -1 & 0 \end{pmatrix}$
    b. $\begin{pmatrix} 3 & 2 \\ 1 & 1 \end{pmatrix} \times \begin{pmatrix} 0 & -1 \\ 0 & -1 \end{pmatrix}$

    c. $\begin{pmatrix} 1 & 2 \\ 3 & 4 \end{pmatrix} \times \begin{pmatrix} -1 & 0 \\ 0 & -1 \end{pmatrix}$
    d. $\begin{pmatrix} 0 & 1 \\ 2 & 1 \end{pmatrix} \times \begin{pmatrix} 0 & -1 \\ -1 & 0 \end{pmatrix}$

    e. $\begin{pmatrix} 4 & 2 \\ 1 & 0 \end{pmatrix} \times \begin{pmatrix} -1 & 0 \\ 0 & 0 \end{pmatrix}$
    f. $\begin{pmatrix} 2 & 3 \\ 1 & 2 \end{pmatrix} \times \begin{pmatrix} 0 & -1 \\ 0 & 0 \end{pmatrix}$

    g. $\begin{pmatrix} 2 & 2 \\ 1 & 5 \end{pmatrix} \begin{pmatrix} 0 & 0 \\ -1 & 0 \end{pmatrix}$
    h. $\begin{pmatrix} 5 & 6 \\ 1 & 1 \end{pmatrix} \times \begin{pmatrix} 0 & 0 \\ 0 & -1 \end{pmatrix}$

8 Finite Systems and Matrices

48. Compute the answers.

a. $\left[\begin{pmatrix} 2 & 3 \\ 4 & -1 \end{pmatrix} + \begin{pmatrix} -1 & -4 \\ -5 & -3 \end{pmatrix}\right] \times \begin{pmatrix} 1 & 2 \\ -1 & -3 \end{pmatrix}$

b. $\begin{pmatrix} 2 & 3 \\ 4 & -1 \end{pmatrix} \times \begin{pmatrix} 1 & 2 \\ -1 & -3 \end{pmatrix} + \begin{pmatrix} -1 & -4 \\ -5 & -3 \end{pmatrix} \times \begin{pmatrix} 1 & 2 \\ -1 & -3 \end{pmatrix}$

c. Did you get the same answer in a and b?

49. Prove: For all 2-by-2 matrices A, B, and C, $(A + B) \times C = (A \times C) + (B \times C)$.

50. Determine the multiplicative identity in the set of 2-by-2 matrices.

8.9 APPLICATIONS OF MATRICES

REVIEW

None

OBJECTIVES

- Consider examples of applications of matrices

There is a variety of applications of matrices in science, particularly in physics. They are also used in business. The following example illustrates a very simple use of matrices in keeping an inventory.

A manufacturer of light bulbs maintains three warehouses storing four kinds of light-bulbs: 25-watt, 60-watt, 75-watt, and 100-watt. He keeps track of his inventory by the use of 2-by-2 matrices. He uses the first row-first column spot to show the number of 25-watt bulbs, the first row-second column spot to show the number of 60-watt bulbs, the second row-first column spot to show the number of 75-watt bulbs, and the second row-second column spot to show the number of 100-watt bulbs. The following equation involving matrices may show the status of the inventory in the three warehouses on a particular date.

$$\underbrace{\begin{pmatrix} 700 & 2000 \\ 10{,}000 & 15{,}000 \end{pmatrix}}_{\text{Warehouse 1}} + \underbrace{\begin{pmatrix} 7000 & 12{,}000 \\ 15{,}000 & 1000 \end{pmatrix}}_{\text{Warehouse 2}} + \underbrace{\begin{pmatrix} 1500 & 300 \\ 17{,}000 & 600 \end{pmatrix}}_{\text{Warehouse 3}}$$

$$= \begin{pmatrix} 9200 & 14{,}300 \\ 42{,}000 & 16{,}600 \end{pmatrix}$$

The matrix on the right of the equality sign gives a manufacturer a quick spot check of his total inventory.

8.9 Applications of Matrices

Consider another example of the use of matrices in a practical situation. Suppose a manufacturer produces three kinds of nails: aluminum, copper, and steel. Each of these is produced in $\frac{1}{4}$ in, $\frac{1}{2}$ in, $\frac{3}{4}$ in, and 1 in lengths. The number of each kind of nail produced in 1 minute is given below by the matrix X:

$$X = \begin{pmatrix} 100 & 200 & 500 & 300 \\ 50 & 20 & 30 & 10 \\ 700 & 600 & 400 & 800 \end{pmatrix} \begin{matrix} \leftarrow \text{aluminum} \\ \leftarrow \text{copper} \\ \leftarrow \text{steel} \end{matrix}$$

with columns labeled $\frac{1}{4}$ in, $\frac{1}{2}$ in, $\frac{3}{4}$ in, 1 in.

From this matrix it follows that 100 aluminum nails of length $\frac{1}{4}$ in are produced in 1 minute, and 30 copper nails of length $\frac{3}{4}$ in are produced in 1 minute.

Now let us assume that the price of the same length nail is the same, whether it is made of aluminum, copper, or steel. The prices in cents are given by the matrix Y below:

$$Y = \begin{pmatrix} 1 \\ 1\frac{1}{2} \\ 2 \\ 2\frac{1}{2} \end{pmatrix} \begin{matrix} \leftarrow \frac{1}{4} \text{ in} \\ \leftarrow \frac{1}{2} \text{ in} \\ \leftarrow \frac{3}{4} \text{ in} \\ \leftarrow 1 \text{ in} \end{matrix}$$

Matrix X is a 3-by-4 matrix (3 rows, 4 columns). Matrix Y is a 4-by-1 matrix (4 rows, 1 column). The following shows how to obtain the product of the two matrices.

$$X \times Y = \begin{pmatrix} 100 & 200 & 500 & 300 \\ 50 & 20 & 30 & 10 \\ 700 & 600 & 400 & 800 \end{pmatrix} \times \begin{pmatrix} 1 \\ 1\frac{1}{2} \\ 2 \\ 2\frac{1}{2} \end{pmatrix}$$

$$\begin{pmatrix} 100 \times 1 + 200 \times 1\frac{1}{2} + 500 \times 2 + 300 \times 2\frac{1}{2} \\ 50 \times 1 + 20 \times 1\frac{1}{2} + 30 \times 2 + 10 \times 2\frac{1}{2} \\ 700 \times 1 + 600 \times 1\frac{1}{2} + 400 \times 2 + 800 \times 2\frac{1}{2} \end{pmatrix} = \begin{pmatrix} 2150 \\ 165 \\ 4400 \end{pmatrix}$$

The resulting product is a 3-by-1 matrix (3 rows, 1 column). By examining the manner in which the three entries were obtained, it is possible to interpret the meaning of the three numbers: 2150 tells the cost in cents of all aluminum nails produced in one minute; 165 is the cost of all copper nails; and 4400 is the cost of all steel nails.

The two examples illustrate some of the simple practical uses of matrices. Applications of matrices in physics are quite numerous, but they are of a more sophisticated nature.

8 Finite Systems and Matrices

Exercises 8.9

1. Show how matrices can be practically applied.
2. Use a matrix to show how a bank could keep track of deposits in two vaults. (*Hint:* Use separate matrices for dollars, cents, gold bullion, stocks, and so on.)

CHAPTER 8 TEST

Compute the answers in S_5:

1. $4 + 2$
2. $3 + 3$
3. $1 + 4$
4. $2 + 1$
5. 3×2
6. 4×2
7. 3×3
8. 1×4
9. $2 - 4$
10. $1 - 4$
11. $1 - 2$
12. $4 - 2$
13. $2 \div 3$
14. $4 \div 3$
15. $1 \div 3$
16. $1 \div 4$
17. Solve the following riddles.
    a. When is 5 the same as 0 but equal to the sum of two numbers?
    b. When is $3^2 = 0$?
18. In S_4, $2x = 0$ has two solutions. What are they?
19. In S_4, what is 3^2 equal to?
20. Why is S_4 not a field?

Solve each equation in S_5:

21. $x + 4 = 0$
22. $x - 2 = 4$
23. $2x = 4$
24. $2x = 1$
25. $x \div 2 = 3$
26. $x \div 4 = 1$
27. $2x + 1 = 4$
28. $3x + 3 = 2$

Compute the answers in S_7:

29. $2 + 6$
30. $3 + 6$
31. $1 + 5$
32. $5 - 2$
33. $1 - 5$
34. $1 - 6$
35. 2×5
36. 3×4
37. 1×6
38. 0×6
39. $2 \div 3$
40. $6 \div 5$

Solve each equation in S_7:

41. $x + 5 = 3$
42. $x + 1 = 6$
43. $2x = 5$
44. $3x = 6$
45. $3 \div x = 2$
46. $5 \div x = 1$
47. $3x + 6 = 2$
48. $2x + 5 = 1$
49. Tell why the following is false.

$$\begin{pmatrix} 5 & 1 \\ 0 & 6 \end{pmatrix} = \begin{pmatrix} \sqrt{25} & \frac{4}{4} \\ 0 & 3^2 \end{pmatrix}$$

8.9 Applications of Matrices

50. A computer makes good use of matrices. Numbers are stored in matrices, so computations can be performed on blocks of data rather than on a single number. Check to see if the following statement has been calculated correctly.

$$\begin{pmatrix} 50 & 11 \\ \frac{2}{3} & 9 \end{pmatrix}^2 = \begin{pmatrix} \sqrt{2500} & \frac{11000}{11000} \\ \frac{36}{99} & \sqrt{82} \end{pmatrix} \times \begin{pmatrix} \sqrt{2500} & \frac{11000}{11000} \\ \frac{36}{99} & \sqrt{82} \end{pmatrix}$$

Tell the replacements of the variables for which the following will be true:

51. $\begin{pmatrix} x & y+1 \\ w & z \\ \frac{1}{2} & \frac{1}{3} \end{pmatrix} = \begin{pmatrix} -4 & -3 \\ -1 & 2 \end{pmatrix}$

52. $\begin{pmatrix} 2x+1 & 3y-1 \\ 4w+3 & 2z-4 \end{pmatrix} = \begin{pmatrix} 1 & -4 \\ -5 & -5 \end{pmatrix}$

Compute the answers:

53. $\begin{pmatrix} -4 & 1 \\ \sqrt{2} & 0 \end{pmatrix} + \begin{pmatrix} -2 & -6 \\ 1 & -3 \end{pmatrix}$

54. $\begin{pmatrix} 2 & -\frac{1}{2} \\ 0 & \frac{1}{3} \end{pmatrix} - \begin{pmatrix} 3 & \frac{1}{4} \\ \frac{1}{3} & -\frac{1}{6} \end{pmatrix}$

55. $-2 \times \begin{pmatrix} 6 & -\frac{1}{2} \\ \frac{1}{3} & \frac{1}{4} \end{pmatrix}$

56. $\begin{pmatrix} 2 & 0 \\ -2 & 5 \end{pmatrix} \times \begin{pmatrix} -1 & 2 \\ 1 & -3 \end{pmatrix}$

57. What is the additive inverse of $\begin{pmatrix} 3 & -6 \\ 0 & -\sqrt{2} \end{pmatrix}$?

58. Give one example to prove that multiplication of 2-by-2 matrices is not commutative.

59. Prove that for each real number r and for each 2-by-2 matrix A, $rA = Ar$.

9

Graphs, Functions, and Linear Programming

9.1 GRAPHING ON A NUMBER LINE

REVIEW

- $x > 3$ is read: x is greater than three
- $x \geq 3$ is read: x is greater than or equal to 3
- If $x < y$, then $x + z < y + z$
- For each *positive* real number z,

  if $x < y$, then $xz < yz$

- For each negative real number z,

  if $x < y$, then $xz > yz$

- $x < y$ means the same as $y > x$

OBJECTIVES

- Define a number line
- Graph inequalities on a number line

Lines are made up of points. By mentally assigning a unique real number to each point on a line and a unique point to each real number, we establish a one-to-one correspondence between the line and the set of real numbers. When this correspondence is set up, we call the line a *number*

9.1 Graphing on a Number Line

line. We can graph subsets of real numbers on the number line. These subsets are described by algebraic sentences.

Figure 9.1 is an example of one such graph.

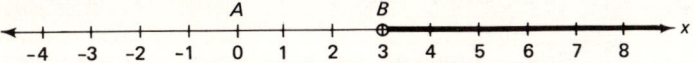

FIGURE 9.1 Graph of $\{x|x > 3\}$

On the graph, the point corresponding to the real number 0 is named by the letter *A* and the point corresponding to the real number 3 by the letter *B*. The number 3 does not belong to the solution set $\{x|x > 3\}$, and this is shown by making a circle around point *B*. The thick line to the right means that every point to the right of *B* belongs to $\{x|x > 3\}$. This is an infinite set which has no greatest member—the members of the set are all the real numbers greater than 3.

The set $\{x|x \geq 3\}$ is different from the set $\{x|x > 3\}$. The former includes the number 3. To indicate this on the graph, a solid dot at the point corresponding to 3 is made. This is shown in Figure 9.2.

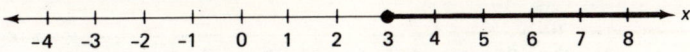

FIGURE 9.2 Graph of $\{x|x \geq 3\}$

In Chapter 6, we learned some algebraic techniques for solving inequalities. We will use these techniques to determine solution sets of inequalities, and then we will graph these solution sets. The examples below illustrate this.

Example 1 Solve and graph the solution set of $2(x + 3) + 1 < 3(x - 1)$.

$$
\begin{aligned}
2(x + 3) + 1 &< 3(x - 1) \\
2x + 6 + 1 &< 3x - 3 \\
2x + 7 - 3x &< 3x - 3 - 3x \\
-x + 7 &< -3 \\
-x + 7 - 7 &< -3 - 7 \\
-x &< -10 \\
-x(-1) &> -10(-1) \\
x &> 10
\end{aligned}
$$

9 Graphs, Functions, and Linear Programming

The following is a graph of this solution set, $\{x|x > 10\}$.

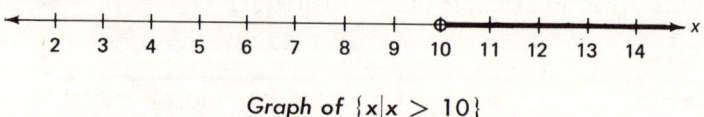

Graph of $\{x|x > 10\}$

Example 2 Solve and graph the solution set of $5(2x - 1) + 3 < 2(3x + 2) - 4$.

$$5(2x - 1) + 3 < 2(3x + 2) - 4$$
$$10x - 5 + 3 < 6x + 4 - 4$$
$$10x - 2 < 6x$$
$$10x - 2 - 6x < 6x - 6x$$
$$4x - 2 < 0$$
$$4x - 2 + 2 < 0 + 2$$
$$4x < 2$$
$$\tfrac{1}{4} \cdot 4x < \tfrac{1}{4} \cdot 2$$
$$x < \tfrac{1}{2}$$

The graph of the solution set, $\{x|x < \tfrac{1}{2}\}$, is as follows:

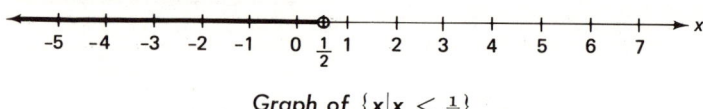

Graph of $\{x|x < \tfrac{1}{2}\}$

Sometimes inequalities are described in such a way that their graphs may consist of two parts. For example:

$$x < -5 \quad \text{or} \quad x > 0$$

is true for any replacement of x by a name of a number which is less than -5 ($x < -5$). It is also true for any replacement which is greater than 0 ($x > 0$). The graph of the solution set of this inequality is shown in Figure 9.3.

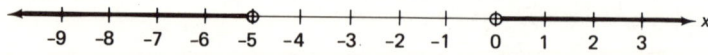

FIGURE 9.3 Graph of $\{x|x < -5 \text{ or } x > 0\}$

9.1 Graphing on a Number Line

Notice that the set $\{x | x < -5 \text{ or } x > 0\}$ is the *union* of the following two sets:

$$\{x | x < -5\}$$

and

$$\{x | x > 0\}$$

that is,

$$\{x | x < -5 \text{ or } x > 0\} = \{x | x < -5\} \cup \{x | x > 0\}$$

The next inequality consists of two parts also, but it has a different form: the two parts are connected with *and*.

$$x > -2 \quad \text{and} \quad x < 4$$

The solution set here consists of all numbers which are between -2 and 4. The graph of this solution set therefore, is as pictured in Figure 9.4.

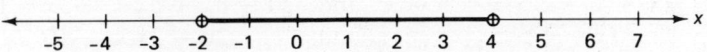

FIGURE 9.4 Graph of $\{x | x > -2 \text{ and } x < 4\}$

The examples below show techniques for solving and graphing inequalities of the type discussed above.

Example 1 Solve and graph $2x + 5 > 3(x - 1)$ or $4(x + 6) > 5x - 1$.

$$\begin{array}{lcl}
2x + 5 > 3(x - 1) & \text{or} & 4(x + 6) > 5x - 1 \\
2x + 5 > 3x - 3 & \text{or} & 4x + 24 > 5x - 1 \\
2x + 5 - 3x > 3x - 3 - 3x & \text{or} & 4x + 24 - 5x > 5x - 1 - 5x \\
-x + 5 > -3 & \text{or} & -x + 24 > -1 \\
-x + 5 - 5 > -3 - 5 & \text{or} & -x + 24 - 24 > -1 - 24 \\
-x > -8 & \text{or} & -x > -25 \\
-x(-1) < -8(-1) & \text{or} & -x(-1) < -25(-1) \\
x < 8 & \text{or} & x < 25
\end{array}$$

Thus, the solution set is $\{x | x < 8 \text{ or } x < 25\}$. This is equivalent to $\{x | x < 25\}$, since every number less than 25 satisfies the inequality

$x < 8$ or $x < 25$. The graph of this solution set is as follows:

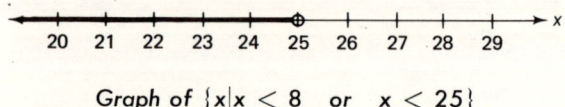

Graph of $\{x|x < 8 \text{ or } x < 25\}$

The solution set here is the union of two sets:

$$\{x|x < 8\}$$

and

$$\{x|x < 25\}$$

that is,

$$\{x|x < 8 \text{ or } x < 25\} = \{x|x < 8\} \cup \{x|x < 25\} = \{x|x < 25\}$$

Example 2 Solve and graph $-2(3 - x) + 2 > x + 5$ and $4(5 - 2x) < -5(x + 3) + 5$.

$-2(3 - x) + 2 > x + 5$	and	$4(5 - 2x) < -5(x + 3) + 5$
$-6 + 2x + 2 > x + 5$	and	$20 - 8x < -5x - 15 + 5$
$2x - 4 > x + 5$	and	$20 - 8x < -5x - 10$
$2x - 4 - x > x + 5 - x$	and	$20 - 8x + 5x < -5x - 10 + 5x$
$x - 4 > 5$	and	$20 - 3x < -10$
$x - 4 + 4 > 5 + 4$	and	$20 - 3x - 20 < -10 - 20$
$x > 9$	and	$-3x < -30$
$x > 9$	and	$-3x(-\frac{1}{3}) > -30(-\frac{1}{3})$
$x > 9$	and	$x > 10$

Thus, the solution set is $\{x|x > 9 \text{ and } x > 10\}$. This is equivalent to $\{x|x > 10\}$, since this set is an intersection of the following two sets:

$$\{x|x > 9\}$$

and

$$\{x|x > 10\}$$

that is,

$$\{x|x > 9 \text{ and } x > 10\} = \{x|x > 9\} \cap \{x|x > 10\} = \{x|x > 10\}$$

Its graph is as follows:

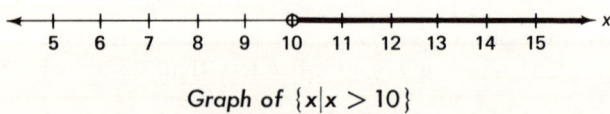

Graph of $\{x|x > 10\}$

9.1 Graphing on a Number Line

Example 3 Solve and graph $3x + 2 > x - 1$ and $x + 5 < 1$.

$$3x + 2 > x - 1 \quad \text{and} \quad x + 5 < 1$$
$$2x + 2 > -1 \quad \text{and} \quad x < -4$$
$$2x > -3 \quad \text{and} \quad x < -4$$
$$x > -\tfrac{3}{2} \quad \text{and} \quad x < -4$$

Thus, the solution set is $\{x \mid x > -\tfrac{3}{2} \text{ and } x < -4\}$. It is the following intersection:

$$\{x \mid x > -\tfrac{3}{2}\} \cap \{x \mid x < -4\}$$

This is the empty set, since no number is greater than $-\tfrac{3}{2}$ *and* less than -4. The two sets are graphed separately on one number line showing that their graphs have no common points.

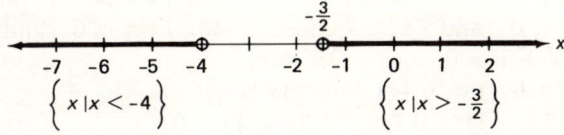

Exercises 9.1

Graph the following sets on the number line:

1. $A = \{x \mid x > 5\}$
2. $B = \{x \mid x \geq 5\}$
3. $C = \{x \mid x < 4\tfrac{1}{2}\}$
4. $D = \{x \mid x \leq 4\tfrac{1}{2}\}$
5. $E = \{x \mid x > -6\}$
6. $F = \{x \mid x \geq -6\}$
7. $G = \{x \mid x < -2\tfrac{1}{3}\}$
8. $H = \{x \mid x \leq -2\tfrac{1}{3}\}$
9. An athletic dot wishes to run on the number line. After the day's exertions, he looks back over his path. If his path is given by the following sets, where did he run?
   First run between $\{x \mid x > 5\}$ and $\{x \mid x \leq 10\}$
   Second run between $\{x \mid x \geq -6\}$ and $\{x \mid x < 2\}$

Graph the following sets on the number line:

10. $\{x \mid x < -2 \text{ or } x > 1\}$
11. $\{x \mid x \leq -2 \text{ or } x \geq 1\}$
12. $\{x \mid x < 1\tfrac{1}{2} \text{ or } x > 4\tfrac{1}{3}\}$
13. $\{x \mid x < 1\tfrac{1}{2} \text{ or } x \geq 4\tfrac{1}{3}\}$
14. $\{x \mid x < 2 \text{ or } x < -1\}$
15. $\{x \mid x \leq 2 \text{ or } x \leq -1\}$
16. $\{x \mid x > -1 \text{ or } x > 0\}$
17. $\{x \mid x \geq -1 \text{ or } x \geq 0\}$

Graph the following sets on the number line:

18. $\{x \mid x < 2 \text{ and } x > -2\}$
19. $\{x \mid x \leq 2 \text{ and } x \geq -2\}$
20. $\{x \mid x > 0 \text{ and } x < 3\tfrac{1}{2}\}$
21. $\{x \mid x \geq 0 \text{ and } x \leq 3\tfrac{1}{2}\}$
22. $\{x \mid x > -2 \text{ and } x > 0\}$
23. $\{x \mid x \geq -2 \text{ and } x \geq 0\}$
24. $\{x \mid x < 1\tfrac{1}{2} \text{ and } x < -5\}$
25. $\{x \mid x \leq 1\tfrac{1}{2} \text{ and } x \leq -5\}$

9 Graphs, Functions, and Linear Programming

Solve each inequality and graph its solution set on the number line:

26. $2x + 6 < x + 4$
27. $3x - 1 > 5x + 3$
28. $3(x + 1) > 2(2x - 1)$
29. $2(2x - 1) < 3(x + 3)$
30. $-3(2 - x) > 2(x + 5)$
31. $-5(2x + 3) - 2 < -9(1 - x)$
32. $3 - 2(x + 3) > 1 - 3(x + 2)$
33. $5(1 - 2x) + 3 < -3(4x + 1) - 6$
34. A rocket is fired from point zero on the number line. Its path is given by the inequality $4(x + 6) < 5x - 1$, where x is the distance above the earth that the rocket coasts without any impulse power. Graph the solution.

Simplify, whenever possible, and graph each solution set on the number line:

35. $\{x | x > -2 \quad \text{or} \quad x < -5\}$
36. $\{x | x > 0 \quad \text{or} \quad x > 2\}$
37. $\{x | x < -1 \quad \text{or} \quad x < 0\}$
38. $\{x | x > -1 \quad \text{and} \quad x < 5\}$
39. $\{x | x < 0 \quad \text{and} \quad x < -3\}$
40. $\{x | x < 0 \quad \text{and} \quad x > 2\}$
41. $\{x | 2x + 1 > 0 \quad \text{or} \quad -3x < 3\}$
42. $\{x | 2(x + 1) < x + 1 \quad \text{or} \quad 3x > 6\}$
43. $\{x | 4(2x - 1) > 0 \quad \text{or} \quad 4(2x - 1) < 0\}$
44. $\{x | 3(x + 1) < -6 \quad \text{and} \quad x > -9\}$
45. $\{x | 3(2x - 1) > 10 \quad \text{and} \quad 3x + 5 < -4\}$
46. $\{x | 3(1 - x) < 0 \quad \text{and} \quad x > -1\}$

9.2 GRAPHING IN A PLANE

REVIEW

- (x, y) is an ordered pair; x is the first member, y the second
- $\{(x, y) | x > 4\}$ is read: the set of all ordered pairs (x, y) such that x is greater than 4

OBJECTIVES

- Establish the rectangular coordinate system
- Graph points on the rectangular coordinate system
- Graph horizontal and vertical lines on the coordinate system
- Graph subsets of the plane on the coordinate system
- Write expressions for the plane in terms of the union of three sets: a line and two half-planes

Consider the set of real numbers, R. It consists of all rational and all irrational numbers. Now imagine that a set of all ordered pairs of real numbers is formed. Call this set $R \times R$ (*read: R cross R*).

9.2 Graphing in a Plane

The set of points corresponding to $R \times R$ is an entire plane. It can be shown graphically as in Figure 9.5. The two perpendicular lines are called the *axes*. In particular, the horizontal line is called the *x-axis* and the vertical line *y-axis*. Once this kind of framework is placed on the plane, it is said that a *rectangular coordinate system* is established. The rectangular coordinate system is also called a *Cartesian coordinate system*.

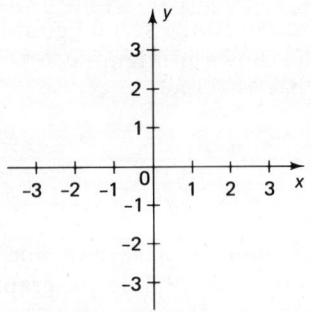

FIGURE 9.5

To locate a point corresponding to a given ordered number pair in a plane, locate the point corresponding to the *first* number on the x-axis, then move up or down, depending on whether the second number is positive or negative, respectively. Several points are shown on the graph in Figure 9.6. Apply the method described above, and verify that the points are correctly placed. For example, the point in the plane corresponding to the ordered pair (3, 5) is found by moving to 3 on the x-axis and then moving up until we are across from 5 on the y-axis.

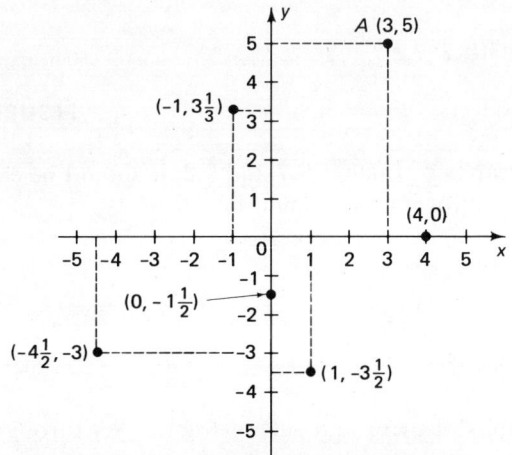

FIGURE 9.6

9 Graphs, Functions, and Linear Programming

The ordered pair (3, 5) is named A in Figure 9.6. The *first coordinate* or *x-coordinate* of point A is 3, and its *second coordinate* or *y-coordinate* is 5.

Consider the set of points described by

$$\{(x, y) | y = 4\}$$

Since x is not specified, it is each real number. Thus, the graph of this set is the set of all points for which the second coordinate is 4. The graph is a line parallel to the x-axis shown in Figure 9.7.

The set of points described by

$$\{(x, y) | x = -1\tfrac{1}{2}\}$$

consists of all points for which the first coordinate is $-1\tfrac{1}{2}$. Since y is not specified, it is each real number. This is a line which is parallel to the y-axis. Its graph is given in Figure 9.8. In the graphs of Figures 9.7 and 9.8, only parts of the graphs are shown. The lines continue on and on, and this fact is indicated by the arrows.

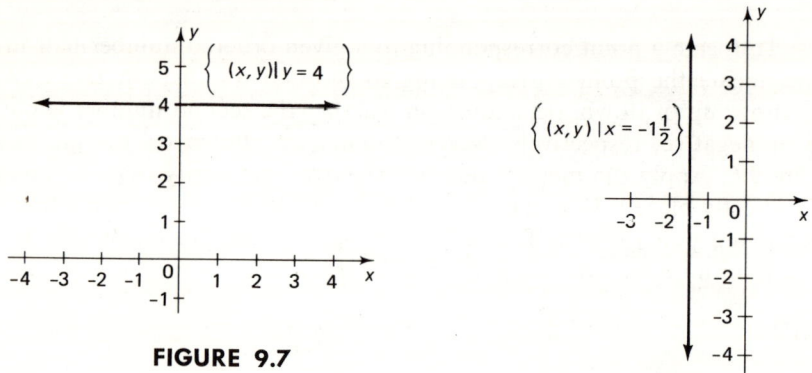

FIGURE 9.7

FIGURE 9.8

With the graphs of Figures 9.7 and 9.8, it should be easy to tell what the graphs of the following sets would be:

$A = \{(x, y) | y > 4\}$
$B = \{(x, y) | y < 4\}$
$C = \{(x, y) | x > -1\tfrac{1}{2}\}$
$D = \{(x, y) | x < -1\tfrac{1}{2}\}$

Examine the graphs of these four sets in Figures 9.9 through 9.12.

9.2 Graphing in a Plane

Notice that the line $y = 4$ is shown by dashed marks in Figure 9.9. This indicates that this line does not belong to the graph of $\{(x, y)|y > 4\}$. The graph extends indefinitely upward as well as to the left and to the right. But no point of the line $y = 4$ and no point below this line belongs to the graph.

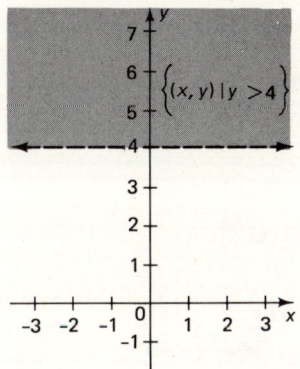
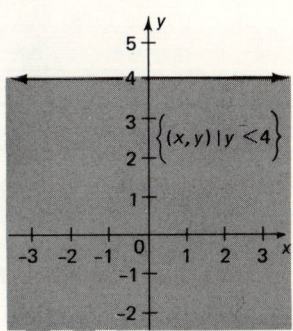

FIGURE 9.9 Graph of $\{(x, y)|y > 4\}$ **FIGURE 9.10** Graph of $\{x, y)|y < 4\}$

In the graph of Figure 9.10, no point of the line $y = 4$ and no point above this line belongs to the graph of $\{(x, y)|y < 4\}$.

Observe that the graph of $\{(x, y)|y = 4\}$ separates the plane into two disjoint subsets. One of these is

$$A = \{(x, y)|y > 4\}$$

and the other is

$$B = \{(x, y)|y < 4\}$$

Describing the set of all points in the plane by the letter P, the following is true:

$$\{(x, y)|y = 4\} \cup \{(x, y)|y > 4\} \cup \{(x, y)|y < 4\} = P$$

Thus, the entire plane has been described by three different sets. Any two of these three sets are disjoint.

In Figure 9.11, the graph extends indefinitely to the right of line $x = -1\frac{1}{2}$ as well as upward and downward, but no point of the line $x = -1\frac{1}{2}$ and no point to the left of this line belongs to the graph of $\{(x, y) | x > -1\frac{1}{2}\}$.

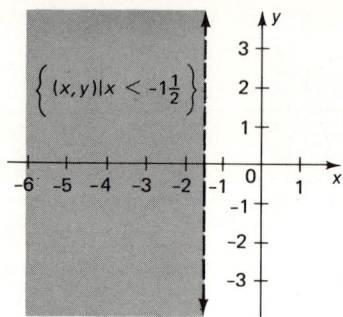

FIGURE 9.11 Graph of $\{(x, y) | x > -1\frac{1}{2}\}$

In Figure 9.12, the graph extends indefinitely to the left of the line $x = -1\frac{1}{2}$ as well as upward and downward, but no point of the line $x = -1\frac{1}{2}$, and no point to the right of this line belongs to the graph of $\{(x, y) | x < -1\frac{1}{2}\}$.

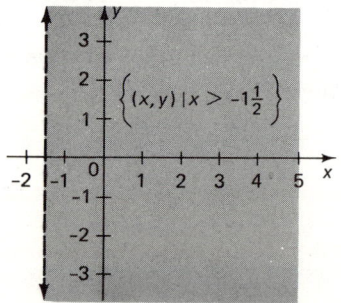

FIGURE 9.12 Graph of $\{(x, y | x < -1\frac{1}{2}\}$

The graph in Figure 9.12 of $\{(x, y) | x = -1\frac{1}{2}\}$ also separates the entire plane into two disjoint subsets. One of these is

$$C = \{(x, y) | x > -1\frac{1}{2}\}$$

9.2 Graphing in a Plane

and the other is

$$D = \{(x, y) | x < -1\tfrac{1}{2}\}$$

Observe that the following is true.

$$\{(x, y) | x = -1\tfrac{1}{2}\} \cup \{(x, y) | x > -1\tfrac{1}{2}\}$$
$$\cup \{(x, y) | x < -1\tfrac{1}{2}\} = P$$

It is also true that any two of these three sets are disjoint sets. For example,

$$C \cap D = \emptyset$$

Exercises 9.2

1. Plot the following points in a coordinate system so that you will determine the secret sign of the "Purple Circle," a subversive math group.
Points:
   a. $(-4, 6)$, $(-3, 6)$, $(-2, 6)$, $(-1, 6)$, $(0, 6)$, $(1, 6)$, $(2, 6)$, $(3, 6)$, $(4, 6)$
   b. $(-3, 5)$, $(-4, 4)$, $(-6, 3)$
   c. $(2, 5)$, $(1, 4)$, $(-1, 3)$

On a coordinate system in a plane locate the following points:

2. $(1, 3)$ 3. $(0, -2)$ 4. $(3\tfrac{1}{2}, 0)$
5. $(-2, 1)$ 6. $(4\tfrac{1}{3}, -1)$ 7. $(-1, -3)$
8. You have just purchased a new super-duper dirt bike. You decide to ride it in the hills on Sunday afternoon. If the main road begins at $(0, 0)$, find the distance you ride if you go to the following coordinates.
   a. $(0, 5)$ b. $(5, 5)$ c. $(5, -5)$ d. $(-5, 0)$ e. $(0, 0)$

The point $(0, 0)$ is called the origin. What is the distance of the following points from the origin?

9. $(5, 0)$ 10. $(-3, 0)$ 11. $(0, 4)$ 12. $(0, -5.6)$

Graph each of the following sets in $R \times R$, where R is the set of real numbers:

13. $\{(x, y) | x = -2\}$ 14. $\{(x, y) | y = 3\tfrac{1}{2}\}$ 15. $\{(x, y) | x > -2\}$
16. $\{(x, y) | x < -2\}$ 17. $\{(x, y) | x \geq -2\}$ 18. $\{(x, y) | y > 3\tfrac{1}{2}\}$
19. $\{(x, y) | y < 3\tfrac{1}{2}\}$ 20. $\{(x, y) | y \leq 3\tfrac{1}{2}\}$

9 Graphs, Functions, and Linear Programming

21. Your new $100,000 house requires some landscaping. If your garden is bounded by the following sets, what shape is it?
$A = \{(x, y) | y \leq 4\}$ $\qquad$ $B = \{(x, y) | x \leq 4\}$
$C = \{(x, y) | y \geq -3\}$ $\qquad$ $D = \{(x, y) | x \geq -3\}$
How many units of area are contained in your yard?
22. What is $\{(x, y) | x > -2\} \cap \{(x, y) | x < -2\}$ equal to?
23. What is $\{(x, y) | y > 3\frac{1}{2}\} \cap \{(x, y) | y \leq 3\frac{1}{2}\}$ equal to?
24. What is $\{(x, y) | y \geq 3\frac{1}{2}\} \cap \{(x, y) | y \leq 3\frac{1}{2}\}$ equal to?
25. What is $\{(x, y) | x > -2\} \cup \{(x, y) | x < -2\} \cup \{(x, y) | x = -2\}$ equal to?

9.3 LINES

REVIEW

- The members of the ordered pair (x, y) are the coordinates of the point corresponding to (x, y)
- In (x, y), x is the first coordinate or abscissa, and y is the second coordinate or ordinate

OBJECTIVES

- Graph lines in the plane
- State two properties of lines
- Define y-intercept and x-intercept

We shall graph some equations and see whether points whose coordinates satisfy these equations form any kind of a pattern. Let us plot some points for the equation

$$\{(x, y) | 2x + 3y = 10\}$$

Six points are plotted in Figure 9.13.
You should verify that the coordinates of each of the six points satisfy the equation $2x + 3y = 10$. It is easy to perceive that these points are on one line. The graph of this line is shown in Figure 9.14.
One could go on graphing such sets as

$$\{(x, y) | 4x + y = 7\}$$
$$\{(x, y) | 2x - 5y = 3\}$$

9.3 Lines

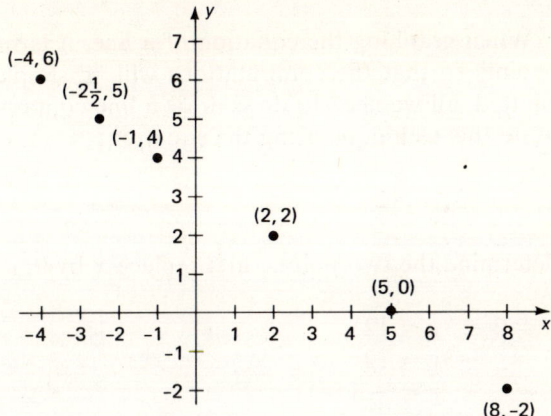

FIGURE 9.13

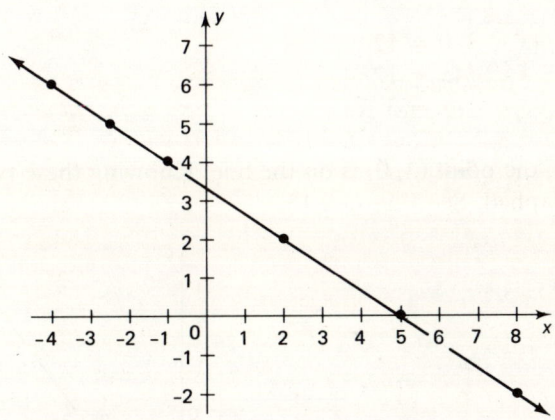

FIGURE 9.14

and generally, sets of the form

$$\{(x, y) \mid ax + by = c\}$$

to verify that in each case the graph would be a line.

We make two observations about lines and state them formally as properties of lines.

LINE PROPERTY 1 The graph of an equation of the form

$$ax + by = c$$

is a line.

LINE PROPERTY 2 A line is determined by two points.

When graphing the equation of a line, it is smart strategy to choose two points so that the computations will be simple. Once the two points are plotted, all we need to do is draw a line containing the two points. We illustrate this technique using the equation:

$$4x - 3y = 12$$

To determine the two points, first replace x by 0:

$$\begin{aligned} 4 \cdot 0 - 3y &= 12 \\ -3y &= 12 \\ y &= -4 \end{aligned}$$

Thus, the point $(0, -4)$ is on the line. Now replace y by 0:

$$\begin{aligned} 4x - 3 \cdot 0 &= 12 \\ 4x &= 12 \\ x &= 3 \end{aligned}$$

Thus, the point $(3, 0)$ is on the line. Knowing these two points, the line can be graphed. See Figure 9.15.

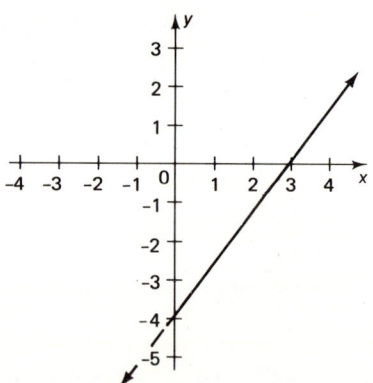

FIGURE 9.15 Graph of $\{(x, y) | 4x - 3y = 12\}$

With many equations, the points at which a line intersects the x-axis and the y-axis are convenient to use for purposes of graphing. The coordinates of those points are given special names.

The second coordinate of the point at which any nonvertical line intersects the y-axis is called the *y-intercept* of the line. It is necessary to specify that a line is nonvertical, since a vertical line has no y-intercept.

9.3 Lines

The first coordinate of the point at which any nonhorizontal line intersects the x-axis is called the *x-intercept* of the line. It is necessary to specify that a line is nonhorizontal, since a horizontal line has no x-intercept.

To determine the y-intercept for any given equation of a line, replace x by 0. For example, in $5x - y = 3$, replacing x by 0 yields:

$$5 \cdot 0 - y = 3$$
$$-y = 3$$
$$y = -3$$

Thus the y-intercept is -3, and the line intersects the y-axis at the point $(0, -3)$. Replacing y by 0 yields:

$$5x - 0 = 3$$
$$5x = 3$$
$$x = \tfrac{3}{5}$$

Thus, the x-intercept is $\tfrac{3}{5}$, and the line intersects the x-axis at the point $(\tfrac{3}{5}, 0)$. For graphing purposes, the last point would not be easily located; therefore, a more convenient point would be sought, preferably one with integer coordinates.

Exercises 9.3

Graph the lines represented by the following equations:

1. $x + y = 7$
2. $x = y$
3. $2x + 4y = 11$
4. $x - 5y = 5$
5. $3x = 2 + 7y$
6. $4y = x - 2$
7. Three moles, Alphonse, Edgar, and Rodney, all frantically burrow to establish a new interhole freeway system. If their burrows are given by the equations below, plot their underground system as a map to help visiting moles.
   Alphonse: $x = y$
   Edgar: $2x + 4y = 11$
   Rodney: $x - 5y = 5$
8. Write the equation of the x-axis.
9. Write the equation of the y-axis.
10. Write a general equation of a horizontal line.
11. Write a general equation of a vertical line.

9.4 SLOPE OF A LINE

REVIEW

- Since the coordinates of the point (1, 2) satisfy the equation $x + y = 3$, this point is on the line given by this equation
- Two points determine a unique line
- $a/b = -a/-b$ for all numbers a and b ($b \neq 0$)

OBJECTIVES

- Solve equations of the form $ax + by = c$ for y, obtaining equations of the form $y = mx + t$
- Determine slopes of lines given by equations in the form $y = mx + t$
- Determine slopes of lines given the coordinates of two points of the lines
- Establish that in an equation of the form $y = mx + t$, m is the slope of the line and t is the y-intercept

The graph of every equation of the form $ax + by = c$ is a line. In this equation, a and b cannot both be 0. If $b \neq 0$, we can solve this equation for y:

$$ax + by = c$$
$$by = -ax + c$$
$$y = -\frac{a}{b}x + \frac{c}{b}$$

Since $-a/b$ and c/b are real numbers, simplification can be achieved by replacing these with more convenient variables. Let us replace $-a/b$ by m and c/b by t, obtaining:

$$y = mx + t$$

We shall now try to discover how different replacements of m in $y = mx + t$ affect the graphs of the equations. On the coordinate system in Figure 9.16, graphs of several equations are shown. You should be able to observe that the replacements for m in $y = mx + t$ have something to do with the "steepness" of the graphs. This "steepness" is determined by m, which is called the *slope*. Thus, the slope of the line described by $y = -2x$ is -2.

9.4 Slope of a Line

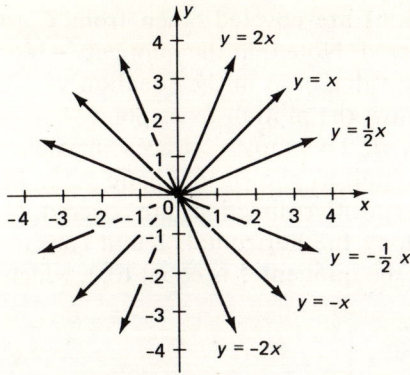

FIGURE 9.16

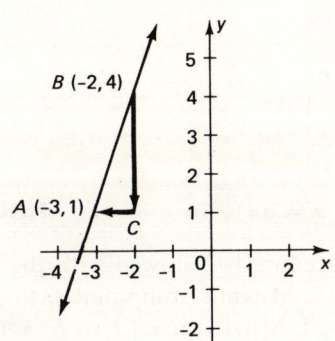

FIGURE 9.17 $y = 3x + 10$

Similarly, the slope of the line $y = 2x$ is 2, of the line $y = \frac{1}{2}x$ is $\frac{1}{2}$, and of the line $y = x$ is 1. In general:

> The *slope* of the line described by the equation $y = mx + b$ is the number m.

We now graph the line $y = 3x + 10$. We determine that points A and B with coordinates $(-3, 1)$ and $(-2, 4)$, respectively, satisfy the equation $y = 3x + 10$. The graph of $y = 3x + 10$ is shown in Figure 9.17.

In Figure 9.17 we graphed the line by first locating two points. Let's now take a look at the idea of the slope of a line in terms of two points on the line. We do this by "moving" from point B to point A, first along the vertical and then along the horizontal. In moving from B to C, -3 units (negative direction) are covered; then from C to A, −1 unit (negative direction) is covered. Note that the quotient −3/−1 is equal to 3. Observe also that 3 is the value of m in the equation $y = 3x + 10$.

Consider now the motion from point A to point B, instead of from point B to point A. This move is traced, first along the vertical and then along the horizontal, as shown in Figure 9.18. Moving from point A to point H, 3 units (positive direction) are covered. Next, moving from point H to point B along the horizontal, 1 unit (positive direction) is covered. Again note that the quotient $\frac{3}{1}$ is equal to 3, which is the slope of the line.

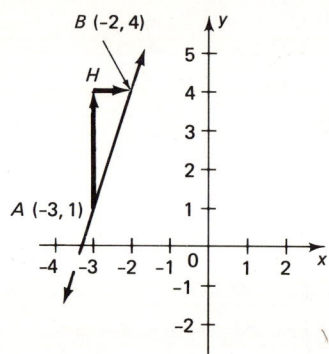 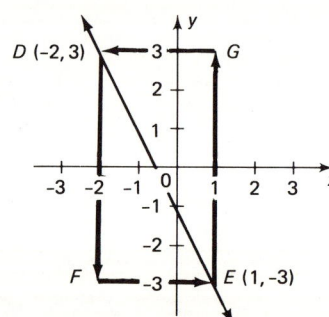

FIGURE 9.18 $y = 3x + 10$ **FIGURE 9.19** $y = -2x - 1$

A similar procedure is followed with the graph of $y = -2x - 1$, as shown in Figure 9.19. Moving from point D to point F, −6 units (negative direction) are covered. Moving from F to E, 3 units (positive direction) are covered. The quotient −6/3 is equal to −2. Observe that −2 is equal to the value of m in the equation $y = -2x - 1$, which is the slope of the line.

Moving from E to G first and then from G to D results in the same slope. From E to G, 6 units (positive direction) are covered, and from G to D, −3 units (negative direction) are covered, thus resulting in the slope of −2.

Actually, it is possible to prove that the slope of a line may be found by considering *any* two points of the line and by taking the quotient of the vertical move to the horizontal move in moving from one point to the other.

Thus, two points at random on the graph of a linear equation can be chosen. Knowing the coordinates of these points, the slope of the graph can be computed. Returning to our previous example, $y = 3x + 10$, suppose the two points are $(-2, 4)$ and $(-3, 1)$. Study the following:

$$\text{slope} = \frac{4 - 1}{-2 - (-3)} = \frac{3}{1} = 3$$

9.4 Slope of a Line

Similarly, for the line $y = -2x - 1$ and the points $(-2, 3)$ and $(1, -3)$ on it, the following is true:

$$\text{slope} = \frac{3 - (-3)}{-2 - 1} = \frac{6}{-3} = -2.$$

In general, if the points (x_1, y_1) and (x_2, y_2) are on a line k, then the slope of k is equal to:

$$\frac{y_2 - y_1}{x_2 - x_1}$$

Note that $\frac{y_2 - y_1}{x_2 - x_1} = \frac{y_1 - y_2}{x_1 - x_2}$, since $y_2 - y_1$ is the additive inverse of $y_1 - y_2$, and $x_2 - x_1$ is the additive inverse of $x_1 - x_2$.

Given an equation of a line in the form $y = mx + t$, what does the number t tell us? To find out, observe that a line intercepts the y-axis at a point whose x-coordinate is 0. Replacing x by 0 in $y = mx + t$, the following is obtained:

$$y = m \cdot 0 + t$$
$$y = 0 + t$$
$$y = t$$

Thus, t is the y-coordinate of the point at which the line intersects the y-axis; that is, t is the y-intercept.

SUMMARY Given an equation of a line in the form

$$y = mx + t$$

m is the slope of the line and t is the y-intercept.

Exercises 9.4

For each of the following equations, (1) give its equivalent equation in the form $y = mx + t$, and (2) tell the slope m and the y-intercept t of the line given by the equation:

1. $x + y = 6$
2. $y + 3x = -4$
3. $y - 2x = 5$
4. $6x + 2y = 22$
5. $4y = 3x - 8$
6. $5x = 2y + 7$
7. $3(2x - y) = 4$
8. $-7(2y - x) = -1$

9 Graphs, Functions, and Linear Programming

9. After cycling through many lands, you intend to cycle through the land of Mathematica. However, visiting Mathematica is a mistake if you can't speak Math. Before any hill can be climbed or descended, you must check whether the slope is within the tolerance of your bike. You know (from painful past experience) that your bike can take a maximum slope of $+4$ or -6. Knowing this fact, can you climb or descend the following hills?
   a. $4y = 3x - 8$
   b. $y + 3x = -4$
   c. $5x = 2y + 7$

Graph each of the following lines on the same coordinate system:

10. $y = 3x + 2$ 11. $y = 2x + 2$ 12. $y = x + 2$
13. $y = -x + 2$ 14. $y = -2x + 2$ 15. $y = -3x + 2$
16. The warlike Mathematicians of Mathematica are having their annual target practice, which consists of throwing eggs at students who got less than a B in math last term. If the plane that the Mathematicians stand on is the x-axis and the students are on the y-axis, calculate where the eggs will land if the trajectories of the eggs are given by the following equations.
    a. Egg 1: $y = 32x + 6$
    b. Egg 2: $y = 6x - 2$
    c. Egg 3: $y = -2x + 16$
    d. Egg 4: $x - 2y = 12$
    e. Egg 5: $x = 17y - 20$
17. What is the y-intercept for each of the lines in exercises 10–15?
18. For each of the lines in exercises 10–15, determine its slope.

Graph each of the following lines on the same coordinate system:

19. $y = 2x + 3$
20. $y = 2x + 1$
21. $y = 2x$
22. $y = 2x - 1$
23. $y = 2x - 2$
24. $y = 2x - 3$
25. How are the lines in exercises 19–24 related to each other?
26. What is the slope of each of the lines in exercises 19–24?
27. For each line in exercises 19–24, determine its y-intercept.
28. If you were one of the unfortunate students victimized in exercise 16, where would you position yourself in order to be relatively safe from eggs?

9.5 LINEAR PROGRAMMING

REVIEW

- To solve a system of inequalities means to find the intersection of the solution sets of all the inequalities
- A *polygonal region* is the union of a polygon and the interior of the polygon
- The symbol $\doteq$ means *is approximately equal to*

OBJECTIVES

- Graph systems of inequalities
- Solve practical problems using linear programming

In Section 9.2 we learned how to graph subsets of the plane described by inequalities. We shall now graph *systems* of inequalities—this consists of graphing the intersection of several sets.

Example 1 Graph the solution set of the following system of inequalities:

$$x \geq 0$$
$$y \geq 0$$
$$x \leq 4$$
$$x + y \leq 8$$

It is not necessary to show the entire graph of each inequality. Examine the graph (shaded portion) in Figure 9.20 to see that the coordinates of each point satisfy *each* of the inequalities. The shaded portion and its boundaries make up a set of points called a *polygonal region*. Since the inequalities that are graphed have the symbol $\geq$ or $\leq$ in them, the boundaries of the polygonal region in the graph belong to the graph of the solution set.

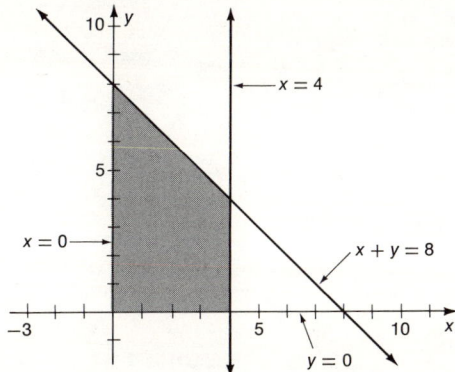

FIGURE 9.20

Now, in Example 2 we will add other conditions to the polygonal region graphed in Example 1.

Example 2 Graph several lines whose equations are of the form $2x + y = k$ on the graph of the system of inequalities of Example 1. Determine the maximum (largest) and the minimum value of k in $2x + y = k$ for which the graphs of $2x + y = k$ intersect the polygonal region of Example 1.

Solution If we express the equation $2x + y = k$ in slope-intercept form, we have:

$$y = -2x + k$$

The following lines of the form $y = -2x + k$ are shown on the graph in Figure 9.21:

$$y = -2x + 14$$
$$y = -2x + 12$$
$$y = -2x + 8$$
$$y = -2x + 4$$
$$y = -2x + 0$$
$$y = -2x + (-4)$$

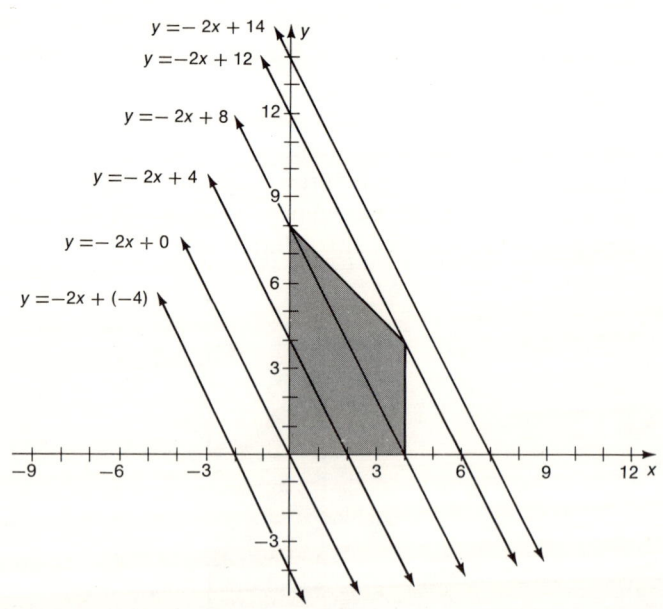

FIGURE 9.21

9.5 Linear Programming

Note that -2 is the slope of each line, and the different values of k give the y-intercepts for the lines. An examination of the graph reveals the following information:

1. The lines $y = -2x + 14$ and $y = -2x + (-4)$ do not intersect the polygonal region.
2. The line $2x + y = 12.01$ would not intersect the region.
3. The line $2x + y = -.01$ would not intersect the region.
4. More generally, no line whose equation is of the form $2x + y = k$, where $k > 12$ or $k < 0$ would intersect the region.

Now we shall answer the two questions we asked at the outset.

What is the maximum value of k in $2x + y = k$ for which the graph of $2x + y = k$ intersects the region? It is 12, since the graph of $2x + y = 12$ intersects the region, and the graph of $2x + y = k$ for every $k > 12$ will not intersect the region.

What is the minimum value of k in $2x + y = k$ for which the graph of $2x + y = k$ intersects the region? It is 0, since the graph of $2x + y = 0$ intersects the region, and the graph of $2x + y = k$ for every $k < 0$ will not intersect the region.

Graphs of equations and inequalities, like those considered in Examples 1 and 2, are useful in solving a variety of problems encountered in business, industry, and government. This technique of problem solving is called *linear programming*. Many of the problems are concerned with decision-making situations, where management is faced with several choices, one of which would lead, say, to the maximum profit. The following example illustrates one such problem.

Example 3 The Econo-Company manufactures two sizes of TV screens, size A and B. The screens are made by two machines. The old machine makes one size A screen in 4 minutes and one size B screen in 1 minute. The new machine makes one size A screen in 2 minutes and one size B screen in 6 minutes. Each machine can operate at most 2 hours a day. If the size A screen sells at a profit of $2 per screen and the size B at a profit of $3 per screen, how many of each size screen should be manufactured per day to bring in the maximum profit?

9 Graphs, Functions, and Linear Programming

Solution Let us summarize the given information.

	Old Machine Time	New Machine Time	Profit
Size A	4 min	2 min	$2
Size B	1 min	6 min	$3
Maximum Time per Day	2 hr	2 hr	

We now write the system of equations and inequalities which represent the essential information. We let the following variables represent the pertinent items involved in the problem:

x = the number of size A screens manufactured per day
y = the number of size B screens manufactured per day
p = the profit in dollars per day's production

$$p = 2x + 3y$$
$$4x + 1y \leq 120$$
$$2x + 6y \leq 120$$

To produce size A and size B screens, the old machine can use at most 2 hours (120 min) per day. We also should keep in mind that the replacement set for x and y is the set of whole numbers.

We seek the values of x and y which satisfy all of the conditions above and yield the maximum profit, p. Let us first graph the system of inequalities:

$$\left.\begin{array}{r}x \geq 0 \\ y \geq 0\end{array}\right\}$$
$$4x + y \leq 120$$
$$2x + 6y \leq 120$$

The solution set of the system is shown in the graph in Figure 9.22. Now let us consider the equation:

$$p = 2x + 3y$$

The equivalent equation in slope-intercept form is:

$$y = -\frac{2}{3}x + \frac{p}{3}$$

9.5 Linear Programming

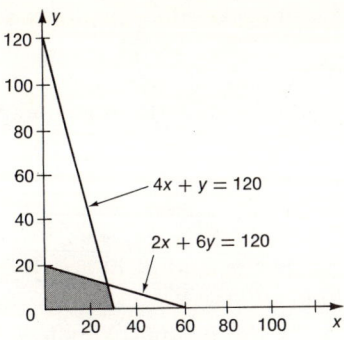

FIGURE 9.22

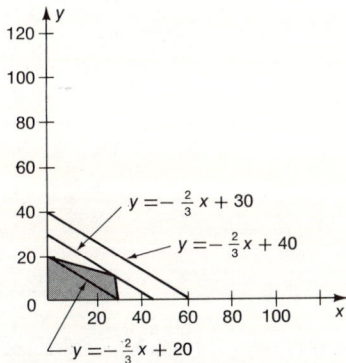

FIGURE 9.23 $y = -\frac{2}{3}x + 40$ $(p = 120)$
$y = -\frac{2}{3}x + 30$ $(p = 90)$
$y = -\frac{2}{3}x + 20$ $(p = 60)$

Replacing p by various numerals will give us graphs which are lines with slope $-\frac{2}{3}$. If p is replaced by 3, the y-intercept is 1, since the equation of this line will be:

$$y = -\frac{2}{3}x + 1$$

The problem calls for selecting the line that intersects the shaded region and has the greatest possible y-intercept. Thus, we are seeking all ordered pairs (x, y) in the intersection of the shaded region and the line $y = -\frac{2}{3}x + p/3$ which will produce the maximum value of p. The graph in Figure 9.23 shows several lines whose equations are of the form $y = -\frac{2}{3}x + p/3$.

9 Graphs, Functions, and Linear Programming

From Figure 9.23 it appears that the line which intersects the shaded region with the greatest y-intercept is $y = -\frac{2}{3}x + 30$. The value of p for this line is 90. The coordinates of the vertex where the region and the line intersect cannot be read with complete accuracy, but they appear to be approximately equal to:

$$(x, y) \doteq (28, 11)$$

We can conclude then that the Econo-Company should manufacture about 28 of size A screens and 11 of size B screens per day, realizing the maximum profit of $90. Substitute 28 for x and 11 for y in $p = 2x + 3y$ to see that p will be approximately 90.

Exercises 9.5

Graph the solution sets of the following systems of inequalities:

1. $x \geq 0$
   $y \geq 0$
   $x \leq 2$
   $y \leq 5$
2. $x \geq 0$
   $y \geq 0$
   $x \leq 3$
   $5x + y \leq 30$
3. $x \geq 0$
   $y \geq 0$
   $x \leq 2$
   $y + 3x \leq 12$
4. $x \geq -4$
   $y \geq -4$
   $x \leq 0$
   $y \leq 0$
   $x + y \geq -4$

5. On the graph for exercise 1, graph several lines whose equations are of the form $x + y = p$. Determine the maximum and the minimum value of p in $x + y = p$ for which the graphs of $x + y = p$ intersect the polygonal region.

6. On the graph for exercise 2, graph several lines whose equations are of the form $2x + y = p$. Determine the maximum and the minimum value of p in $2x + y = p$ for which the graphs of $2x + y = p$ intersect the polygonal region.

7. On the graph for exercise 3, graph several lines whose equations are of the form $x + y = p$. Determine the maximum and the minimum values of p in $x + y = p$ for which the graphs of $x + y = p$ intersect the polygonal region.

8. On the graph for exercise 4, graph several lines whose equations are of the form $2x + y = p$. Determine the maximum and the minimum value of p in $2x + y = p$ for which the graphs of $2x + y = p$ intersect the polygonal region.

9.5 Linear Programming

9. Your old great-great-granduncle Scrooge decided to leave his estate partially to you and partially to the Happy Home for Fallen Cats and Estranged Dogs. Your plot of land is given by this solution set of inequalities:

$$x \geq 0$$
$$y \geq 0$$
$$x \leq 3$$
$$x + 2y \leq 9$$

   What is the shape of your polygonal region?

10. After cultivating the plot of land left to you by old Scrooge, several of the fallen cats and estranged dogs escape and run straight across your beautiful garden. If there are seven dogs and cats, and their path is given by the equation $y = x + K$, determine how many of the beasties would run over your garden if K differs by 2 each time—for example, $\{-6, -4, -2, 0, 2, 4\}$, starting with $K = -6$. Then find the maximum and the minimum value of K that allows any animal to trample upon your garden.

11. Using x for the number of size 1 buttons and y for size 2 buttons, write a mathematical sentence which fits the following conditions: The total cost of x size 1 buttons and y size 2 buttons is at most $8.00, if the price of size 1 buttons is 10¢ and of size 2 buttons 8¢.

12. For exercise 11, let c be the total cost of x size 1 buttons and y size 2 buttons. Write an equation for the total cost of these buttons in terms of x and y.

13. J. J. Gobblesfield, the well-known horseradish manufacturer, wishes to display his knowledge of math by showing his costs in the form of a mathematical expression. Unfortunately, he knows nothing about math, so he hires you to do the job. Given the following information, prepare an expression for J.J.

    $x = $ number of ordinary-size horseradish bottles
    $y = $ number of super-duper (hot) horseradish bottles
    Total cost of producing one day's supply of horseradish (after all benefits are accounted for) is $120 or more.
    The price of ordinary-size bottles is $2 each; the super-duper size cost $4 each.

14. The Bolt Company manufactures two kinds of handmade, special purpose bolts, zero-bolts and one-bolts. Boltmaker A makes zero-bolts at the rate of 1 per 2 min and one-bolts at 1 per 4 min. Boltmaker B makes zero-bolts at the rate of 1 per 3 min and one-bolts at 1 per 1 min.

9 Graphs, Functions, and Linear Programming

The profit on each zero-bolt is $3 and on each one-bolt is $4. The maximum time devoted to the production of bolts by each maker is 3 hr per day.

	Maker A Time	Maker B Time	Profit
Zero-Bolt	2 min	3 min	$3
One-Bolt	4 min	1 min	$4
Maximum Time	3 hr	3 hr	

a. Using the following variables:

x = for the number of zero-bolts manufactured per day
y = for the number of one-bolts manufactured per day
p = for the profit in cents per day's production

write an equation showing the relation between p, and x and y
b. Write a sentence showing that it takes boltmaker A at most 3 hr (180 min) to produce x zero-bolts and y one-bolts.
c. Write a sentence showing that it takes boltmaker B at most 3 hr to produce x zero-bolts and y one-bolts.
d. Graph the sentences in parts b and c. What is the replacement set for x and y?
e. Graph several lines which have the slope indicated by the equation in part a.
f. Find the point of intersection which gives the maximum value of p.
g. State the approximate maximum profit.

9.6 RELATIONS AND FUNCTIONS

REVIEW

- In the ordered pair (a, b), a is the first member, and b is the second member

OBJECTIVES

- Define a relation
- Define the domain of a relation
- Define the range of a relation
- Define a function
- Determine which relations are functions and which are not by the use of the vertical line test

9.6 Relations and Functions

Among the most central concepts in mathematics are the concepts of relation and function. We define a relation first.

A *relation* is a set of ordered pairs. Keep in mind that the elements of a pair do not have to be numbers. For example,

{(Jones, 5 ft 8 in), (Adams, 6 ft)}

is a relation because it is a set of ordered pairs. This relation consists of two ordered pairs.

The set of all first elements of the ordered pairs in a relation is called the *domain* of the relation. For example, the domain of the relation

{(Jones, 5 ft 8 in), (Adams, 6 ft)}

is:

{Jones, Adams}

It is the set of all first elements of the ordered pairs that belong to the relation.

The set of all second elements of the ordered pairs in a relation is called the *range* of the relation. For example, for the relation above, the range is:

{5 ft 8 in, 6 ft}

This is the set of all second elements of the ordered pairs that belong to the relation.

Of special interest in mathematics are relations which have a property stated below.

A relation in which no two ordered pairs have the same first element is called a *function*.

For example, the relation

{(1, 5), (−2, 7), (1, 9)}

is not a function because two of its ordered pairs have the same first elements. These pairs are (1, 5) and (1, 9). They share the same first element, the number 1.

However, the relation

{(0, 3), (5, 3), (6, −5)}

is a function because no two pairs have the same first element.

It is easy to decide whether a relation is a function by examining its graph. If at least one vertical line intersects the graph of the relation at more than one point, then the relation is not a function. This occurs because the relation contains two or more ordered pairs with the same first element. See Figure 9.24.

Note that in Figure 9.24 the given vertical line intersects the graph of the relation at two points, (3, 2) and (3, −2). The element 3 of the domain is paired with two different elements of the range: 2 and −2. Thus the relation whose graph is shown is *not* a function. This idea can be generalized as follows:

> VERTICAL LINE TEST FOR A FUNCTION A relation is a function if no vertical line intersects or touches the graph of the relation at more than one point.

Now consider the graph of $y = 2x - 1$ given in Figure 9.25. An examination of the graph of $y = 2x - 1$ suggests that no vertical line would intersect it at more than one point. Thus, the relation

$$\{(x, y) | y = 2x - 1\}$$

is a function.

Since the graph of every linear equation is a line, it is reasonable to conclude that the solution set of every linear equation is a function, except for linear equations whose graphs are vertical lines. Solution sets of the latter equations are relations that are not functions. This can be seen in Figure 9.26.

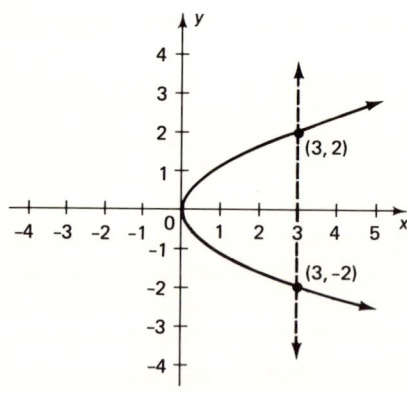

FIGURE 9.24

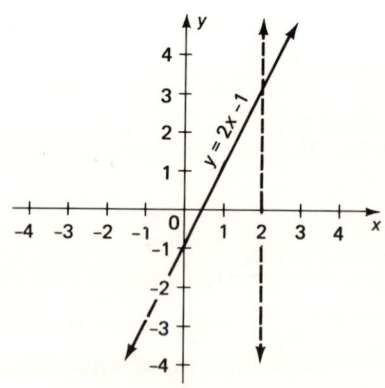

FIGURE 9.25

9.6 Relations and Functions

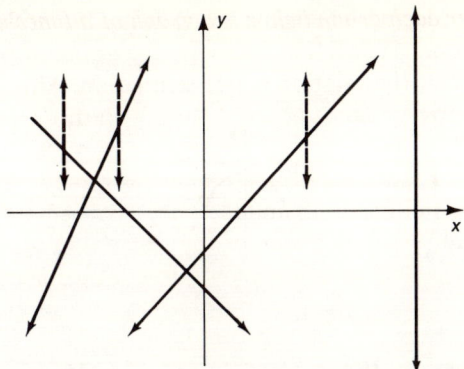

FIGURE 9.26

Note that the vertical line graphed on the right in Figure 9.26 fails the vertical line test; that is, a vertical line intersects it at an infinite number of points. Equations of vertical lines are of the form $x = a$, where a is a real number. Thus, the solution set of every linear equation, except those of the form $x = a$, is a function.

Exercises 9.6

1. After intercepting the following message, you remember Section 9.6 about Domain and Range. The message then becomes clear. Read it. (Meet, 16) (at, or) (the, more) (old, will) (tree, come)

Which of the following relations are also functions?

2. $\{(1, 3), (1, 2)\}$
3. $\{(3, 1), (2, 1)\}$
4. $\{(-1, 0), (0, -1), (2, 0)\}$
5. $\{(-2, 6), (3, 6), (-2, 5)\}$
6. $\{(0, 0), (1, 1), (2, 2)\}$
7. $\{(1, 2), (2, 4), (3, 2)\}$
8. $\{(-1, 1), (-2, 4), (1, 1), (2, 4)\}$
9. $\{(1, -1), (4, -2), (1, 1), (4, 2)\}$
10. After breaking into the Math Department storeroom, you come across the safe in which the final exam answers are kept. Slippery Sam, the archvillain safecracker, has told you that the combination is a *function* of the stereoisotopic properties of the lock. Determine which of the following will open the safe.
    a. $(2, -\frac{3}{2})$ $(6, 100)$ $(2, 2)$ b. $(1, \frac{2}{3})$ $(\frac{21}{20}, \frac{101}{9})$ $(2, 4)$
    c. $(6, 24)$ $(\frac{36}{6}, 9)$ $(\frac{108}{18}, 2)$

9 Graphs, Functions, and Linear Programming

Determine whether each graph below is a graph of a function. Use the vertical line test.

11.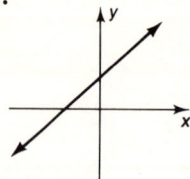

12.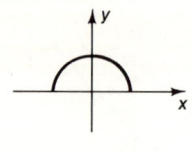

13.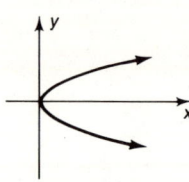

14.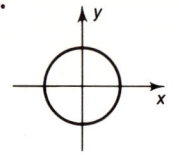

15.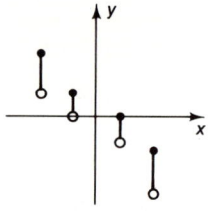

16.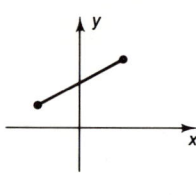

17.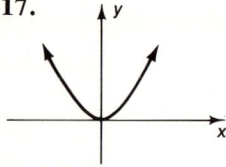

18.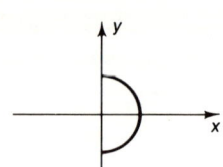

19.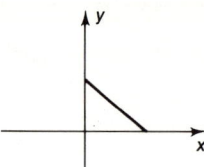

20.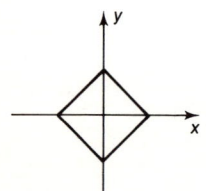

21.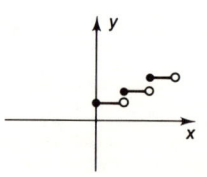

22.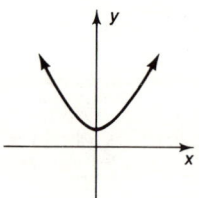

23.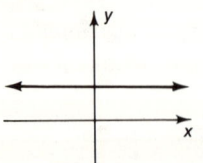

24.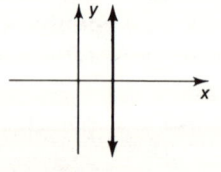

25.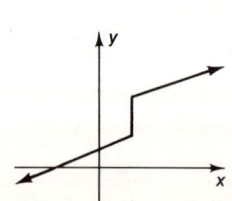

9.6 Relations and Functions

26.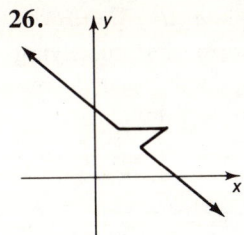
27.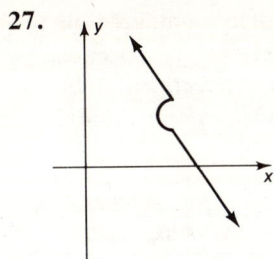
28.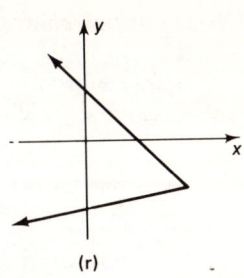
(r)

CHAPTER 9 TEST

Graph each of the following sets on a number line:

1. $A = \{x|x < 2\}$
2. $B = \{x|x \geq -1\}$
3. $C = \{x|x < -3 \text{ or } x > 0\}$
4. $D = \{x|x \leq 0 \text{ or } x \geq 0\}$
5. $E = \{x|x < 4 \text{ and } x > 1\}$
6. $F = \{x|x \leq 0 \text{ and } x \geq -2\}$

Solve each inequality, and graph its solution set on the number line:

7. $4x + 1 < 3x - 6$
8. $2(3x - 2) > 3(3x + 1)$
9. $\{x|3(2x + 1) > 0 \text{ or } 5(x - 2) < 0\}$
10. $\{x|5(x + 1) > 0 \text{ and } x < 2\}$
11. The Mad Bomber, alias Arnold the Termite, is zooming along the number line in an attempt to blow up whatever he finds at the end. If Arnold's path is given by the inequality $3x + 2 > x + 7$, graph his path. What area will be free from his attacks?

Graph each of the following in $R \times R$, where R is the set of real numbers:

12. $\{(x, y)|x = -1\}$
13. $\{(x, y)|x > -1\}$
14. $\{(x, y)|x < -1\}$
15. What is $\{(x, y)|x = -1\} \cup \{(x, y)|x > -1\} \cup \{(x, y)|x \leq -1\}$ equal to?

For each of the following equations, (1) write its equivalent equation in the form $y = mx + t$, and (2) determine the slope and the y-intercept of the line given by the equation:

16. $x + y = 4$
17. $y - 7x = 3$
18. $4x = 5y - 1$
19. $6(2y + x) = -5$
20. Visitors from outer space at last have decided to land and are on their way through the atmosphere. You just happen to be looking at the heavens with your high-powered telescope. Using the knowledge of

equations gained in Chapter 9, you are able to calculate the equation of their motion: $16(\frac{3}{2}y + 5x) = \frac{144}{3}$. Because you know that the flying saucer will land on its y-intercept, plot the results on this special observers' map. Show on which point the flying saucer will land.

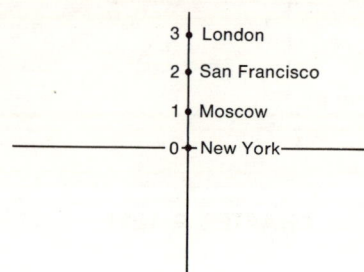

21. Graph each of the lines in exercise 16–19 on a coordinate system.

Graph the solution set of each of the following systems of inequalities:

22. $x \geq 0$
$y \geq 0$
$x \leq 3$
$y \leq 4$

23. $x \geq 0$
$y \geq 0$
$x \leq 4$
$y \leq 6$

24. $x \geq 0$
$y \geq 0$
$x \leq 3$
$x + y \leq 6$

25. $x \geq 0$
$y \geq 0$
$x \leq 3$
$y \leq 2x + 1$

26. A manufacturer stores bicycles in his warehouses in Detroit and in Chicago. The Detroit warehouse has 600 bicycles and the Chicago warehouse 800 bicycles. The manufacturer received an order from Atlanta for 500 bicycles and from St. Louis for 400 bicycles. The profit on each bicycle shipped from Chicago to Atlanta is $30 and from Detroit to Atlanta is $20. The profit on each bicycle shipped from Chicago to St. Louis is $25 and from Detroit to St. Louis is $15. What is the maximum profit the manufacturer can make?

Classify each of the following as (1) a relation but not a function (2) a function.

27. $\{(9, 0), (2, 0)\}$
28. $\{(1, 8), (2, -1), (3, 0) (1, 6)\}$
29. $\{(3, 1), (-1, 0), (6, 0)\}$
30. $\{(x, y) | y = -3\}$
31. $\{(x, y) | x = 5\}$
32. $\{(x, y) | x + y = 6\}$
33. $\{(x, y) | 2x = y\}$

9.6 Relations and Functions

After catching Arnold of exercise 11, you train him to run his vertical line over the following graphs to see whether or not they are functions. Arnold, being very cooperative, does this without blowing up a single one.

34.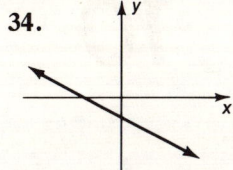

35.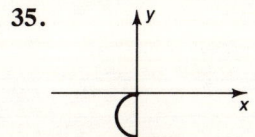

36.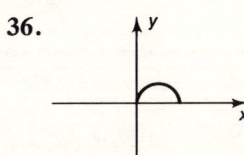

37.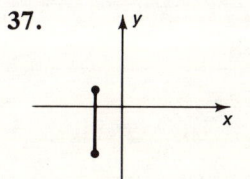

38.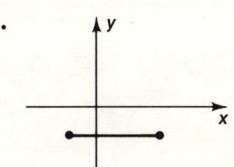

39.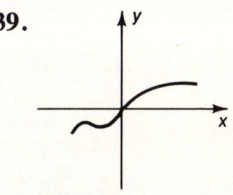

10

Logic

10.1 STATEMENTS AND THEIR TRUTH-VALUES

REVIEW

None

OBJECTIVES

- Consider open statements
- Consider true statements
- Consider false statements
- Given a statement, judge it to be true, false, or an open statement
- Know the concept of a variable
- Know the concept of a replacement set

Have you ever accused someone of being illogical? If you did, what did you mean by this accusation? Is it that the person came to an unwarranted (invalid) conclusion, or is it that the person's conclusion did not agree with yours?

10.1 Statements and Their Truth-Values

Logic is concerned with the study of drawing valid conclusions. If you hear an ad:

All beautiful women use brand X soap
Crickett uses brand X soap

do you conclude that Crickett is a beautiful woman? Is the conclusion that follows from this ad valid?

In this chapter we shall systematically examine various kinds of statements so that you can think clearly and reach valid conclusions.

Let's consider the following three statements:

He was the Cy Young Award recipient in 1972.
George Washington was the first president of the United States.
Carol Koslow wrote *Gone with the Wind*.

The first statement is neither true nor false. We can replace *he* by a name of a man, and then we would have either a true or a false statement. Replacing *he* by Morris Nichi would give us a false statement, for example. But replacing *he* by Steve Carlton would result in a true statement:

Steve Carlton was the Cy Young Award recipient in 1972.

The second statement is true—it is a historic fact.
The third statement is false, since Margaret Mitchell wrote *Gone with the Wind*.

The situation in mathematics is similar to that illustrated by these three examples. We have these three kinds of statements in mathematics. Here are analogous examples:

$x + 5 = 8$
$3 + 4 = 7$
$9 - 3 = 5$

The first statement is neither true nor false. There are many replacements for x that would result in false statements. But there is only one replacement that will give a true statement: 3. The letter x is called a *variable*. Usually, when we use a variable in statements, we specify the *replacement set* for it. For example, for the statement $x + 5 = 8$, we may say that the replacement set is the set of whole numbers. Statements which have one or more variables in them are sometimes called *open statements*.

The second statement is true. The third statement is false. We say, in somewhat fancy language, that

the truth-value of $3 + 4 = 7$ is true (T for short)
the truth-value of $9 - 3 = 5$ is false (F for short)

Exercises 10.1

Determine which of the following are true, which are false, and which are open statements:

1. $2^3 = 6$
2. $\sqrt{4} = 2$
3. $6 - x = 2$
4. $n + 1 = 4$
5. $100\% = 1$
6. $\sqrt{-4} = -2$
7. $a + a = 2a$
8. $\sqrt{\frac{1}{16}} = \frac{1}{4}$
9. $25\% = .25$
10. $10^3 = 1000$
11. $\frac{0}{4} = 0$
12. $.4 = \frac{2}{5}$
13. $2 \cdot 0 = 2$
14. $\frac{9}{m} = 2$
15. $x + y = y + x$
16. $\left(\frac{1}{2}\right)^2 = 1$
17. $2 + 0 = 0$
18. $\frac{4k}{8} = \frac{1}{2}m$
19. You and several companions have been locked for 8 hr in a wine cellar. Your friends are becoming affected by their confinement. When you hear the shout, "The doors are open!" you can't be sure that the statement is accurate. Then you hear a second shout, "$x^2/z^3 = 33K^8$." This statement is obviously not the product of an incoherent mind, but are the doors open or not?
20. The police arrest you for being drunk and disorderly. You maintain that you are completely sober and that, to prove it, you will answer some mathematical questions. The answers you gave are shown below. Are you sober?
    a. $2^3 = 6$
    b. $a + a + b = 2a$
    c. $10^3 = 1000$
    d. $3 = \frac{9}{2}$
    e. $6^3 = 2^4$

10.2 QUANTIFIERS

REVIEW

- Letters of the alphabet are used as variables
- Open statements are statements that contain one or more variables

OBJECTIVES

- Use the universal quantifier
- Use the existential quantifier
- Determine the nonpermissible replacements of variables in open statements

Pick up any algebra book. Leaf through it. Did you notice lots of open statements? You probably did. In algebra we are concerned with the study of all kinds of open statements. Many of the techniques taught in algebra serve the purpose of identifying those numbers which in place of variables will result in true statements. Identifying those numbers comes under the heading of solving equations or inequalities.

We said that open statements are neither true nor false. We can make them into true statements by prefixing some phrases. For example, we can say:

for each real number x, $x + x = 2x$

This is a true statement, even though it has a variable in it. We can prove that this is true for each real number.

The phrase *for each* is called a *universal quantifier*. It states that something is true for *all* members of a given replacement set.

Sometimes replacing a variable by some member of a replacement set will result in a meaningless symbol, such as replacing x by 2 in

$$\frac{x-2}{x-2} = 1$$

Notice that we can replace x by any real number other than 2, and a true statement will result. In such cases we still use a universal quantifier, but indicate the exception:

for each real number $x(x \neq 2)$, $\frac{x-2}{x-2} = 1$.

10 Logic

Or we might say:

for each permissible value of x, $\dfrac{x-2}{x-2} = 1$

In this case, 2 is a nonpermissible value of x.

To save writing, a symbol for the universal quantifier was invented: $\forall$. The statement:

for each real number x, $x + x = 2x$

would be written as follows:

$\forall_x, x + x = 2x$

Of course, it is ethically sound to state in advance what the replacement set for x is. Sometimes it is stated right along with the quantifier:

$\forall_{x \in R}, x + x = 2x$

We also work with statements that are true for one or just a few replacements of a variable. For example,

$x^2 = 4$

is true for two real numbers, 2 and -2. In such cases, we can state the following:

for some real numbers x, $x^2 = 4$

This is a true statement. Instead of saying *for some*, one of the following phrases will be used:

there exists
there is at least one

Each of these is called an *existential quantifier*. An abbreviation for this quantifier has also been invented: $\exists$. Thus, the statement:

for some real numbers x, $x^2 = 4$

would be written as

$\exists_x, x^2 = 4$

10.3 Negating Statements

Exercises 10.2

Prefix each of the following statements with the appropriate quantifier, $\forall$ or $\exists$:

1. $x + 5x = 6x$
2. $x + 5x = 3x$
3. $n + 1 = 4$
4. $\dfrac{r + 2}{r + 3} = \dfrac{2}{3}$
5. $y(y - 1) = 0$
6. $ab = ba$
7. $3(x + y) = 3x + 3y$
8. $\dfrac{6w + 3z}{3} = 2w + z$
9. $x^2 = x$

Determine the nonpermissible replacements of the variables:

10. $\dfrac{m - 3}{m - 3} = 1$
11. $\dfrac{2x - 6}{2x - 6} = 1$
12. $\dfrac{y}{y} = 1$
13. $\dfrac{3(a - 2)}{a - 2} = 3$
14. $\dfrac{(a - 2)(a - 1)}{a - 1} = a - 2$
15. $\dfrac{(n - 10)(n - 2)}{n - 10} = n - 2$

16. Many numbers have split personalities. What are the two faces of x when it is in the form $x^2 = 1$?
17. You may have heard the statement, "A square peg will not fit into a round hole." Rephrase the above statement to describe the permissible replacement and the nonpermissible replacement. The mathematical expression is $x/(x - 1)$.
18. There are two replacements for x in $x^2 = 9$ which result in true statements. Give these two replacements.
19. Why is 0 a permissible replacement for x in $x/(x - 1)$? What is the resulting number?
20. Why is 1 not a permissible replacement for x in $x/(x - 1)$?

10.3 NEGATING STATEMENTS

REVIEW

- Meanings of symbols:
  $=$ is equal to
  $\neq$ is not equal to
  $<$ is less than
  $\not<$ is not less than
  $>$ is greater than
  $\not>$ is not greater than

OBJECTIVES

- Negate statements in two ways
- Make a truth-table for a negation

10 Logic

In our day-to-day affairs, we often deal with opposites. What is the opposite of cold, of love, of justice, of revolt? Similarly, we can ask about the opposite of a statement, called *negating* a statement. Consider the following true statement:

Spring follows winter

This statement can be negated in two ways:

1. Prefix the statement with the phrase *it is not true that*; this gives us the statement:

    <u>It is not true that</u> spring follows winter.

2. Change the verb *follows* to *does not follow*; this gives the statement:

    Spring <u>does not follow</u> winter.

Each of these two derived statements is called a *negation* of the original statement. Observe that the original statement is true, but its negations are false.

Given a false statement

$3 = 5$ (3 is equal to 5)

its negation can also be stated in two ways:

1. It is not true that $3 = 5$.
2. $3 \neq 5$.

Notice that in $3 \neq 5$, the relation is *is not equal to*, which is a negation of the relation *is equal to*. The statement $3 \neq 5$ is true. Thus, $3 = 5$ is a false statement that has a negation that is true.

To generalize the discussion above, a variable to replace statements is needed. For example, in the statement

for every whole number n, $2n$ is an even number

n is a *numerical variable* that can be replaced by the name of any whole number. For example, replacing n by 3 results in the statement:

$2 \cdot 3$ is an even number

which is true. Of course, it is impossible to make all replacements for n because the *replacement set*, the set of whole numbers, is an infinite set.

10.3 Negating Statements

The situation with statements is quite similar: to say something about many statements, a letter that can be replaced by any statement meeting the specific conditions can be used. For example, the letter p can be used as a *statement variable*.

For every statement p, the negation of p will be denoted by the symbol

$\sim p$ *Read:* not p

Using this symbol, the observations made above concerning negations can be generalized in a concise form as follows:

> For every statement p,
> if p is true, then $\sim p$ is false
> if p is false, then $\sim p$ is true.

This can be stated in a still more concise form by using a *truth-table*. The truth-table for negation is shown in Table 10.1.

Table 10.1 Truth-Table for Negation

p	$\sim p$
T	F
F	T

Some people have misconceptions about how to negate quantified statements. In the case of a universally quantified statement, it is necessary to produce only one case as an exception and the statement is negated.

Example 1   All students at Valley College are girls.

Negation: At least one student at Valley College is not a girl.

To negate an existentially quantified statement, it is necessary to assert that the statement is false for all cases.

Example 2   At least one student at Valley College is a boy.

Negation   No student at Valley College is a boy.

Up to this point, we have considered only *simple statements*—statements that contain only one subject and one verb. In mathematics, as well as in everyday use of language, we form statements which consist of two or more simple statements. The most common statements of this kind are those in which two simple statements are combined with the connectives *and* and *or*. They will be considered in the next section.

Exercises 10.3

In our odyssey through this textbook, we have come across many strange diseases. We now meet another: "negativitis." The afflicted person cannot avoid putting all statements in a negative form. Imagine you have the disease, and put the following statements in a negative form:

1. War is bitter.
2. This is a good college.
3. Each triangle has three sides.
4. Not every whole number is even.
5. $21 = 13$
6. Reverse the following symptoms of negativitis.
   a. Love is not good.   b. Math is not interesting.
   c. Students are not concerned.
7. In exercise 1, one form of a negation of *war is bitter* is *war is not bitter*.
   a. What is a negation of *war is not bitter*?
   b. Did you obtain the original statement, that is, *war is bitter*?
   c. Does it follow that a negation of a negation of *war is bitter* is the statement itself?
8. Do you think this is true for every statement, that is, for every statement p, $\sim\sim p = p$?
9. What symbol is a negation of the symbol $=$ (is equal to)?

The symbol $<$ means *is less than*, and the symbol $>$ means *is greater than*. The negation of $<$ is $\not<$ (*is not less than*). Give a negation of each of the following statements:

10. $5 < 7$
11. $8 < 3$
12. $2 \not< 5$
13. $0 \not< 2$
14. $10 > 6$
15. $1 > 5$
16. $3 \not> 1$
17. $3 \not> 7$
18. For each statement and its negation in exercises 10–17 tell their truth-values.

Write a negation of each of the following statements.

19. All students in this course are A students.
20. At least one student in this course is an A student.

10.4 CONJUNCTIONS AND DISJUNCTIONS

REVIEW

None

OBJECTIVES

- Define a conjunction
- Construct the truth-table for a conjunction
- Define a disjunction
- Construct the truth-table for a disjunction
- Construct truth-tables for more complicated compound statements

Two simple statements can be combined into what is called a *compound statement*. One way to form a compound statement is to insert the connective *and* between the two simple statements. Here is an example of such a compound statement:

The sun is shining *and* the farmers are pleased

This compound statement consists of two simple statements:

1. The sun is shining.

2. Farmers are pleased.

Any statement of the form:

p and q

where p and q are simple statements is called a *conjunction*.

Here are some examples of mathematical conjunctions.

$5 = 3 + 2$ and $3 > 1$
$4 < 3$ and $5 > 2$
$9 = 4 + 5$ and $1 > 2$
$1 = 2 + 0$ and $4 = 5 + 1$

The two simple statements in the first of these conjunctions are:

1. $5 = 3 + 2$

2. $3 > 1$

The symbol ∧ is usually used in place of *and*. Thus, a conjunction is a statement of the form $p \land q$ (*read: p and q*).

The next question to be resolved concerns the truth-values of various conjunctions. For the four statements above, the truth-values are as follows:

$(5 = 3 + 2) \land (3 > 1)$ T
$(4 < 3) \land (5 > 2)$ F
$(9 = 4 + 5) \land (1 > 2)$ F
$(1 = 2 + 0) \land (4 = 5 + 1)$ F

It is a matter of a somewhat arbitrary agreement that a conjunction is true only when each of the two simple statements is true. Table 10.2 is a truth-table for a conjunction.

Table 10.2   *Truth-Table for Conjunction*

p	q	$p \land q$
T	T	T
T	F	F
F	T	F
F	F	F

Another compound statement form frequently used in mathematics is obtained by combining two simple statements with *or*. The connective *or* is abbreviated as ∨.

> A statement of the form $p \lor q$ (*read: p or q*) is called a *disjunction*.

Using the same simple statements as those above, four disjunctions can be formed. To each disjunction thus formed, a truth-value is assigned:

$(5 = 3 + 2) \lor (3 > 1)$ T
$(4 < 3) \lor (5 > 2)$ T
$(9 = 4 + 5) \lor (1 > 2)$ T
$(1 = 2 + 0) \lor (4 = 5 + 1)$ F

Again, it is a matter of a somewhat arbitrary agreement that a disjunction is false only when each of the two simple statements is false. This is summarized in the truth-table in Table 10.3.

10.4 Conjunctions and Disjunctions

Table 10.3 *Truth-Table for Disjunction*

p	q	$p \vee q$
T	T	T
T	F	T
F	T	T
F	F	F

As stated above, there is some degree of arbitrariness in assigning truth-values to compound statements. However, there are also logical reasons for assigning one truth-value in preference to the other to a given statement. For example, note that if we should assign the truth-value false to $p \vee q$ when only one of p and q is false, then the truth-table for $p \vee q$ would be the same as the truth-table for $p \wedge q$. Thus, from the point of view of logic, a disjunction would be indistinguishable from a conjunction.

With the truth-tables for a conjunction and a disjunction, we can construct truth-tables for more complex statements. Consider the compound statement

not p or q ($\sim p \vee q$)

Table 10.4 is a truth-table for this statement.

Table 10.4 *Truth-Table for $\sim p \vee q$*

p	q	$\sim p$	$\sim p \vee q$
T	T	F	T
T	F	F	F
F	T	T	T
F	F	T	T

Another compound statement is:

not (p or not q)

abbreviated:

$\sim(p \vee \sim q)$

The truth-table for this statement is Table 10.5.

10   Logic

Table 10.5   Truth-Table for $\sim(p \lor \sim q)$

p	q	$\sim q$	$p \lor \sim q$	$\sim(p \lor \sim q)$
T	T	F	T	F
T	F	T	T	F
F	T	F	F	T
F	F	T	T	F

Exercises 10.4

1. Victims of the dread disease "digititis" replace as many words as possible by a digit or a symbol. This centuries-old disease is somewhat typical of math teachers, who get so used to writing symbols that they eventually start saying them. Translate the following statement into English:

   My students $\land$ I do $\sim$ argue $\lor$ fight; it's just that they can$\sim$ understand me.

Determine the truth-value of each of the following conjunctions:

2. $(5 > 4) \land (3 = 2 + 1)$
3. $(1 < 0) \land (0 < 1)$
4. $(\frac{1}{2} \neq \frac{2}{4}) \land (5 \not< 6)$
5. $(5 > 1) \land (3 \neq 2 + 1)$
6. $(0 \neq 0) \land (5 \neq 5)$
7. $(1 < 0) \land (1 > 0)$

Replace p by the moon is shining and q by I am sad, and then write in words the statements you obtain from the following:

   Example $\sim(p \land \sim q)$
   Solution   It is not true that *the moon is shining and I am not sad.*

8. $p \land q$
9. $\sim p \land q$ (Note: $\sim p \land q$ means $(\sim p) \land q$; it does *not* mean $\sim(p \land q)$.)
10. $p \land \sim q$
11. $\sim p \land \sim q$
12. $\sim(p \land q)$
13. $\sim(\sim p \land q)$
14. $\sim(\sim \land \sim q)$

Replace p by $5 > 3$ and q by $2 < 1$; then determine the truth-values of the statements you obtain from the following:

15. $p \land q$
16. $\sim p \land q$
17. $p \land \sim q$
18. $\sim p \land \sim q$
19. $\sim(p \land q)$
20. $\sim(\sim p \land q)$
21. $\sim(p \land \sim q)$
22. $\sim(\sim p \land \sim q)$
23. $\sim[\sim(p \land \sim q)]$
24. $\sim[\sim(\sim p \land \sim q)]$

10.5 Implications

Determine the truth-value of each of the following disjunctions:

25. $(5 > 4) \lor (3 = 2 + 1)$ **26.** $(1 < 0) \lor (0 < 1)$
27. $(\frac{1}{2} \neq \frac{2}{4}) \lor (5 \not< 6)$ **28.** $(5 > 1) \lor (3 \neq 2 + 1)$
29. $(0 \neq 0) \lor (5 \neq 5)$

Replace p by $5 > 3$ and q by $2 < 1$; then determine the truth-values of the statements you obtain from the following:

30. $p \lor q$ **31.** $\sim p \lor q$ **32.** $p \lor \sim q$
33. $\sim p \lor \sim q$ **34.** $\sim(p \lor q)$ **35.** $\sim(\sim p \lor q)$
36. $\sim(p \lor \sim q)$ **37.** $\sim(\sim p \lor \sim q)$ **38.** $\sim[\sim(p \lor \sim q)]$
39. $\sim[\sim(\sim p \lor \sim q)]$
40. If $A = 1, B = 2, C = 3, D = 4, \ldots, Z = 26$, check the truth of the following statements.
 a. $(B < N) \land (J > A)$ b. $(C/D < Z/B) \land (S > T)$
 c. $(E > Z) \lor (B = A + C)$ d. $(D < Z) \lor (A > D)$

10.5 IMPLICATIONS

REVIEW	OBJECTIVES
None	<ul><li>Define an implication</li><li>Identify the antecedent and the consequent in an implication</li><li>Construct a truth-table for implication</li><li>Rationalize the truth-values of an implication</li><li>Define logically equivalent statements</li><li>Define a biconditional</li><li>Establish logical equivalence of statements</li><li>Use De Morgan's laws</li></ul>

Many of the theorems in mathematics, particularly in geometry, are stated in the form *if . . . , then . . . .*

10 Logic

For example,

If two sides of a triangle are congruent, *then* the two angles opposite these sides are congruent

is such a statement. The part immediately following *if*:

two sides of a triangle are congruent

is called an *antecedent* or *hypothesis*, and the part immediately following *then*:

the two angles opposite these sides are congruent

is called a *consequent* or *conclusion*. If we denote the antecedent in the statement above by p and the consequent by q, then the statement above is abbreviated:

if p, then q

which is further abbreviated as

$p \to q$ *Read:* if p, then q; or read: p implies q

A statement of the form $p \to q$ is called an *implication* or a *conditional*.
The truth-values of $p \to q$ are determined as follows:

The truth-value of $p \to q$ is F only in the case where the truth-value of p is T and the truth-value of q is F

The complete truth-table for $p \to q$ is shown in Table 10.6.

Table 10.6   *Truth-Table for Implication*

p	q	$p \to q$
T	T	T
T	F	F
F	T	T
F	F	T

The last two entries in the table cause some concern. Why, for example, should the statement "If I am a donkey, then the moon is made of

10.5 Implications

blue cheese" be true? Two observations can be made concerning such statements.

First, such statements are usually not dealt with in mathematics, since there is no sensible relationship between the antecedent and the consequent.

Second, it needs to be recognized that neither the absence of any relationship between the antecedent and the consequent nor the meaninglessness of the statement has any bearing on the truth-value of the implication. The only prerequisite for determining the truth-value of the implication is the knowledge of the truth-values of the antecedent and the consequent.

Thus, according to the truth-table for an implication, the following statement is true:

If I am a donkey, *then* the moon is made of blue cheese

The following argument provides some justification for the choice of the truth-values in the truth-table for implication. There is a choice of either the truth-value T or F for $p \rightarrow q$, when p is false. Suppose the truth-value F were chosen. Then the entries in the last column in the truth-table for implication would be T, F, F, F in this order. But this is precisely the table for $p \wedge q$. Thus, $p \rightarrow q$ would be logically indistinguishable from $p \wedge q$. This is one reason for assigning the truth-value T to $p \rightarrow q$, whenever p is false.

To rationalize this choice in another way, consider the following example. Suppose I think of a number, which is not known to anybody (but me) at this time. I make the following statement about this number:

If this number is 3, then the square of the number is 9

It is easy to see that the truth-value of this statement is T.

Now I shall disclose the number: it is 4. Thus, the truth-value of the antecedent in the implication above is F and of the consequent is also F. But we have agreed that the truth-value of the implication is T.

Now, suppose I have not thought of 4. Instead, I thought of the number -3. This makes the antecedent false and the consequent true, since $(-3)^2$ is equal to 9. But we have agreed that the original implication was true.

It is important to realize that the above is merely an example which makes plausible the choice of the truth-values assigned to $p \rightarrow q$.

In logic, frequently a question is raised as to whether two different statements are *logically equivalent*. Statements p and q are said to be *logically equivalent* or simply *equivalent* if and only if they have the same truth-value for all truth-values of their component parts.

For example, the truth-table in Table 10.7 shows that $\sim(\sim p \land q)$ and $p \lor \sim q$ are logically equivalent. In the last two columns the same truth-values occur for each statement.

Table 10.7 Truth-Table Showing $\sim(\sim p \land q)$ and $p \lor \sim q$ Equivalent

p	q	$\sim p$	$\sim q$	$\sim p \land q$	$\sim(\sim p \land q)$	$p \lor \sim q$
T	T	F	F	F	T	T
T	F	F	T	F	T	T
F	T	T	F	T	F	F
F	F	T	T	F	T	T

Many theorems in mathematics are stated in the form *p if and only if q*, abbreviated $p \leftrightarrow q$. This is a short form of stating the following conjunction:

if *p*, then *q*
and
if *q*, then *p*

A statement of this form is called a *biconditional*.

Exercises 10.5

1. Construct a truth-table to prove that $p \rightarrow q$ and $\sim p \lor q$ have the same truth-values. Are these two statements logically equivalent?

Write the following sentences in the form: If . . . , then . . . .

2. (Mary is a freshman) → (She is taking algebra)
3. A triangle has three sides.
4. $(3x = 12) \rightarrow (x = 4)$
5. A parallelogram has opposite sides parallel.
6. The absolute value of a negative number is the additive inverse of that number.
7. Translate the following examples of "digititis" into English.
    a. John can walk → he can go
    b. $(3x/2 = 9/2) \rightarrow (x = 3)$
    c. You can decipher this → you understand the principle
8. Make a truth-table to show that $\sim(p \land q)$ and $(\sim p) \lor (\sim q)$ are logically equivalent.
9. Make a truth-table to show that $\sim(p \lor q)$ and $(\sim p) \land (\sim p)$ are logically equivalent.

10.6 Converses and Inverses

The following two biconditionals are called De Morgan's laws:

$$\sim(p \wedge q) \leftrightarrow (\sim p) \vee (\sim q)$$
$$\sim(p \vee q) \leftrightarrow (\sim p) \wedge (\sim q)$$

Using these laws, give the logical equivalent of each of the following:

10. It is not true that $(5 = 4$ and $4 > 1)$.
11. It is not true that $(5 = 4$ or $4 > 1)$.
12. Overheard at Mafia headquarters:
    A: "If we give in, then you must give in!"
    B: "No, if we give in, then you must give in!"
    (A muffled exchange of blows followed.) Assuming you are an FBI spy, how could you communicate this exchange to headquarters using a math code?

10.6 CONVERSES AND INVERSES

REVIEW

- The symbol $\sim p$ denotes the negation of p; it is read "not p"
- An implication is a statement of the form "if p, then q," abbreviated $p \rightarrow q$

OBJECTIVES

- Write converses of implications
- Define a converse of an implication
- Write inverses of implications
- Define an inverse of an implication
- Construct truth-tables for a converse and an inverse
- Observe that a converse or an inverse of a true implication is not necessarily true

Consider the following implication:

If it is snow, then it melts

This is a true implication. Now let's form a statement related to it:

If it melts, then it is snow

This, of course, is false. It could be butter, couldn't it?

Here is a mathematical implication:

if $4 > 1$, then $4 + 2 > 1 + 2$

A related statement formed in the same manner as the one above is:

if $4 + 2 > 1 + 2$, then $4 > 1$

Both of these implications are true.

These derived implications are called converses of the original implications. In general:

a *converse* of $p \to q$ is $q \to p$

The first statement above has proved that the converse of a true implication is not necessarily true. Table 10.8 shows a complete truth-table for a converse.

The third row of the table shows that when p has the truth-value F and q has the truth-value T, then $p \to q$ is true, but its converse $q \to p$ is false.

Table 10.8 *Truth-Table for Converse of Implication*

p	q	$p \to q$	$q \to p$
T	T	T	T
T	F	F	T
F	T	T	F
F	F	T	T

Now we shall derive an implication from a given implication in a different manner. From the original implication:

If it is snow, then it melts

we derive the following implication:

If it is not snow, then it does not melt

This is an inverse of the original implication. In general:

an *inverse* of $p \to q$ is $\sim p \to \sim q$

Did you notice that the inverse of the above true implication is false? This example proves that an inverse of a true implication is not necessarily true.

The word *inverse* is used in mathematics with several different meanings. For example, there is an inverse of a function, a multiplicative inverse,

10.6 Converses and Inverses

and an additive inverse of a number. In each instance, the appropriate definition assigns the intended meaning to *inverse*.

Exercises 10.6

Give a converse of each of the following statements. (Hint: if the statement is not in the form if . . . , then . . . . you might want to change it to this form first.)

1. If two finite sets are matching sets, then each set has the same number of elements.
2. Two equivalent equations have the same solution set.
3. In an isosceles triangle at least two sides have the same measure.
4. Everything that is considered to be geometry is also mathematics.
5. A triangle that is a scalene triangle is also a nonequilateral triangle.
6. Make a truth-table for $\sim p \rightarrow \sim q$ (inverse of $p \rightarrow q$). State for what truth-values of p and of q, $p \rightarrow q$ is true, but $\sim p \rightarrow \sim q$ is false.

Given an example of an implication from arithmetic, so that:

7. $p \rightarrow q$ is true, but $\sim p \rightarrow \sim q$ is false
8. $p \rightarrow q$ is false, but $\sim p \rightarrow \sim q$ is true
9. Both $p \rightarrow q$ and $\sim p \rightarrow \sim q$ are true
10. You have met the disease "digititis," and by now you should be familiar with all its symptoms. As a concerned student, you should try to help the afflicted teacher by speaking with him in "digitalis," a language of digits and symbols, instead of words. Put the following statement into digitalis.
    If it is a hard problem, then I will do it. If it is not a hard problem, then I will not do it.
11. In answer to your previous question (exercise 10), the teacher replies: $q \rightarrow p$; $\sim q \rightarrow \sim p$. What is the relation between your original statement and the teacher's reply?
12. Using the truth-table you constructed for exercise 6, show that it is not possible to have a false implication whose inverse is also false.

Give an example of an implication from geometry, so that:

13. $p \rightarrow q$ is true, but $\sim p \rightarrow \sim q$ is false
14. $p \rightarrow q$ is false, but $\sim p \rightarrow \sim q$ is true
15. Both $p \rightarrow q$ and $\sim p \rightarrow \sim q$ are true
16. Pick several statements from a current newspaper or article. Construct a converse and an inverse, and test their validity.
17. Copy the truth-table you constructed for exercise 6. Now add another column for a converse of an inverse of $p \rightarrow q$, that is $\sim q \rightarrow \sim p$. Write the truth-values for this last column. How do these truth-values compare with the truth-values of $p \rightarrow q$?

10.7 CONTRAPOSITIVES AND TAUTOLOGIES

REVIEW

- The symbol $\vee$ is read "or"; a statement of the form $p \vee q$ is a disjunction
- The symbol $\wedge$ is read "and"; a statement of the form $p \wedge q$ is a conjunction

OBJECTIVES

- Define a contrapositive of a statement
- Form contrapositives of given statements
- Define a tautology
- Construct truth-tables for tautologies
- Construct truth-tables for various laws of logic

In exercise 17 of Section 10.6, it was established that $p \to q$ and $\sim q \to \sim p$ are logically equivalent for all statements p and q:

$\sim q \to \sim p$ is called a *contrapositive* of $p \to q$

To know that these are logically equivalent is very useful. It permits us to replace a statement, whose form we might not like for one reason or another, with its contrapositive (which we might like better). We know that, if the original statement was true, we still have a true statement. If the original statement was false, we still have a false statement.

There is another reason for the importance of this finding. Frequently, it is extremely difficult to prove a theorem directly. In such cases, it may be considerably easier to prove a contrapositive of a theorem. This is the same as proving a theorem, since an implication and its contrapositive are logically equivalent. For example, instead of proving the theorem:

If a quadrilateral is a parallelogram, then its opposite sides are congruent

its contrapositive may be proved. The contrapositive is:

If the opposite sides of a quadrilateral are not congruent, then it is not a parallelogram.

In considering compound statements, we sometimes find that a statement is true no matter what the truth-value of the component statements are. Such a compound statement is called a *tautology*. One example of a tautology is any compound statement of the form $p \vee \sim p$. The truth-table in Table 10.9 shows that $p \vee \sim p$ is true for all truth-values of its component statements.

10.7 Contrapositives and Tautologies

Table 10.9 Truth-Table for $p \lor \sim p$

p	$\sim p$	$p \lor \sim p$
T	F	T
F	T	T

The tautology $p \lor \sim p$ is called *the law of the excluded middle*.

Exercises 10.7

Construct the truth-tables to show that for all truth-values of p and q each of the following is true (tv means truth-value):

1. $tv(p \land \sim p) = F$
2. $tv(p \lor \sim p) = T$
3. If $tv(q) = T$, then $tv(p \land q) = tv(p)$
4. If $tv(q) = F$, then $tv(p \land q) = F$
5. If $tv(q) = T$, then $tv(p \lor q) = T$
6. If $tv(q) = F$, then $tv(p \lor q) = tv(p)$

Write a contrapositive of each of the following:

7. $(3 = 4) \rightarrow (5 = 6)$
8. $(3 = 4) \rightarrow (0 = 157)$
9. $(3 = 4) \rightarrow (9 = 2 \times 4\frac{1}{2})$
10. $(2 = 1 + 1) \rightarrow (5 = 6)$
11. $(2 = 1 + 1) \rightarrow (4 = 3 + 1)$
12. $(2 < 3) \rightarrow (2 > 6)$
13. $(1 > 5) \rightarrow (2 > 6)$
14. $(1 > 5) \rightarrow (6 > 2)$
15. For each implication in exercises 7–14, determine its truth-value and the truth-value of its contrapositive. Is the truth-value of a contrapositive of an implication the same as that of the implication in each case?
16. The dreaded Math Murderer struck again! This time Sherlock Holmes and Dr. Watson were called in. Holmes theorized that the guilty party was the mysterious Mr. Snow, who was driven insane by teaching logic. Holmes came to this conclusion because the following note was found at the scene of the murder. (It was written by the victim [who taught contrapositives] just before he died.) NOTSNOW NOTMR. Watson asked Holmes how he knew Mr. Snow was the murderer. Holmes replied, "Elementary, my dear Watson, because..." Finish the statement.

Construct truth-tables to show that each of the following is a tautology:

17. $p \leftrightarrow \sim \sim p$ (law of double negation)

18. $[p \wedge (p \rightarrow q)] \rightarrow q$ (law of detachment)

Exercise 18 is completed as a sample of how to handle compound statements involving three simple statements.

p	q	$p \rightarrow q$	$p \wedge (p \rightarrow q)$	$[p \wedge (p \rightarrow q)] \rightarrow q$
T	T	T	T	T
T	F	F	F	T
F	T	T	F	T
F	F	T	F	T

Since $[p \wedge (p \rightarrow q)] \rightarrow q$ is true for all truth-values of p and q, as is seen from the last column, it is a tautology.

19. $p \wedge q \leftrightarrow q \wedge p$ ⎱ (commutative laws)
20. $p \vee q \leftrightarrow q \vee p$ ⎰
21. $(p \vee q) \vee r \leftrightarrow p \vee (q \vee r)$ ⎱ (associative laws)
22. $(p \wedge q) \wedge r \leftrightarrow p \wedge (q \wedge r)$ ⎰
23. $p \wedge (q \vee r) \leftrightarrow (p \wedge q) \vee (p \wedge r)$ ⎱ (distributive laws)
24. $p \vee (q \wedge r) \leftrightarrow (p \vee q) \wedge (p \vee r)$ ⎰
25. $(p \rightarrow q) \leftrightarrow {\sim}p \vee q$ (law of equivalence of implication and disjunction)
26. ${\sim}(p \rightarrow q) \leftrightarrow p \wedge {\sim}q$ (law of equivalence of negation of implication and conjunction)
27. $(p \leftrightarrow q) \leftrightarrow [(p \rightarrow q) \wedge (q \rightarrow p)]$ (law of equivalence of biconditional and conjunction)
28. ${\sim}(p \wedge q) \leftrightarrow {\sim}p \vee {\sim}q$ ⎱ (De Morgan's law)
29. ${\sim}(p \vee q) \leftrightarrow {\sim}p \wedge {\sim}q$ ⎰
30. Lie detector tests have often been accused of being inaccurate and unreliable. A subject was tested with the following.

a.
	p q	$p \rightarrow q$	$q \rightarrow p$
i.	T T	T	T
ii.	T F	F	F
iii.	F T	T	F
iv.	F F	T	T

b.
	p	${\sim}p$	$p \vee {\sim}p$
i.	T	F	F
ii.	F	T	T

Determine the accuracy of the lie detector if the subject's answers are indicated by p and q and the lie detector's conclusion is indicated by the: next two columns; the next column.

10.8 VALIDITY OF ARGUMENTS

REVIEW

- $\sim p$ is a negation of p; it is read "not p"

OBJECTIVES

- Establish valid and invalid forms of arguments

Most people agree that it is desirable to be logical in their thinking, yet very few people follow their convictions. People do argue in invalid ways. Let us consider some forms of arguments and see which of them are valid and which are not.

Example 1   GIVEN   If Mary passed Mathematics 215, then Mary graduated.
                  GIVEN   Mary passed Mathematics 215.
                  CONCLUSION   Mary graduated.

Let us show the form of this argument by assigning these abbreviations:

p: Mary passed Mathematics 215
q: Mary graduated

The argument has this form:

$$[(p \rightarrow q) \wedge p] \rightarrow q$$

This argument will be considered valid, if the form above is a tautology. We know it is from exercise 18 of Section 10.7. We called this the *law of detachment*. Another name for this form of argument is *modus ponens*. Notice that in this form of argument, we assume that $p \rightarrow q$ is true and if we know that p is true, then the conclusion q follows.

Example 2   GIVEN   If you had an average of above 90, then you got an A.
                  GIVEN   You got an A.
                  CONCLUSION   You had an average of above 90.

We use these abbreviations:

p: You had an average of above 90.
q: You got an A

10 Logic

The form of the argument is:

$$[(p \rightarrow q) \wedge q] \rightarrow p$$

For the argument to be valid, this must be a tautology. We construct a truth-table to see, Table 10.10.

Table 10.10 Truth-Table for $[(p \rightarrow q) \wedge q] \rightarrow p$

p	q	$p \rightarrow q$	$(p \rightarrow q) \wedge q$	$[(p \rightarrow q) \wedge q] \rightarrow p$
T	T	T	T	T
T	F	F	F	T
F	T	T	T	F
F	F	T	F	T

Since the last column does not have all Ts, $[(p \rightarrow q) \wedge q] \rightarrow p$ is not a tautology. Thus, this form of argument is not valid. Notice that this argument is equivalent to assuming that a converse of a true implication is true. This was proved to be false in Section 10.6.

This line of unreasoning is frequently used by advertisers, who say:

If you want to be beautiful, use *Cutie* soap

What they want you to assume is:

If I use *Cutie* soap, I will be beautiful.

Example 3   GIVEN   If you went fishing Sunday, you were not ready for the test on Monday.
               GIVEN   You did not go fishing Sunday.
               CONCLUSION   You were ready for the test on Monday.

Let p: You went fishing Sunday
     q: You were not ready for the test on Monday

The form of this argument is:

$$[(p \rightarrow q) \wedge \sim p] \rightarrow \sim q$$

We construct the truth-table for this argument in Table 10.11.

10.8 Validity of Arguments

Table 10.11 Truth-Table for $[(p \rightarrow q) \wedge \sim p] \rightarrow \sim q$

p	q	$p \rightarrow q$	$\sim p$	$\sim q$	$(p \rightarrow q) \wedge \sim p$	$[(p \rightarrow q) \wedge \sim p] \rightarrow \sim q$
T	T	T	F	F	F	T
T	F	F	F	T	F	T
F	T	T	T	F	T	F
F	F	T	T	T	T	T

The last column is not all Ts. Thus, the argument is not valid. Did you notice that this is arguing on the basis that an inverse of a true implication is true? This was proved to be false in Section 10.6.

This form is also used by advertisers in the hope that people will fall for their line of argument by assuming the inverse. It may go like this:

If you visit Spa regularly, you will be healthy

The advertiser hopes that you will think as follows:

If you don't visit Spa regularly, you will not be healthy

Example 4   GIVEN   If you swallowed too many oysters, then you got sick.
    GIVEN   You didn't get sick.
    CONCLUSION   You didn't swallow too many oysters.

Let p: You swallowed too many oysters
    q: You got sick
The form of this argument is:

$[(p \rightarrow q) \wedge \sim q] \rightarrow \sim p$

Observe that this argument is based on the contrapositive. You have shown in exercise 17 of Section 10.6 that an implication and its contrapositive are logically equivalent. Thus, this form of argument is valid. You may wish to construct a truth-table for this form to verify that it is a tautology.

Example 5   GIVEN   If you study mathematics, then you are intelligent.
    GIVEN   If you are intelligent, then you are successful.
    CONCLUSION   If you study mathematics, then you are successful.

10 Logic

Let p: You study mathematics
q: You are intelligent
r: You are successful

The form of this argument is:

$$[(p \to q) \land (q \to r)] \to (p \to r)$$

This form is quite frequently used in mathematics in proving theorems. We have a chain of implications, each of which is true, and we finally arrive at a conclusion. We thus arrive at the final implication that the first statement leads to the final conclusion.

We construct a truth-table for this argument in Table 10.12.

Table 10.12 Truth-Table for $[(p \to q) \land (q \to r)] \to (p \to r)$

p	q	r	$p \to q$	$q \to r$	$(p \to q) \land (q \to r)$	$p \to r$	$[(p \to q) \land (q \to r)] \to (p \to r)$
T	T	T	T	T	T	T	T
T	T	F	T	F	F	F	T
T	F	F	F	T	F	F	T
F	T	T	T	T	T	T	T
F	F	T	T	T	T	T	T
F	F	F	T	T	T	T	T
T	F	T	F	T	F	T	T
F	T	F	T	F	F	T	T

Thus, this is a valid argument. Of course, the conjunction, which serves as the antecedent here, can consist of as many implications as we wish to have.

Now comes the tricky part: we have shown that every argument of the form above is valid. **Its validity does not depend on the truth-value of the implications.** For example, the following argument is valid:

GIVEN If a polygon has three sides, then it is a square.
GIVEN If it is a square, then it has twenty sides.
CONCLUSION If a polygon has three sides, then it has twenty sides.

Each of the given implications is false, and the conclusion is false as well. But the argument is valid, since it is a tautology, as shown in Table 10.12. **Do not confuse truth with validity!**

10.8 Validity of Arguments

Exercises 10.8

Determine whether each argument is valid or not valid:

1. GIVEN If Gary ran a mile in 3:51, then he bettered the world record.
   GIVEN Gary ran a mile in 3:51.
   CONCLUSION Gary bettered the world record.
2. GIVEN If it snows, schools are dismissed.
   GIVEN Schools are not dismissed.
   CONCLUSION It does not snow.
3. GIVEN If Susan is in the band, then she travels to all football games.
   GIVEN Susan travels to all football games.
   CONCLUSION Susan is in the band.
4. GIVEN If Peggy is a cheerleader, then she attends all basketball games.
   GIVEN Peggy is not a cheerleader.
   CONCLUSION Peggy does not attend all basketball games.
5. GIVEN If Kevin smokes, then he will get cancer.
   GIVEN If Kevin will get cancer, then he will die soon.
   CONCLUSION If Kevin smokes, then he will die soon.
6. GIVEN If Carol graduated from high school, then she went to college.
   GIVEN If Carol went to college, then she received a college degree.
   CONCLUSION If Carol graduated from high school, then she received a college degree.
7. GIVEN If Dave sold the most newspapers, then he won a trip to the Bahamas.
   GIVEN If Dave is not the best newspaper boy, then he did not win a trip to the Bahamas.
   CONCLUSION If Dave sold the most newspapers, then he is the best newspaper boy.
8. GIVEN If the tulips were in bloom, then it was springtime.
   GIVEN If it was not springtime, then the birds did not migrate north.
   CONCLUSION If the tulips were in bloom, then the birds did not migrate north.
9. GIVEN If $4 = 3 + 1$, then $4 \cdot 5 = (3 + 1) \cdot 5$.
   GIVEN If $4 \cdot 5 \neq (3 + 1) \cdot 5$, then $4 \neq 3 + 1$.
   CONCLUSION If $4 = 3 + 1$, then $4 \neq 3 + 1$.
10. GIVEN If a triangle has five sides, then a square has eight sides.
    GIVEN If a square has eight sides, then a rectangle has ten sides.
    CONCLUSION If a triangle has five sides, then a rectangle has ten sides.
11. Given the statement $[(p \to q) \land (q \to r)] \to (p \to r)$, find a situation to fit the reasoning.

12. Is the following valid?
GIVEN   If I have a green nose with pink spots, then I am normal.
GIVEN   If I am normal, I have six ears and four eyes.
CONCLUSION   If I have a green nose with pink spots, I have six ears and four eyes.

13. Can the following be valid?
GIVEN   If a vehicle has oars, it is a car.
GIVEN   If it is a car, then it has wings.
CONCLUSION   If a vehicle has oars, then it has wings.

CHAPTER 10 TEST

Determine which of the following are true, which are false, and which are open statements:

1. $5^2 = 10$
2. $200\% = 2$
3. $x + 4 = 4$
4. $x^2 = x \cdot x$
5. $0 \cdot 7 = 7$
6. $(\frac{1}{3})^2 = \frac{1}{9}$
7. You apply to America's most exclusive country club and are told the club is closed. Dismayed, you look up the club's federal license number and see it is: $100 = 2.3x40y$. Are they really closed?
8. If 66 players enter the All-American Celebrity Tennis Tournament, how many matches must be played to determine the victor if each person plays at least one match, with the winner advancing to the next round?

Prefix each of the following statements with the appropriate quantifier, $\forall$ or $\exists$:

9. $x^3 = x \cdot x \cdot x$
10. $2x + 6 = 8$
11. $\dfrac{x}{x} = 1$

12. What is the nonpermissible replacement of the variable in $(2x - 3)/(2x - 3) = 1$?

Write a negation of each of the following statements:

13. This was a cold winter.
14. $7 = 5 + 2$
15. $3 < 9$
16. $5 \not> 3$
17. $8 \neq 9$
18. All squares are quadrilaterals.
19. At least one triangle is a right triangle.

Determine the truth-value of each of the following compound statements:

20. $(2 < 4) \vee (4 = 2^2)$
21. $(5 > 6) \vee (3 = 2 + 1)$
22. $(1 = 2) \vee (5 \not< 7)$
23. $(4 > 1) \wedge (1 < 4)$
24. $(3 > 4) \wedge (3 < 4)$
25. $(2 \neq 1 + 1) \wedge (9 \neq 3^2)$

10.8 Validity of Arguments

26. GIVEN p: Martha is a freshman this year
               q: Martha will be a sophomore next year
    Write the two statements of the forms: $p \rightarrow q$, $\sim p \vee q$. Are they logically equivalent?
27. Write the
    a. inverse   b. converse   c. contrapositive
    of the following statement: If four sides of a quadrilateral are congruent, then the quadrilateral is a rhombus.
28. State the
    a. inverse   b. converse   c. contrapositive

Determine whether each of the following arguments is valid:

29. GIVEN   If A is a right triangle, then it has one right angle.
    GIVEN   If A is a triangle with one right angle, then it can be an isosceles triangle.
    CONCLUSION   If A is a triangle with one right angle, then it can be an isosceles triangle.
30. GIVEN   If x is an integer, then its square is an integer.
    GIVEN   If the square of x is an integer, then it is positive.
    CONCLUSION   If x is an integer, then it is positive.

11

Probability

11.1 INTRODUCTION TO PROBABILITY

REVIEW

None

OBJECTIVES

- Know the historical beginnings of the study of probability
- Distinguish between the a priori and the statistical probability
- Give an example of an experiment
- Define a sample space of a given experiment
- Define an event
- Give the sample space for the experiment of tossing a coin
- List all events for the experiment of tossing a coin
- State the probabilities for the events resulting from tossing a coin
- State probability principle 1

11.1 Introduction to Probability

The study of probability was motivated by games of chance. The year 1654 is regarded by historians as the year in which the systematic study of probability began. It is ascribed to Blaise Pascal and Pierre de Fermat, who discussed the ideas involved in games of dice and other games of chance. Their discussion was provoked by a professional gambler, Chevalier de Méré, who raised some questions about the dice games with Blaise Pascal. So began a new branch of mathematics, probability. Today probability is a well-developed field of mathematics.

All the mathematics in this book up to this point has been mathematics in which one can be sure of the answers. For example, after solving the equation $3x = 24$, we can be *certain* that the solution is 8 because $3 \cdot 8 = 24$. Probability deals with phenomena of *uncertainty*. In such cases, we ask about the *likelihood* of something taking place. For example,

When tossing a coin, how *likely* is it that it will fall heads up?

Learning to answer this question will not provide a way to be *certain* that the answer is correct. It will not be possible to verify the answer.

The question asked above falls into the category of *a priori probability*. (A priori is Latin for "from the previous.") It can be answered in advance of any experimentation. The answer depends only on the nature of the coin—it can land either heads or tails, no other possibility exists. Furthermore, each outcome is equally likely to occur. Of course, we are speaking of an unbiased coin.

There is another kind of probability, *statistical probability*. An example of a question asked in statistical probability is the following:

If I am 19 years old, what are my chances of living to be 65?

This question cannot be answered without some data. Cases must be studied and projections made on the basis of experience. Insurance companies are concerned with this kind of probability, and life insurance premiums are established and revised on the basis of the study of many individual cases.

In this book we shall be concerned with a priori probability. To study it, we first need a few basic terms.

An activity such as tossing a coin or rolling a die is called an *experiment*. In the experiment of tossing a coin, one of two things will happen: the coin will land either "heads" or "tails." Thus, the set of all *outcomes* is {H, T}, where H stands for "heads" and T for "tails." The set of all possible outcomes of a given experiment is called the *sample space* of the experiment.

11 Probability

Since all of the possible outcomes of the experiment of tossing a coin are heads (H) and tails (T), the sample space of this experiment is {H, T}. There are four subsets of the sample space {H, T}. They are:

{H}   {T}   [H, T}   ∅

Each subset of the sample space of an experiment is called an *event*.

Thus, the experiment of tossing a coin consists of four events. They are the four subsets of {H, T}. These events may be described as follows:

{H}   the event of the coin landing "heads"
{T}   the event of the coin landing "tails"
{H, T}   the event of the coin landing either "heads" or "tails"
∅   the event of the coin landing both "heads" and "tails"

We shall agree on a method of assigning numbers which will be measures of the likelihood of the occurrence of the given event. To do this, observe the following characteristics of the experiment of tossing a coin.

1. There are *two* possible outcomes.

2. Each of the two outcomes has the *same likelihood* of occurring.

The following is a very important principle which will serve as a guide in assigning numbers that tell the likelihood of various events.

> PROBABILITY PRINCIPLE 1   If there are two possible outcomes and each is equally likely to occur, then the probability of each outcome is $\frac{1}{2}$.

On the basis of this principle, it can now be concluded that, in the experiment of tossing a coin, the probability of H (heads) is $\frac{1}{2}$ and the probability of T (tails) is $\frac{1}{2}$. To say that the probability of H is $\frac{1}{2}$, we shall write

$P(H) = \frac{1}{2}$

Similarly, to say that the probability of T is $\frac{1}{2}$, we shall write

$P(T) = \frac{1}{2}$

Exercises 11.1

1. Briefly relate the historical beginnings of probability.
2. Give an example of a question which falls into the a priori probability.

11.2 Multiple Tosses of a Coin

3. Give an example of a question which falls into the statistical probability.
4. Give an example of an experiment.
5. Define a sample space.
6. Define an event.
7. What is the sample space for the experiment of tossing a coin?
8. List all the events for the experiment of tossing a coin.
9. It has been discovered by many corrupt mathematicians who have a passion for gambling that weighting one side of a coin will influence its falling action. What does this do to the probability?
10. Analyze your previous day by characterizing the probability of the events that took place (was each high or low?).
11. What is $P(H)$ equal to? $P(T)$?

11.2 MULTIPLE TOSSES OF A COIN

REVIEW

- When tossing a coin, $P(H) = \frac{1}{2}$ and $P(T) = \frac{1}{2}$

OBJECTIVES

- List the sample space for tossing a coin twice
- State probability principle 2
- Tell the probability of each outcome when tossing a coin twice
- List the sample space for tossing a coin three times
- Tell the probability of each outcome when tossing a coin three times

The sample space of the experiment *tossing a coin once* is {H, T}. Now consider the experiment *tossing a coin twice*. All possible outcomes of this experiment are the following:

HH, HT, TH, TT

Thus, the sample space of this experiment consists of four elements:

{HH, HT, TH, TT}

Probability principle 1 needs to be extended so that it will apply to the experiment of tossing a coin more than one time.

> PROBABILITY PRINCIPLE 2   If there are n possible outcomes and each is equally likely to occur, then the probability of each outcome is $1/n$.

Since, when tossing a coin twice, there are four possible outcomes, n is equal to 4. It follows that, according to probability principle 2, the probability of each of the four outcomes is $\frac{1}{4}$. Thus, for example, $P(HH) = \frac{1}{4}$.

What would be the sample space for the experiment of tossing a coin *three times*? Here it is:

$$\{HHH, HHT, HTT, HTH, THH, THT, TTH, TTT\}$$

The sample space in this case consists of eight elements. Thus, according to probability principle 2, the probability of each of the eight outcomes is $\frac{1}{8}$.

Exercises 11.2

1. Thinking you know all about probability, you bravely make the statement that you can predict which way a coin will fall. The first time it works, but what happens to your chance of success each time you flip
   a. once?   b. twice?   c. three times?

What is the sample space for the experiment of tossing a coin:

2. Once?   3. Twice?   4. Three times?
5. Give the sample space for tossing a coin four times.

Below is a table summarizing the number of outcomes as they relate to the number of tosses of a coin. These can be shown in a way which would suggest a pattern pointing to the relationship between the number of outcomes and the number of tosses. Examine the pattern and, assuming that the pattern continues, predict the number of outcomes when a coin is tossed:

6. 4 times   7. 5 times   8. 6 times   9. n times

Number of Tosses	Number of Outcomes
1	2
2	4
3	8

11.3 Sample Spaces for Various Experiments

10. a. When tossing a coin once, what is $P(T)$ equal to?
    b. Is $P(H) = P(T)$ true?
    c. Is $P(H) + P(T) = 1$ true?
11. a. When tossing a coin twice, what is $P(TT)$ equal to?
    b. Is $P(HH) = P(TT)$ true?
    c. What is $P(HH) + P(HT) + P(TH) + P(TT)$ equal to?

What is the probability of each outcome when a coin is tossed:

12. 3 times? 13. 4 times? 14. 5 times? 15. 6 times?
16. k times?
17. Bill the Breaker, the well-known boxer, is famous for always putting his opponents on the floor several times during a match. If Charlie the Cruncher, his opponent, is knocked down 5 times, what is the probability of his landing face down all 5 times?

11.3 SAMPLE SPACES FOR VARIOUS EXPERIMENTS

REVIEW

- $E \subseteq S$ means that set E is a subset of set S; that is, each element of E is also in S
- $P(1)$ means the probability of obtaining the outcome 1, as for example in rolling a die
- $(1, 2)$ is an ordered pair of numbers in which 1 is the first member and 2 is the second member

OBJECTIVES

- List the sample space for the experiment of rolling a die
- Tell the probability of each outcome when rolling a die
- Assign the probability 0 to an impossible outcome
- List the sample space for the experiment of rolling a die twice
- State probability principle 3
- Tell the probabilities of various events when rolling a die twice

The use of dice is common in games of chance. When rolling a die, there are six possible outcomes. Assuming a nonbiased die, each outcome has the same probability. It is $\frac{1}{6}$. For example, $P(1) = \frac{1}{6}$ and $P(5) = \frac{1}{6}$. The sample space for this experiment is:

$\{1, 2, 3, 4, 5, 6\}$

We assign the probability 0 to an outcome which is impossible. Thus, for example, $P(7) = 0$. Stated another way, the event $\{7\} = \emptyset$.

Let's now consider the experiment of rolling a die twice. If the outcome on the first roll is 2 and on the second is 5, we write (2, 5) to denote this. Table 11.1 gives a record of all possible outcomes of the experiment of rolling a die twice. In each pair of numbers, the first number gives the outcome on the first roll and the second number the outcome on the second roll. The total number of outcomes for this experiment is 36.

Table 11.1 All Possible Outcomes of Rolling a Die Twice

	\multicolumn{6}{c}{Second Roll}
First Roll	1	2	3	4	5	6
1	(1, 1)	(1, 2)	(1, 3)	(1, 4)	(1, 5)	(1, 6)
2	(2, 1)	(2, 2)	(2, 3)	(2, 4)	(2, 5)	(2, 6)
3	(3, 1)	(3, 2)	(3, 3)	(3, 4)	(3, 5)	(3, 6)
4	(4, 1)	(4, 2)	(4, 3)	(4, 4)	(4, 5)	(4, 6)
5	(5, 1)	(5, 2)	(5, 3)	(5, 4)	(5, 5)	(5, 6)
6	(6, 1)	(6, 2)	(6, 3)	(6, 4)	(6, 5)	(6, 6)

Consider some events in connection with this experiment. The event of obtaining 1 on the first roll is:

$$\{(1, 1), (1, 2), (1, 3), (1, 4), (1, 5), (1, 6)\}$$

This event has 6 elements. Since 1 can be obtained in 6 different ways out of 36 possible outcomes, the probability of obtaining 1 on the first roll is $\frac{6}{36}$, which is the same as $\frac{1}{6}$.

The conclusion above is based on the following general principle.

PROBABILITY PRINCIPLE 3 Denote a finite sample space of an experiment by S. Let E be an event such that $E \subseteq S$. Let the number of elements in S be $n(S)$ and the number of elements in E be $n(E)$. Then:

$$P(E) = \frac{n(E)}{n(S)}$$

Example 1 The event of obtaining 1 on the first roll of a die is

$$E = \{(1, 1), (1, 2), (1, 3), (1, 4), (1, 5), (1, 6)\}$$

11.3 Sample Spaces for Various Experiments

The sample space for rolling a die twice has 36 elements. Thus:

$$n(E) = 6 \quad n(S) = 36$$

Therefore, $P(E) = \frac{6}{36}$ or $\frac{1}{6}$.

Let us examine two extreme cases to which probability principle 3 applies. One case is when an event and a sample space are the same set; that is, $E = S$. For example, for the experiment of tossing a coin once, consider the event of obtaining heads or tails on one toss. This event is $E = \{H, T\}$, and the sample space for this experiment is $S = \{H, T\}$. Therefore, $E = S$, and:

$$P(E) = \frac{n(E)}{n(S)} = \frac{2}{2} = 1$$

It is *generally* true that, if $E = S$, then:

$$P(E) = \frac{n(E)}{n(S)} = 1$$

If the probability of an event is 1, the event is *certain* to occur.

The second extreme case is when $E = \emptyset$. For example, the event of obtaining both heads and tails on one toss of a coin contains no elements. Thus, in this case, $E = \emptyset$ and $S = \{H, T\}$. The probability of this event is:

$$P(E) = \frac{n(E)}{n(S)} = \frac{0}{2} = 0$$

In general, if $E = \emptyset$, then $n(E) = 0$. Thus:

$$P(E) = \frac{n(E)}{n(S)} = \frac{0}{n(S)} = 0$$

The following examples give some instances of applications of probability principle 3.

Example 2 A hat contains four slips of paper, one with a letter M on it, one with A, one with T, and one with H. What is the probability of drawing a slip with the letter M on it?

The event in this case is $E = \{M\}$ and the sample space is $S = \{M, A, T, H\}$. Thus, $n(E) = 1$ and $n(S) = 4$. Therefore:

$$P(E) = \frac{n(E)}{n(S)} = \frac{1}{4}$$

Example 3 In the experiment of rolling a die once, what is the probability that the face landing up will have an odd number of dots?

$$E = \{1, 3, 5\}$$
$$S = \{1, 2, 3, 4, 5, 6\}$$
$$n(E) = 3$$
$$n(S) = 6$$

Thus, $P(E) = n(E)/n(S) = \frac{3}{6} = \frac{1}{2}$.

Example 4 In the experiment of rolling a die once, what is the probability that the face landing up will have more than 4 dots?

$$E = \{5, 6\}$$
$$S = \{1, 2, 3, 4, 5, 6\}$$
$$n(E) = 2$$
$$n(S) = 6$$

Thus, $P(E) = n(E)/n(S) = \frac{2}{6}$ or $\frac{1}{3}$.

Example 5 In the experiment of rolling a die twice, what is the probability of obtaining less than 3 dots on the second roll?

The event for this case is:

$$E = \{(1, 2), (2, 2), (3, 2), (4, 2), (5, 2), (6, 2),$$
$$(1, 1), (2, 1), (3, 1), (4, 1), (5, 1), (6, 1)\}$$

Since E has 12 elements and the sample space S has 36 elements, we have

$$P(E) = \frac{n(E)}{n(S)} = \frac{12}{36} \text{ or } \frac{1}{3}$$

Exercises 11.3

1. A hat contains four slips of paper of different colors: red, green, black, and blue. What is the probability that, in one drawing, a blue slip will be obtained?

In the experiment of rolling a die once, what is the probability that the face landing up will have:

2. An even number of dots?
3. Fewer than 7 dots?
4. Fewer than 4 dots?
5. More than 6 dots?
6. More than 1 dot?
7. Less than 1 dot?

11.3 Sample Spaces for Various Experiments

Consider the experiment of tossing a coin twice. What is the probability that a head will come up:

8. Once?   **9.** Twice?   **10.** Not at all?

A box contains three red marbles and eight green marbles. What is the probability that, in one drawing:

11. A red marble will come up?
12. A green marble will come up?

A box contains 50 light bulbs, 10 of which are 50-watt, 15 are 75-watt, and the remaining are 100-watt. What is the probability that, on one drawing:

13. A 50-watt light bulb will come up?
14. A 75-watt light bulb will come up?
15. A 100-watt light bulb will come up?

Consider the experiment of rolling a die twice. What is the probability of the following coming up?

16. 6 dots on the first roll
17. Fewer than 2 dots on the second roll
18. Fewer than 6 dots on the first roll
19. More than 0 dots on the first roll
20. Fewer than 7 dots on the second roll
21. The sum of the numbers of dots on both rolls will be 6
22. The sum of the numbers of dots on both rolls will be at least 10
23. The sum of the numbers of dots on both rolls will be at most 10
24. The product of the numbers of dots on both rolls will be 4
25. The difference of the numbers on both rolls will be 0
26. Your mean friend sends you a bag full of snakes, A, B, C, and D. You eagerly open your present to see what was sent you. If snake C is poisonous, what is the probability that you will not see your friend to thank him for the gift?
27. You are sailing in your luxury boat when you see a deadly sea mine rolling toward you. If on its axis of rotation there are ten prongs and only numbers 3 and 4 are explosive, what is the probability that your boat will be blown up?

Each numeral from 0 to 9 inclusive is written on a separate slip of paper, and the slips are placed in a box. One slip is drawn from the box. What is the probability that the numeral on the slip drawn names:

28. An odd number?   **29.** An even number?
30. A prime number?   **31.** A composite number?
32. A number divisible by 3?   **33.** A number divisible by 4?
34. A number divisible by 1?   **35.** A number divisible by 9?

Consider the experiment of rolling a die three times.

★36. How many elements does the sample space have?

★37. What is the probability that the sum of the numbers of dots on the three rolls will be 18?

★38. What is the probability that the sum of the numbers of dots on the three rolls will be greater than 10?

★39. What is the probability that the sum of the numbers of dots on the three rolls will be less than 9?

★40. What is the probability that the number of dots on each of the three rolls will be the same?

★41. A regular octahedron is a symmetrical solid with eight faces. Some of the faces of an octahedron are painted blue and the rest are painted yellow. If, when the octahedron is rolled, the probability of a blue face landing on the bottom is three times the probability of a yellow face, then how many faces are painted blue?

42. If in one pocket you have a $1 bill, a $5 bill, a $10 bill, and a $20 bill, and in the other pocket you have a penny, a nickel, a dime, and a quarter, what is the probability that in any one try of reaching into both pockets you will pull out $10.05?

11.4 OCCURRENCE OF MORE THAN ONE EVENT

REVIEW

- Given $S = \{1, 2, 3, 4, 5, 6\}$ and its subset $E = \{1, 2\}$, the complement of E in S, denoted by $\bar{E}$, is $\{3, 4, 5, 6\}$. In general, if $A \subseteq B$, then the complement of A in B, denoted by $\bar{A}$, is the set of all elements in B that are not in A
- If $A \subseteq B$, then $A \cup \bar{A} = B$

OBJECTIVES

- Relate probability and complement of a set
- State probability principle 4
- Solve problems about probabilities of two events
- Solve problems about probabilities of at least one event

11.4 Occurrence of More Than One Event

The concept of probability and the concept of complement of a set are related. To see how, let's consider the experiment of rolling a die. Its sample space is:

$$S = \{1, 2, 3, 4, 5, 6\}$$

Example 1 Let us suppose that the event E consists of outcomes of the die landing with less than 3 dots up. Then:

$$E = \{1, 2\}$$

Thus, E consists of 2 elements. The *complement* of the set E in S is $\{3, 4, 5, 6\}$; that is,

$$\bar{E} = \{3, 4, 5, 6\}$$

Since $n(S) = 6$ and $n(E) = 2$, we have $P(E) = n(E)/n(S) = \frac{2}{6}$ or $\frac{1}{3}$. And since $n(\bar{E}) = 4$, we have $n(\bar{E})/n(S) = \frac{4}{6}$ or $\frac{2}{3}$. Observe that $P(E) + P(\bar{E}) = \frac{2}{3} + \frac{1}{3} = 1$. Observe also that $E \cup \bar{E} = S$.

This example leads us to another probability principle.

PROBABILITY PRINCIPLE 4 Given the sample space S and an event E. Let $\bar{E}$ be the complement of E in S. Then $P(E) + P(\bar{E}) = 1$.

Next we shall consider the probability that two events will take place. In statistical probability, such a question may be

What is the probability that Mr. Jones, now 23 years old, *and* Mrs. Jones, now 21 years old, will *both* live to be 65?

To generalize this question, let E_i and E_j be two events occurring in an experiment with a sample space S. What is the probability that *both* E_i and E_j will occur?

To gain insight into this situation, consider the experiment of rolling a die twice. The sample space for this experiment was given in Table 11.1. It has 36 elements.

Example 2 Consider the following two events:

E_1 = the event of obtaining a 2 on the *first* roll
E_2 = the event of obtaining a 4 on the *second* roll

The events E_1 and E_2 are marked in Table 11.2.

11 Probability

Table 11.2 Events E_1 and E_2 When a Die Is Rolled Twice

		Second Roll					
		1	2	3	4	5	6
First Roll	1	(1, 1)	(1, 2)	(1, 3)	(1, 4)	(1, 5)	(1, 6)
	2	(2, 1)	(2, 2)	(2, 3)	(2, 4)	(2, 5)	(2, 6) ← E_1
	3	(3, 1)	(3, 2)	(3, 3)	(3, 4)	(3, 5)	(3, 6)
	4	(4, 1)	(4, 2)	(4, 3)	(4, 4)	(4, 5)	(4, 6)
	5	(5, 1)	(5, 2)	(5, 3)	(5, 4)	(5, 5)	(5, 6)
	6	(6, 1)	(6, 2)	(6, 3)	(6, 4)	(6, 5)	(6, 6)
					↑ E_2		

Observe in Table 11.2 that $n(E_1) = 6$ and $n(E_2) = 6$. From the table one can see that the probability of the event

E_3 = the event of both E_1 and E_2 occurring

consists of one element, namely (2, 4). Thus, $E_3 = \{(2, 4)\}$. Since

$n(E_3) = 1$ and $n(S) = 36$

the probability of E_3, $P(E_3) = n(E_3)/n(S) = \frac{1}{36}$. Observe further that $E_1 \cap E_2 = E_3$ since

$E_1 = \{(2, 1), (2, 2), (2, 3), (2, 4), (2, 5), (2, 6)\}$
$E_2 = \{(1, 4), (2, 4), (3, 4), (4, 4), (5, 4), (6, 4)\}$
$E_3 = \{(2, 4)\}$

Thus, $P(E_3) = P(E_1 \cap E_2)$.

Example 3 Consider the following events:

E_4 = the event of obtaining a 3 on the *first* roll
E_5 = the event of obtaining a 5 on the *first* roll

The events E_4 and E_5 are marked in Table 11.3.

$E_4 = \{(3, 1), (3, 2), (3, 3), (3, 4), (3, 5), (3, 6)\}$
$E_5 = \{(5, 1), (5, 2), (5, 3), (5, 4), (5, 5), (5, 6)\}$

Let E_6 be the event of obtaining both 3 and 5 on a roll of a die. Since it is impossible for this to occur, the probability of E_6 is 0; that is, $P(E_6) = 0$. Observe that

$E_6 = E_4 \cap E_5 = \emptyset$

11.4 Occurrence of More Than One Event

Table 11.3 Events E_4 and E_5 When a Die Is Rolled Twice

		\multicolumn{6}{c}{Second Roll}
		1	2	3	4	5	6
First Roll	1	(1, 1)	(1, 2)	(1, 3)	(1, 4)	(1, 5)	(1, 6)
	2	(2, 1)	(2, 2)	(2, 3)	(2, 4)	(2, 5)	(2, 6)
	3	(3, 1)	(3, 2)	(3, 3)	(3, 4)	(3, 5)	(3, 6)	← E_4
	4	(4, 1)	(4, 2)	(4, 3)	(4, 4)	(4, 5)	(4, 6)
	5	(5, 1)	(5, 2)	(5, 3)	(5, 4)	(5, 5)	(5, 6)	← E_5
	6	(6, 1)	(6, 2)	(6, 3)	(6, 4)	(6, 5)	(6, 6)

thus, $n(E_6) = 0$.

$$P(E_6) = \frac{n(E_6)}{n(S)} = \frac{0}{36} = 0$$

It is true again that $P(E_6) = P(E_4 \cap E_5)$.

The two examples above suggest another probability principle.

PROBABILITY PRINCIPLE 5   Let E_i and E_j be two events in the sample space S. Let E be the event that both E_i and E_j will occur. Then $P(E) = P(E_i \cap E_j)$.

The following question from statistical probability exemplifies another type of question for which probability theory has an answer:

One hundred babies were born during the month of May in Fertile Town. What is the probability that *at least* 98 of them will live to be 1 year old?

To simplify the problem, let's consider a less complicated situation falling into the same category. We shall use the two events, E_1 and E_2, marked in Table 11.2, and ask the question:

What is the probability that *at least* one of E_1 and E_2 will occur?

Example 4   The two events are:

E_1 = the event of obtaining a 2 on the *first* roll
E_2 = the event of obtaining a 4 on the *second* roll

11 Probability

Asking if at least one of the two events occurs is the same as asking if either E_1 or E_2 occurs. We are using the inclusive *or*, which means that both events may also occur. Let E_3 be the event of E_1 or E_2 occurring. Then E_3 is the following set:

$$E_3 = \{(2, 1), (2, 2), (2, 3), (2, 4), (2, 5), (2, 6),\\ (1, 4), (3, 4), (4, 4), (5, 4), (6, 4)\}$$

It consists of 11 elements. Thus, the probability of E_3 is:

$$P(E_3) = \frac{n(E_3)}{n(S)} = \frac{11}{36}$$

Observe also that:

$$E_3 = E_1 \cup E_2$$

Thus,

$$P(E_3) = P(E_1 \cup E_2)$$

Example 5 Consider events E_4 and E_5, marked in Table 11.3.

E_4 = the event of obtaining a 3 on the *first* roll
E_5 = the event of obtaining a 5 on the *first* roll

Let E_6 be the event of at least one of E_4 and E_5 occurring:

$$E_6 = \{(3, 1), (3, 2), (3, 3), (3, 4), (3, 5), (3, 6),\\ (4, 1), (4, 2), (4, 3), (4, 4), (4, 5), (4, 6)\}$$

E_6 consists of 12 elements; that is, $n(E_6) = 12$. Thus,

$$P(E_6) = \frac{n(E_6)}{n(S)} = \frac{12}{36} = \frac{1}{3}$$

Now observe that

$$E_6 = E_4 \cup E_5$$

Thus, again,

$$P(E_6) = P(E_4 \cup E_5)$$

These two examples suggest probability principle 6:

PROBABILITY PRINCIPLE 6 Let E_i and E_j be two events in the sample space S. Let E be the event that E_i or E_j will occur. Then $P(E) = P(E_i \cup E_j)$.

Example 6 Consider the experiment of rolling a die twice. The sample space for this experiment is given in Table 11.1 in Section 11.3.

11.4 Occurrence of More Than One Event

Let:

E_1 be the event of obtaining 1 on the first roll
E_2 be the event of obtaining a 5 on the first roll
E_3 be the event of obtaining a 2 on the first roll
E_4 be the event of obtaining a sum of 3 on both rolls

The following shows the elements of each of the events involved in this problem:

$E_1 = \{(1, 1), (1, 2), (1, 3), (1, 4), (1, 5), (1, 6)\}$
$E_2 = \{(5, 1), (5, 2), (5, 3), (5, 4), (5, 5), (5, 6)\}$
$E_3 = \{(2, 1), (2, 2), (2, 3), (2, 4), (2, 5), (2, 6)\}$
$E_4 = \{(1, 2), (2, 1)\}$

a. Determine the probability of E_1 or E_2.
   $P(E_1 \text{ or } E_2) = P(E_1 \cup E_2)$.
   From the listing of the sets E_1 and E_2, we see that $E_1 \cup E_2$ has 12 elements. Thus, $P(E_1 \cup E_2) = \frac{12}{36} = \frac{1}{3}$.

b. Determine the probability of E_1 or E_3.
   $P(E_1 \text{ or } E_3) = P(E_1 \cup E_3)$.
   $E_1 \cup E_3$ has 12 elements. Thus, $P(E_1 \cup E_3) = \frac{12}{36} = \frac{1}{3}$.

c. Determine the probability of E_4.
   Since E_4 consists of two elements, $P(E_4) = \frac{2}{36} = \frac{1}{18}$.

d. Determine the probability of E_1 or E_4.
   $P(E_1 \text{ or } E_4) = P(E_1 \cup E_4)$.
   Since the union of E_1 and E_4 has 7 elements,
   $P(E_1 \cup E_4) = \frac{7}{36}$.

e. Determine the probability of E_1 and E_2.
   $P(E_1 \text{ and } E_2) = P(E_1 \cap E_2)$. Note from the listing of the sets $E_1 \cap E_2 = \emptyset$. Thus, $P(E_1 \cap E_2) = 0$.

f. Determine the probability of E_3 and E_4.
   $P(E_3 \text{ and } E_4) = P(E_3 \cap E_4)$. Since $E_3 \cap E_4 = \{(2, 1)\}$, it consists of one element. Thus, $P(E_3 \cap E_4) = \frac{1}{36}$.

Exercises 11.4

Consider the experiment of rolling a die twice. The sample space for this experiment is given in Table 11.1 in Section 11.3. Let E be the event that the face with 3 or 4 dots will land up on the second roll of the die:

1. How many elements does E consist of?
2. What is $P(E)$ equal to?
3. Describe $\bar{E}$.
4. How many elements does $\bar{E}$ consist of?
5. What is $P(\bar{E})$ equal to?

6. What is $P(E) + P(\bar{E})$ equal to?
7. Marvelous Malbo the mind reader claims that he can tell which card in a 52-card deck is pulled out. What is the probability that he *guesses* correctly?

Tell which of the following are true and which are false:

8. $P(E_1 \cup E_2) = P(E_1) + P(E_2)$ if and only if E_1 and E_2 are disjoint sets.
9. If E is an event in a finite sample space S, then $P(E \cup \bar{E}) = 0$.
10. If E is an event in a finite sample space S, then $P(E \cap \bar{E}) = 0$.
11. If E is an event in a finite sample space S, and $P(E) = \frac{1}{3}$, then $P(\bar{E}) = \frac{1}{3}$.
12. If E_1 and E_2 are events in a finite sample space S, and $E_1 \cup E_2 = S$, then $P(E_1 \cup E_2) = 1$.
13. If E_1 and E_2 are events in a finite sample space S, and $E_1 \cup E_2 = S$, and $E_1 \cap E_2 \neq \emptyset$, then $P(E_1) + P(E_2) = 1$.
14. If E_1 and E_2 are events in a finite sample space S, and $P(E_1) + P(E_2) > 1$, then $E_1 \cap E_2 \neq \emptyset$.
15. If E_1 and E_2 are events in a finite sample space S, $P(E_1) = \frac{1}{4}$, $P(E_2) = \frac{1}{3}$, and $P(E_1 \cap E_2) = \frac{1}{12}$, then $P(E_1 \cup E_2) = \frac{1}{2}$.
16. If E_1 and E_2 are events in a finite sample space S, $P(E_1) = 1$, and $P(E_2) = 1$, then $E_1 \cap E_2 = \emptyset$.
17. If E_1 and E_2 are events in a finite sample space S, $P(E_1) = 0$, and $P(E_2) = 1$, then $P(E_1 \cup E_2) = 1$.
18. If E_1 and E_2 are disjoint events in a finite sample space S, then $P(E_1 \cap E_2) = 0$.
19. If E_1 and E_2 are events in a finite sample space S, $P(E_1) = 0$, and $P(E_2) = 0$, then $P(\bar{E}_1) + P(\bar{E}_2) = 2$.
20. Four women start to play poker for high stakes. To decide who will deal, each draws a card. What is the probability that they will all draw aces?

A marble is drawn from a box containing 6 red marbles, 4 green marbles, and 7 blue marbles. What is the probability that the marble is:

21. Red or green?
22. Red or blue?
23. Green or blue?
24. Both red and green?
25. Red or green or blue?

One card is drawn from a deck of 52 cards. What is the probability of drawing:

26. An ace or a king?
27. A heart or a spade?
28. A red card or a black card?
29. An ace or a diamond?

11.5 Some Counting Problems

Each numeral from 1 through 25 is written on a separate slip of paper, and the slips are placed in a box. What is the probability that the numeral on the slip names a number that is:

30. Even or divisible by 3?
31. Odd or prime?
32. Odd or composite?
33. Prime or divisible by 8?
34. Divisible by 3 or by 5?
35. Divisible by 2 or by 6?

The following statements refer to the experiment of rolling a die twice. Let x_1 refer to the number of spots that appear on the first roll and x_2 to the number of spots that appear on the second roll. Find the probability of each of the following:

36. $x_1 < 4$ or $x_2 \geq 5$
37. $x_1 \geq 2$ or $x_2 \leq 1$
38. $x_1 + x_2 = 12$
39. $x_1 + x_2 = 2$
40. $x_1 + x_2 = 1$
41. $x_1 + x_2 \neq 12$
42. $x_1 + x_2 \neq 1$
43. $x_1 + x_2 \neq 4$
44. $x_1 + x_2 < 10$
45. $x_1 < 1$ or $x_2 > 6$

★46. The faces of each of two regular tetrahedra (a solid with four faces) are numbered 1, 2, 3, and 4. If two such tetrahedra are tossed, what is the probability that the sum of 4 or 5 will come up?

★47. A box of coins contains 5 dimes and 4 nickels. If a clerk draws three coins, what is the probability that she will be able to make change for a quarter?

11.5 SOME COUNTING PROBLEMS

REVIEW

None

OBJECTIVES

- Tell the number of ways two tasks can be performed, knowing the number of ways each task can be performed
- Tell the number of ways three tasks can be performed, knowing the number of ways each task can be performed
- Generalize the first two objectives to any number of tasks
- Analyze counting problems of different kinds
- Compute the number of k-member subsets formed from an n-member set
- Use the term *combinations* and a shorthand notation for it

11 Probability

Counting is the basic process of mathematics. However, there is a special kind of counting we frequently do when answering a question of the type:

In how many ways can we ... ?

Let's consider a specific question of this kind:

There are two different routes from Nowhere to Here and three different routes from Here to There. In how many different ways can you travel from Nowhere to There?

A diagram will help:

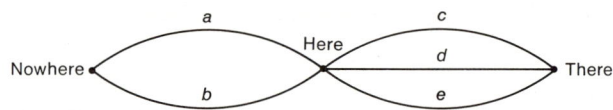

We can choose route a from Nowhere to Here. Then we have three different choices of routes from Here to There. Or we can choose route b from Nowhere to Here. This is also followed by three choices of routes from Here to There. Here are all combinations of the routes:

a—c b—c
a—d b—d
a—e b—e

Thus, all together there are six different routes. With each of the two choices from Nowhere to Here, there are three choices from Here to There: $2 \cdot 3 = 6$.

Let's consider another problem: we are to form 3-member committees from three groups of individuals. The first group consists of two Democrats, the second of three Republicans, and the third of four Independents. Two committees will be considered different if they differ in at least one individual. Each committee is to consist of one individual from each party.

We will give short names to the individuals in the groups: D_1 and D_2 are the two Democrats; R_1, R_2, and R_3 are the three Republicans; and I_1, I_2, I_3, and I_4 are the four Independents. We can count the total number of committees by arranging them quite systematically, as done below:

D_1—R_1—I_1 D_2—R_1—I_1
D_1—R_1—I_2 D_2—R_1—I_2
D_1—R_1—I_3 D_2—R_1—I_3
D_1—R_1—I_4 D_2—R_1—I_4

11.5 Some Counting Problems

$D_1 - R_2 - I_1$ $\quad\quad$ $D_2 - R_2 - I_1$
$D_1 - R_2 - I_2$ $\quad\quad$ $D_2 - R_2 - I_2$
$D_1 - R_2 - I_3$ $\quad\quad$ $D_2 - R_2 - I_3$
$D_1 - R_2 - I_4$ $\quad\quad$ $D_2 - R_2 - I_4$

$D_1 - R_3 - I_1$ $\quad\quad$ $D_2 - R_3 - I_1$
$D_1 - R_3 - I_2$ $\quad\quad$ $D_2 - R_3 - I_2$
$D_1 - R_3 - I_3$ $\quad\quad$ $D_2 - R_3 - I_3$
$D_1 - R_3 - I_4$ $\quad\quad$ $D_2 - R_3 - I_4$

However, as in most of mathematics, we can save a lot of writing by reasoning. Think: a Democrat can be chosen in 2 ways; with each of these two choices, there are 3 choices of Republicans—a total of $2 \cdot 3$. With each of these $2 \cdot 3$ choices, there are 4 choices of Independents—a total of $2 \cdot 3 \cdot 4$ or 24 different committees.

The two situations considered above suggest the following generalizations:

> If one task can be performed in k ways and another in m ways, then the first and the second tasks can be performed in $k \cdot m$ different ways.

> If one task can be performed in k ways, a second in m ways, and a third in n ways, then the first, second, and third tasks can be performed in $k \cdot m \cdot n$ different ways.

You probably can see a way to generalize this to any number of tasks. You may also have realized that it does not matter in which order we choose to arrange the tasks. For example, we could have decided to choose an Independent first, then a Democrat, then a Republican—the count would be the same. Note that $D_1 - R_1 - I_1$ and $I_1 - D_1 - R_1$ are the same committee.

One must be cautious not to follow a formula blindly in reasoning out answers to counting problems. Each problem should be analyzed first. Let's consider another example.

Example 1 How many license plates can be made of the form

letter-letter-digit-digit-digit-digit

Solution There are 26 choices for the first letter, 26 choices for the second letter, and 10 choices for each of the four digits. Thus, together there are

$26 \cdot 26 \cdot 10 \cdot 10 \cdot 10 \cdot 10$

choices. This is 6,760,000.

11 Probability

Quite often counting problems are of the type where we are trying to find out how many subsets with a given number of members can be formed from a set with a given number of members (the former set having at most as many members as the latter set). For example, we may ask:

how many 2-member subsets can be formed from a 5-member set, $U = \{a, b, c, d, e\}$?

It is customary to call the k-member subsets *combinations* and phrase the question in the form:

How many combinations of 2 objects can be formed from the set of 5 objects?

Let's list all of these 2-member subsets:

$\{a, b\}$ $\{b, c\}$ $\{c, d\}$ $\{d, e\}$
$\{a, c\}$ $\{b, d\}$ $\{c, e\}$
$\{a, d\}$ $\{b, e\}$
$\{a, e\}$

Thus, there are 10 combinations of 2 objects from a set of 5 objects.

In problems of this kind, listing of all subsets is impractical when there are many members in the set. Mathematicians thus derived a formula for finding the number of combinations of k objects from a set of n objects. To state this formula, we need a shorthand notation for products of consecutive numbers. This notation looks like an exclamation mark, but is read *factorial*:

$2! = 1 \cdot 2$ *Read:* "two factorial"
$3! = 1 \cdot 2 \cdot 3$
$4! = 1 \cdot 2 \cdot 3 \cdot 4$
  and so on

In general,

$n! = 1 \cdot 2 \cdot 3 \cdot \ldots \cdot (n-1) \cdot n$

Now for the formula:

The number of k-member subsets formed from an n-member set is

$$\frac{n!}{k!(n-k)!}$$

Where k is at most equal to n.

11.5 Some Counting Problems

Now suppose $k = n$. Then the formula becomes:

$$\frac{n!}{n!\ 0!}$$

All is well if we define 0! to be 1, which we do, since the number of n combinations from n objects is 1. Thus:

$$\frac{n!}{n!\ 0!} = \frac{n!}{n! \cdot 1} = \frac{n!}{n!} = 1$$

Example 1 Using the formula, compute the number of combinations of 4 objects that can be formed from a 10-member set.

Solution Here $n = 10$, $k = 4$.

$$\frac{10!}{4!\ (10-4)!} = \frac{10!}{4!\ 6!}$$
$$= \frac{\cancel{1 \cdot 2 \cdot 3 \cdot 4 \cdot 5 \cdot 6} \cdot 7 \cdot 8 \cdot 9 \cdot 10}{1 \cdot 2 \cdot 3 \cdot 4 \cdot \cancel{1 \cdot 2 \cdot 3 \cdot 4 \cdot 5 \cdot 6}}$$
$$= \frac{7 \cdot \cancel{8}\overset{}{} \cdot \cancel{9}\overset{3}{} \cdot 10}{1 \cdot \cancel{2} \cdot \cancel{3} \cdot \cancel{4}}$$
$$= 7 \cdot 3 \cdot 10$$
$$= 210$$

The first cancelling above suggests a shortcut that would save the necessity of writing out the entire products. Try to discover this shortcut.

A handy notation for the number of 4-member subsets formed from a 10-member set is:

$C(10, 4)$

And, in general, the number of k-member subsets from an n-member set is:

$C(n, k)$

Exercises 11.5

1. There are 5 different routes from A to B and 9 different routes from B to C. How many different routes are there from A to C?
2. There are 16 different routes from X to Y and m different routes from Y to Z. How many different routes are there from X to Z?
3. You are going bald and have only 10 hairs left on your head—3 gray, 2 black, and 5 brown. In how many ways can you style these hairs?

11 Probability

4. There are 5 girls and 8 boys from whom to form 2-member committees. Each committee must consist of one boy and one girl. How many such committees can be formed?
5. There are 10 couples and 15 children. Groups of three individuals each are to be formed, such that each group will have a woman, a man, and a child. How many such groups can be formed?
6. There are 10 couples, 12 boys, and 14 girls. Groups of four individuals each are to be formed, such that each group has a woman, a man, a boy and a girl. How many such groups can be formed?
7. You are a racing driver who buys tires from Badpoor Tires, Inc. If they provide you with 7 perfect tires, 5 quarter-bald tires, 10 half-bald tires, and 3 completely bald tires, how many ways can the tires be arranged?
8. How many license plates can be made of the form letter-letter-letter-digit-digit?
9. How many license plates can be made of the form given in exercise 8, if no letter and no digit is repeated?
10. How many different 5-member committees can be formed from a class of 30 students?
★11. How many different 5-card poker hands can be dealt from a deck of 52 cards?
★12. How many different bridge hands can be dealt from a deck of 52 cards? (There are 13 cards in a bridge hand.)
13. You are arranging a meeting between the local opposing football teams. It would be desirable to make sure that no two players on opposing teams sit together. If the rows are 4, 6, and 10 seats long, how many different seating arrangements can be made?

11.6 ODDS AND MATHEMATICAL EXPECTATION

REVIEW

- In rolling an ordinary die the probability of each outcome is $\frac{1}{6}$
- If there are n outcomes and each outcome is equally likely to occur, then the probability of each outcome is $1/n$

OBJECTIVES

- Know the meaning of the term *odds in favor* or *against*
- Define mathematical expectation
- Solve problems involving bets and mathematical expectation

Suppose that Mark claims that the odds in favor of Playball Community College's winning the basketball tournament are 10 to 1. He is

11.6 Odds and Mathematical Expectation

willing to bet on it, and Susan is ready to take him up on this bet. If Mark bets $10, Susan puts up $1 against that bet. If Mark wins, he gets the $1 Susan lost and his bet of $10 is returned to him. The *payoff rate* here is 10 for 1.

Now consider the odds in rolling an ordinary die. What are the odds for rolling a 6? We know that the probability of rolling a 6 is $\frac{1}{6}$. The probability of failing to roll a 6 is therefore $\frac{5}{6}$. The ratio of the probability of success to the probability of failure is $\frac{1}{6}$ to $\frac{5}{6}$, which is the same as 1 to 5. Thus, we say that the odds in *favor* of rolling a 6 are 1 to 5, and the odds *against* rolling a 6 are 5 to 1.

The concept of odds in games of chance is related to the concept of *mathematical expectation*. Mathematical expectation is the product of the probability of an event and the amount of, say, money one receives in case the event does take place.

Example 1 The player will receive 30 cents each time a 5 comes up on the roll of a die. For any other outcome he receives nothing. What is the mathematical expectation of this event?

Solution The mathematical expectation here is equal to:

$$\tfrac{1}{6} \cdot 0 + \tfrac{1}{6} \cdot 0 + \tfrac{1}{6} \cdot 0 + \tfrac{1}{6} \cdot 0 + \tfrac{1}{6} \cdot 30 + \tfrac{1}{6} \cdot 0$$

The first number in each of the products above is the probability of each of the outcomes of rolling of a die. The second number is the payoff for each outcome. Note that for each outcome, except 5, the payoff is 0. The sum above is equal to 5. Thus, the mathematical expectation of this game is 5 cents.

What is the practical meaning of the mathematical expectation of the game considered in Example 1? The mathematical expectation tells you what you can expect in the long run. Thus, in Example 1, if you bet a nickel for each roll of the die in the long run you are going to break even. The expectation is that you will win a nickel in the long run.

This conclusion is reasonable, since it is expected that a 5 will come up once in six rolls of a die and during these six rolls you would have bet 30 cents. Thus, you would have received your money back and neither won nor lost.

Example 2 The bet that a 5 comes up on a roll of a die is 6 cents and the payoff is 30 cents. What is the long range expectation of wins or losses?

Solution In Example 1 we saw that a 5-cent bet would result in a break-even situation. Thus, we would expect to lose 1 cent for each 6-cent bet. Thus, the house would have an edge of $\frac{1}{6}$ or $16\frac{2}{3}\%$ in this game—rather terrible odds for the players.

11 Probability

Example 3 A rather common bet in a roulette game is $1 that a particular number will come up. Suppose you bet $1 on number 13. What is the mathematical expectation of your bet?

First, you should know that there are 38 possible outcomes in a roulette game. Each has a probability of $\frac{1}{38}$. Second, you should know that you will receive $36 if the number 13 comes up. For any other outcome you receive nothing.

Solution In computing the mathematical expectation, we find that all addends are 0, except the addend that is the product of the probability of 13 coming up and the amount of payoff. This product is

$$\tfrac{1}{38} \cdot 36 \quad \text{or} \quad \tfrac{36}{38}$$

That's in dollars. Since you bet $1 (that is, $\frac{38}{38}$), you are expected to lose $\$\frac{2}{38}$ for every $1 you bet. This is a little more than a 5-cent loss on each dollar. Thus, the house in the long run wins more than a nickel on each $1 bet—pretty good odds in favor of the house.

To consider mathematical expectation in most general terms, let us suppose that we have n outcomes with probabilities $p_1, p_2, p_3, \ldots, p_n$. Furthermore, let us suppose that the corresponding payoffs are $A_1, A_2, A_3, \ldots, A_n$. Let the mathematical expectation of this game be M. Then:

$$M = p_1 A_1 + p_2 A_2 + p_3 A_3 + \ldots + p_n A_n$$

Exercises 11.6

1. A player bets $1 that a high number (4, 5, or 6) will come up on a roll of a die. He wins $2 if one of these numbers comes up. What is the mathematical expectation of this game? What is its meaning in terms of long-range wins or losses?
2. You bet with Slick Sam that the 49ers will win the Super Bowl. The odds against the 49ers are 1:50. If you bet $50,000, what will you get if they win?
3. A player bets $1 that 1 or 2 will come up on a roll of a die. He wins $2.50 if one of these numbers comes up. What is the mathematical expectation of this game? What is its meaning in terms of long-range wins or losses?
4. Overheard in the Math Department: "We have great expectations for the top student in our probability class. If we invested $100 in him at a probability of $\frac{1}{2}$, we should soon be rich!" What expectations did that enterprising teacher have?

11.6 Odds and Mathematical Expectation

5. You bet $1 in a game of tossing two coins. You get $3 if two heads come up. What is the mathematical expectation of this game? What does it mean for you in the long run?
6. The payoff for each of the 38 outcomes in a roulette game is $36. What should be your bet in order for it to be a break-even game?
7. After breaking the bank of Monte Carlo, you decide to attack Las Vegas. The game is betting on picking a card from a 52-card deck. If you bet $1 that the ace of spades will come up, what can you expect mathematically? (You will win $50 if your card comes up.)
8. What is the effect of betting $10 instead of $1 in exercise 7?
9. Based on exercises 7 and 8, what can you conclude about betting heavily on a certain constant probability?
10. You are betting on any one of four chosen numbers in a roulette game. What should be your bet in order for this to be a break-even game?
11. A game is played by rolling three dice. You place a bet of $5 that a low sum (3 through 9) will come up. You get $10 if a low sum comes up. What is the winning situation in this game?
12. A player is paid $10 if a king is drawn on one draw from a deck of 52 ordinary cards. What is the mathematical expectation of this game?

CHAPTER 11 TEST

1. What is the meaning of the term *sample space*?
2. Give the sample space for tossing a coin three times. How many elements does it have?

Give each of the following for the experiment of tossing a coin three times:

3. $P(HHH)$
4. $P(HTT$ or $HHT)$
5. $P(HHH$ and $TTT)$
6. $P(HHH$ or TTT or $HTT)$
7. How many elements are there in the sample space for the experiment of rolling a die twice?

Tell each of the following for the experiment of rolling a die twice:

8. $P(1, 1)$
9. $P((1, 1)$ or $(6, 6))$
10. $P((1, 1)$ and $(6, 6))$
11. Probability that 1 will occur on the first roll
12. Probability that 1 or 6 will occur on the first roll

11 Probability

13. Probability that 1 will occur on the first roll and 1 or 6 will occur on the second roll
14. Probability that anyone of 1 through 6 will occur on the first roll
15. Probability that 7 will occur on the first roll
16. Probability that the sum of the numbers of dots will be 8
17. Probability that the sum of the numbers of dots will be less than 5
18. Probability that the sum of the numbers of dots will be less than 2
19. Probability that the sum of the numbers of dots will be even
20. If $S = \{1, 2, 3, 4, 5, 6\}$ and $A = \{1, 3, 5\}$ then what is $\overline{A}$ in S equal to?
21. Given the universal set U, what is $\overline{U}$ equal to?
22. What is $\overline{\emptyset}$ in U equal to?
23. Here is the long-range weather forecast for July: It will rain on July 7, 11, and 23. What is the probability that it will rain in July?
24. One card is drawn from a deck of 52 cards. What is the probability of drawing a king or a heart?
25. How many license plates can be made of the form
    letter-letter-digit-digit-digit?
26. If Big Brother wishes to give a code number of the following form to every person
    letter-digit, letter, digit-digit-digit-digit, letter,
    how many people will he be able to encode?
27. How many different 3-member committees can be formed from a class of 20 students?
28. You bet $15 on a horse in a field of 20. If this is a handicap race where all have an equal chance of winning, what is the mathematical expectation of your horse winning?
★29. You bet $1 that any one of the numbers 12 through 23 will come up in the game of roulette. If you win, you get $3. What is the mathematical expectation in this game? What percent edge does the house have?
★30. A player is paid $5 if a red card is drawn on one draw from an ordinary deck of cards. What is the mathematical expectation of this game?

12

Statistics

12.1 INTRODUCTION TO STATISTICS

REVIEW

None

OBJECTIVES

- Define descriptive statistics
- Give an example of the use of descriptive statistics
- Define inferential statistics
- Give an example of the use of inferential statistics

Statistics is a mathematical subject with many uses. These uses fall into two categories: discriptive statistics and inferential statistics.

In descriptive statistics we use numbers to describe some quantitative characteristics of a situation. Here is an example of descriptive statistics:

> The percentile ranks of the four students who took the Easy Algebra 1 Test were 97, 91, 83, 72.

12 Statistics

In inferential statistics we use data obtained from samples to make inferences about the entire population from which the samples were drawn. Here is an example of inferential statistics:

A random sample of fourth grade students in Friendly City was given a mathematics attitude inventory. On the basis of the results obtained for this sample, the experimenter inferred that the majority of all fourth graders in Friendly City has a favorable attitude toward mathematics.

The use of inferential statistics is a rather complicated matter. A great deal of study is necessary to explore this side of statistics to a satisfactory degree. For this reason, in this chapter we will study the more basic part of statistics, descriptive statistics. The basic concepts of descriptive statistics are needed before one can make statistical inferences. Thus, if you are or become interested in continuing into the study of inferential statistics, this chapter will be of help.

Exercises 12.1

1. What purpose is served by descriptive statistics?
2. Give an example of the use of descriptive statistics.
3. What purpose is served by inferential statistics?
4. Give an example of the use of inferential statistics.
5. Can you think of an application of statistics in a men's magazine?
6. What classification do the statistics in exercise 5 fall under?

12.2 MEAN, MEDIAN, MODE

REVIEW

- The *sum* of a set of numbers is the number obtained by adding all the numbers

OBJECTIVES

- Give examples of raw data
- Define the mean of a set of numbers
- Compute means
- Define the median of a set of numbers
- Find the median of a given set of numbers
- Define the mode of a set of numbers
- Find the mode of a given set of numbers

12.2 Mean, Median, Mode

Ten students have taken a mathematics test on which a perfect (maximum possible) score was 70. Here are the scores made by these students, given in descending order:

70, 70, 65, 60, 59, 54, 51, 47, 42, 30

These are the *raw data*. We will use these raw data to compute some *measures of central tendency*, which describe and give meaning to the raw data.

If we add these scores and divide the sum by the number of scores, 10 in this case, we obtain what is called the *mean*. The mean is sometimes called the *average* or the *arithmetic mean*. The sum of the scores given above is 548. Since there are 10 scores, the mean is 548 ÷ 10 or 54.8.

We need to be aware of the limitations of the mean. It does not tell us anything about the spread of the scores. For instance, if each of the students had obtained a score of 54.8, the mean of these scores would also be 54.8. Explain the reason for this.

We illustrate this limitation of the mean with another example. Consider the three numbers: 1, 12, and 20. Since the sum is 33, the mean is 33 ÷ 3 or 11. Note that the mean of 11, 11, and 11 is also 11. So is the mean of 11, 12, and 13. Thus, the mean does not give us much information about the nature of the distribution of the scores.

Another measure of central tendency is the *median*. The median is the middle score. If the number of scores is odd and they are arranged in order, then there is a middle score. For example, in a set of 11 numbers, the median would be the sixth number from the top or the bottom. In our first example, however, we had an even number of scores, namely, 10. In such cases it is customary to take the median to be the mean of the two middle scores. We take the fifth and sixth scores and compute their mean:

59 + 54 = 113
113 ÷ 2 = 56.5

Thus, the median for our distribution is 56.5. Note that it is different from the mean.

Observe that the median doesn't tell us much about the nature of the distribution either. For example, the median of 1, 2, and 500 is 2; 2 is also the median of 1, 2, and 3. And the two distributions are certainly different!

The third frequently used measure of central tendency is the *mode*. The mode is the score that occurs most frequently. For our distribution, the mode is 70. This is the only score that occurs twice. Each of the remaining scores occurs only once. If a distribution is such that no score occurs most frequently, then we simply conclude that the distribution has no mode.

12 Statistics

Exercises 12.2

Using the following set of five numbers: 3, 7, 15, 20, 20

1. Compute the mean.
2. What is the difference of the mean and the least number?
3. What is the difference of the mean and the greatest number?
4. What is the median of this set of numbers?
5. Which is greater, the mean or the median?
6. What is the mode of this set of numbers?
7. The upper measurements of the models in a certain men's magazine were 36, 40, 44, 50, 38, 36, 36. How do you describe these data? Determine the arithmetic mean of the models.
8. In the previous exercise, which measurement was the median?
9. In exercise 7, what is the mode of the models?
10. Give an example of three numbers for which the mean, the median, and the mode are the same. Now give another set of three numbers for which the three measures are the same as for the first set of numbers.
★11. Can the mean of a set of numbers be greater than the greatest number in the set?
★12. Can the mean of a set of numbers be less than the least number in the set?
13. You casually walk around the gymnasium watching the wrestling matches (and listening to the grunts and groans). You see numbers 7, 11, 5, 4, and 8 wrestling on the floor. Based upon your mathematical knowledge, which wrestler is the meanest?

12.3 STANDARD DEVIATION

REVIEW

- The square root of a number is the number which when multiplied by itself results in the original number. For example, $\sqrt{64} = 8$ because $8 \cdot 8 = 64$.

OBJECTIVES

- Define standard deviation of a set of numbers (scores)
- Compute the standard deviation of a given set of numbers
- Define variance

We have observed that completely different sets of numbers can have the same measures of central tendency: mean, median, and mode. Thus, these three measures do not give a very accurate picture of distributions. We shall look now at another measure. This one falls into the category of *measures of dispersion*, since it indicates the extent of deviation of numbers

12.3 Standard Deviation

from the mean. We will explore the measure of dispersion called *standard deviation*, denoted by the Greek letter σ (*read:* "sigma").

We shall illustrate by an example how to compute the standard deviation for a given set of numbers.

Example 1 Find the standard deviation, σ, for the following set of scores: 93, 87, 84, 80, 71

1. Compute the mean. It is 83.
2. Find the difference of each score and the mean:

$$93 - 83 = 10$$
$$87 - 83 = 4$$
$$84 - 83 = 1$$
$$80 - 83 = -3$$
$$71 - 83 = -12$$

3. Find the sum of the squares of these differences:

$$100 + 16 + 1 + 9 + 144 = 270$$

4. Divide this sum by the number of scores:

$$\frac{270}{5} = 54$$

This is called the *variance*.

5. Now compute the square root of the variance:

$$\sqrt{54} \doteq 7.3$$

This is the *standard deviation* of the given set of five scores.

Let us now review the steps and develop a general formula for computing the standard deviation for any set of data. To represent the sum of a set of numbers, we shall use the summation symbol Σ. For example, to represent the sum of our five scores above, we would write:

$$\sum_{}^{5} X = 93 + 87 + 84 + 80 + 71$$

The 5 above the summation sign is used to indicate that we have 5 scores. In general, we shall use n above the summation sign to indicate that a distribution has n scores. Using M for the mean of a distribution, we can now denote the sum of the differences of the scores and the mean as:

$$\sum_{}^{n} (X - M)$$

and the sum of the squares of the differences as:

$$\sum_{}^{n} (X - M)^2$$

12 Statistics

Assuming that the distribution has n scores, we found the variance by dividing the last sum by the number of scores:

$$\frac{\sum_{n}^{n}(X - M)^2}{n}$$

And, finally, to find the standard deviation, we took the square root of the last number:

$$\sigma = \sqrt{\frac{\sum_{n}^{n}(X - M)^2}{n}}$$

The last expression is a formula for the standard deviation. It states that the standard deviation is equal to the square root of the mean of the squares of the deviations from the mean.

Exercises 12.3

The following are scores on a test obtained by 20 students:

    99, 96, 93, 90, 88, 83, 80, 79, 78, 78, 75, 73, 73, 73, 72, 70, 67, 62, 58, 53

1. Compute the mean of this distribution.
2. What is the median?
3. What is the mode?
4. Compute the standard deviation.
5. How many scores fall within one standard deviation of the mean?
6. What percent of the scores are within one standard deviation of the mean?
7. How many scores are within two standard deviations of the mean?
8. What percent of the scores are within two standard deviations of the mean?
9. You are the judge for the high school javelin-throwing contest (a dangerous occupation). If the distances thrown are 102, 58, 83, 49, and 70 ft, compute the mean of this distribution.
10. What is the median distance of the high-flying throws in exercise 9?
11. Now that you have the mean and median of the energetic efforts, compute the mode and the variance.
12. Now compute the standard deviation from the variance.

12.4 PERCENTILE RANK

REVIEW

- $1\% = .01$ or $\frac{1}{100}$. This is a definition of *percent*

OBJECTIVES

- Interpret the meaning of percentile rank
- Compute the percentile ranks for individuals whose scores are given
- Develop a formula for computing percentile ranks

One of the most frequently used statistics is *percentile rank*. For example, you took a standardized test in trigonometry and your score placed you in the 85th percentile. What does this mean? It means that approximately 85 percent of the students in the population for which the norms were developed scored below the score you obtained. Usually it is assumed that the population used for norms is rather large. We shall consider an example to illustrate how a percentile rank is determined.

Example 1 Susan took a reading test. She obtained a raw score of 48. A large number of students of Susan's age took the same test. Of these students, 60 percent scored lower than Susan, 8 percent obtained the score of 48 (the same as Susan's) and 32 percent had a score higher than Susan's. What is Susan's percentile rank.

We compute Susan's percentile rank in the following way:

1. Take the percent scoring lower than Susan: 60.
2. Take one-half of the percent scoring the same as Susan: $\frac{1}{2} \cdot 8 = 4$.
3. Add the second number to the first: $60 + 4 = 64$.

Thus, we conclude that Susan's percentile rank is 64.

Now let's consider an example in which the data are given in terms of the number of students obtaining a particular score on a test. This will illustrate a procedure for computing percentile ranks when results are in terms of numbers of students.

Example 2 There were 250 students who took a trigonometry test. Ted obtained a raw score of 63. Of all the students, 180 obtained a score below Ted's (below 63). But 20 students obtained the score of 63 (the same as Ted's). What is Ted's percentile rank?

12 Statistics

Here are the steps we would follow to compute Ted's percentile rank:

1. Take the number of students scoring lower than Ted: 180.
2. Add one-half of the number of students obtaining the same score as Ted: $180 + \frac{1}{2} \cdot 20 = 190$.
3. Divide this number by the total number of students taking the test: $\frac{190}{250}$.
4. Multiply the result by 100: $\frac{190}{250} \cdot 100 = 76$.

Thus, Ted's percentile rank is 76.

The last procedure can be generalized by providing a formula for computing an individual's percentile rank. We shall use the following code:

T_S = the raw score of a particular individual
L = the number of raw scores below T_S
E = the number of raw scores equal to T_S
k = the total number of raw scores

Using these designations, we state the formula for the percentile rank of T_S, designated by P_T, which reflects the steps followed in the last example:

$$P_T = \frac{L + \frac{1}{2} \cdot E}{k} \cdot 100$$

To apply this formula to Ted's case, we would make the following substitutions:

$$L = 180 \quad E = 20 \quad k = 250$$

$$P_T = \frac{180 + \frac{1}{2} \cdot 20}{250} \cdot 100$$

$$= \frac{180 + 10}{250} \cdot 100$$

$$= \frac{19\cancel{0}}{25\cancel{0}} \cdot 100$$

$$= \frac{19}{\cancel{25}} \cdot \cancel{100}^{4}$$

$$= 19 \cdot 4$$

$$= 76$$

12.4 Percentile Rank

Thus, the percentile rank for the raw score of 63 for this particular distribution is 76.

Exercises 12.4

1. Sam is 19 years old. He is 6 ft 3 in tall. His height is reported as being in the 92nd percentile for the 19-year olds. What does this mean?
2. After running down 4 pedestrians, going the wrong way down a one-way street, running 3 stop signs and 2 red lights, you are told you failed your driving test. The instructor calmly said, "Well, you are in the first percentile." What did he mean?
3. Carolyn obtained the raw score of 52 on a spelling test. The group with which she competed consisted of a very large number of children. Forty percent of all children scored below 52, and 20 percent obtained the score of 52. What was Carolyn's percentile rank?
4. Of the players in a tennis tournament, 60 percent placed less than 85 percent of their first serves in the court, 30 percent placed 85 percent of their first serves in, and 10 percent placed over 85 percent of their first serves in. If Arnold the Ace placed 85 percent of his first serves in, what percentile is he in?
5. Mark took a general mathematics test. He obtained the score of 21. Fifteen percent of all those taking this test scored lower than Mark. Six percent obtained the score of 21. What was Mark's percentile rank?
6. On an algebra test, Beverly obtained the score of 65. Of a large group of students taking the test, 8 percent scored higher than Beverly, and 2 percent obtained the score of 65. What was Beverly's percentile rank?
7. Bulging University administered an English test to 2000 freshmen. Karl obtained the score of 45. Of the 2000 students, 1400 obtained scores below 45, and 150 students had the score of 45. What is Karl's percentile rank?
8. Poverty Country has a population of 4000 people. Mr. Walking has an annual income of $1400. Of the 4000 people, 3900 have an income higher than that of Mr. Walking, and 6 people have an income of $1400. What is Mr. Walking's percentile rank?
9. If only 10 percent of the golfers make holes-in-one in the $1,000,000 celebrity golf tournament, what percentile ranking are they in?
10. Of the 13 people who took part in the bullfrog croaking contest, 2 did better than you, 3 did the same, 7 did worse, and 1 was chased home by an offended bullfrog. What is your percentile rank?

12.5 REPRESENTING DATA BY GRAPHS

REVIEW

- 1 kilogram ≐ 2.20 pounds
- 1 centimeter ≐ .39 inch

OBJECTIVES

- Make a bar graph from a given set of data
- Change the bar graph to a histogram
- Interpret histograms

Many different kinds of graphs are used to represent data. There are circle graphs, picture graphs, bar graphs, and others. In using graphs, we assume that one can scan some important characteristics of a distribution by taking a quick glance at the graph. To illustrate this, we shall use what is called a *bar graph* and show how it can be developed into a more general graph, called a *histogram*.

We first present data collected on the weights of men at Scales Community College. In Table 12.1, the left-hand column gives the weights in kilograms, and the right-hand column gives the numbers of students of this weight.

Table 12.1 *Weight Data from Scales Community College*

Weight in kg.	Frequency
50	2
55	4
60	7
65	9
70	12
75	15
80	10
85	5
90	1

Note that weights are given in 5-kilogram intervals and that the total number of students is 65. We shall now picture these data using a bar graph.

A quick glance at Figure 12.1 reveals the following facts:

1. The most frequently occurring weight is 75 kilograms.
2. The least frequently occurring weight is 90 kilograms.
3. All given weights are represented by different numbers of students.
4. The smallest weight is 50 kilograms, and the largest weight is 90 kilograms.

12.5 Representing Data by Graphs

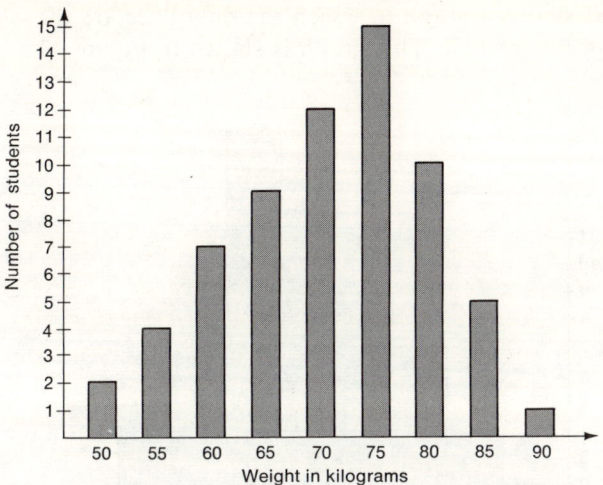

FIGURE 12.1

We are now going to "doctor up" the graph in Figure 12.1. We divide each interval between the 5-kilogram weights in half and connect the boundaries of these newly created intervals by dashed lines. Notice that the 5-kilogram multiples now become the midpoints of the newly created intervals. This is shown in Figure 12.2.

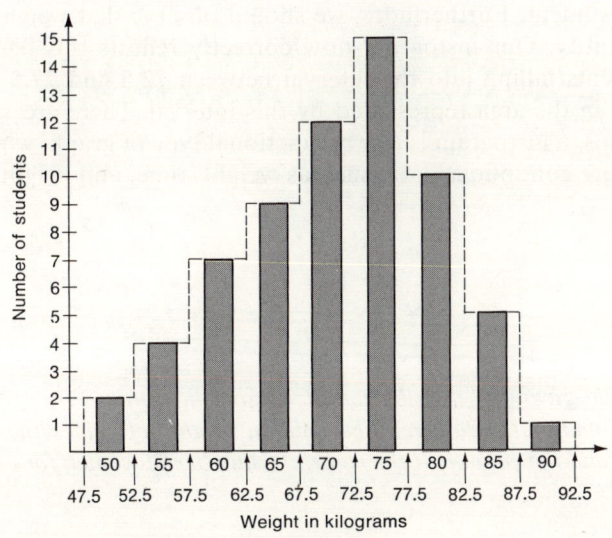

FIGURE 12.2

The next step is a graph in which the only lines are those shown by dashed lines in Figure 12.2. This graph is shown in Figure 12.3.

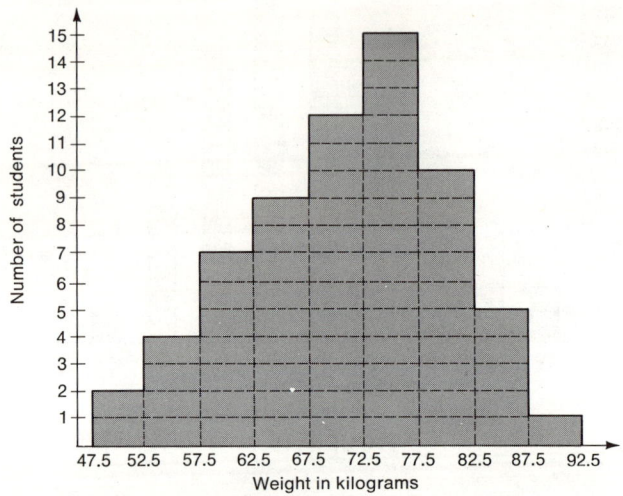

FIGURE 12.3

Another change from Figure 12.2 was made in Figure 12.3. We drew dashed lines to show frequencies. We can use these dashed lines to think of numbers of students in terms of areas. The area of each little rectangle represents one student. Furthermore, we should observe that weight is a continuous quantity. Our histogram now correctly reflects this fact. For example, students falling into the interval between 72.5 and 77.5 kilograms are pictured in the area represented by this interval. There are 15 of these students. Thus, a histogram is a very functional type of graph, which can be used to picture continuous data, such as weight, time, and height.

Exercises 12.5

The next table gives the distribution of heights in centimeters of 62 women at Height Community College. The column at the left gives the heights in centimeters, and the column at the right, the number of women for each height.

12.5 Representing Data by Graphs

Height in cm	Frequency
140	4
145	6
150	10
155	12
160	11
165	7
170	6
175	3
180	2
185	1

1. Make a bar graph showing these data.
2. Create new intervals and draw dashed lines as done in Figure 12.2.
3. Make a histogram showing the frequencies in terms of areas.
4. Interpret the newly created intervals in terms of heights.
5. In the table below the number of people in Vegetable City who eat artichokes and the number of artichokes they can eat are listed.

Number of People	Number of Artichokes
2	10
4	12
6	14
8	12
10	10
12	8

Make a bar graph of the above data for the local gourmet magazine.
6. Put on your surgeon's hat, pick up your scalpel, and "doctor up" the bar graph obtained from exercise 5.
7. Drop your surgeon's role and become a historian by turning the graph of exercise 6 into a histogram.

CHAPTER 12 TEST

1. What is the difference between descriptive and inferential statistics?

Using the following set of six numbers:

   5, 6, 12, 15, 15, 27

2. Compute the mean.
3. What is the difference of the mean and the least number?
4. What is the difference of the mean and the greatest number?
5. What is the median of this set of numbers?
6. Which is greater, the mean or the median?
7. What is the mode of this set of numbers?
8. In Big Brother's society of 1984, criminals are classified by number. In the maximum security section, the following prisoners are kept: 5, 7, 11, 6, 2, 4, 3, 10. Who is the meanest prisoner?

The following are scores on a test obtained by 12 students:

   99, 91, 90, 90, 88, 83, 80, 80, 80, 72, 61, 46

9. Compute the mean of this distribution.
10. What is the median?
11. What is the mode?
12. Compute the standard deviation.
13. Barbara obtained the raw score of 65 on an English test. Of the students who took the same test, 75 percent scored below 65, and 10 percent obtained the score of 65. What was Barbara's percentile rank?
14. At a California university, 5000 freshmen took a basic mathematics test. Paul obtained the score of 57. Of the 5000 students, 2000 obtained scores below 57 and 200 students had the score of 57. What was Paul's percentile rank?

A mathematics test was administered to a group of students. The table below gives the times students took to complete the test.

Time in minutes	Frequency
55	5
50	8
45	14
40	10
35	6
30	4
25	3
20	1

12.5 Representing Data by Graphs

15. Make a bar graph showing these data.
16. Create new intervals and draw dashed lines as done in Figure 12.2.
17. Make a histogram showing the frequencies in terms of areas.

Answers to Selected Problems

CHAPTER 1

Exercises 1.1

1-6. Answers will vary.
7. Each sack lasts one cow one year. To order enough vitamin supplement to last 50 cows for one year, each cow must have one sack. Therefore, 50 sacks are needed.

Exercises 1.2

1.
$$\begin{array}{r} 2 + 4 + 6 + \ldots + 100 \\ 100 + 98 + 96 + \ldots + 2 \\ \hline 102 + 102 + 102 + \ldots + 102 \end{array}$$
$$\frac{100 \times 102}{2} = 5100$$

3. $1^3 + 2^3 = 1 + 8 = 9 = 3^2$
$1^3 + 2^3 + 3^3 = 9 + 27 = 36 = 6^2$
$1^3 + 2^3 + 3^3 + 4^3 = 36 + 64 = 100 = 10^2$
$1^3 + 2^3 + 3^3 + 4^3 + 5^3 = 225 \quad\quad = 15^2$
or
$$1^3 + 2^3 + \ldots + n^3 = \left(\frac{n(n+1)}{2}\right)^2$$

5. *Pattern*   a. 89...91   b. 89...91
    81   19 nines $(n-1)$ nines
    891
    8991
    89991
    etc.

7. $\$2 + \$4 + \$6 + \$8 + \cdots + \$40 = \$420$. The second prize is obviously better. We see that $420 = 20 \times 21$, but $21 = 20 + 1$. If we let n equal the number of terms, then the sum of consecutive even terms is $n(n+1)$.

Exercises 1.3

1. Ways of writing names for numbers—i.e., the grammar
3. 0, 1, 2, 3, 4, 5, 6, 7, 8, 9

Answers

5. a. 3.6×10^7 b. 2.21×10^4 c. 6.29×10^3 d. 3.98×10^{11}
7. In 2 seconds your aunt has fallen 10 ft. In 2 seconds your plane has dived 20 ft. You start diving 2 sec after she falls out of the plane. In that 2 sec, she falls 10 ft. You now dive and in 2 sec go 20 ft. But in that 2 sec, your aunt falls another 10 ft, making the two distances equal. You are then able to reach out and pull your aunt back into the plane.
9. Answers will vary.

Exercises 1.4

1. $3 \times 10^1 + 7 \times 10^0$
3. $5 \times 10^3 + 1 \times 10^2 + 3 \times 10^1 + 7 \times 10^0$
5. $6 \times 10^5 + 0 \times 10^4 + 1 \times 10^3 + 9 \times 10^2 + 8 \times 10^1 + 4 \times 10^0$
7. $4 \times 10^3 + 8 \times 10^2 + 3 \times 10^1 + 5 \times 10^0 + 3 \times 10^{-1} + 9 \times 10^{-2} + 6 \times 10^{-3}$
9. a. 3600   b. b. 2,930,000   c. 0.121   d. 936
11. $3 \times 5^2 + 1 \times 5^1 + 2 \times 5^0$
13. $4 \times 5^1 + 0 \times 5^0 + 3 \times 5^{-1}$
15. $1 \times 5^4 + 1 \times 5^3 + 4 \times 5^2 + 0 \times 5^1 + 0 \times 5^0 + 0 \times 5^{-1} + 0 \times 5^{-2} + 1 \times 5^{-3} + 3 \times 5^{-4}$
17. $100000_{\text{five}} = 3125$
    $1000000_{\text{five}} = 15625$
19. Length $= 2 \times 5^1 + 0 \times 5^0 = 10 + 0 = 10$ in.
    Height $= 3 \times 5^1 + 4 \times 5^0 + 15 + 4 = 19$ in.
    Width $= 4 \times 5^1 + 3 \times 5^0 = 45 + 3 = 48$ in.
21. ... 4^3 4^2 4^1 4^0 4^{-1} 4^{-2} 4^{-3}
    ... □ □ □ □ · □ □ □ ...

Exercises 1.5

1. $6 + (4 + 5) = (6 + 4) + 5 = 10 + 5$
3. $4 + (6 + 2) = (4 + 6) + 2 = 10 + 2$
5. 104_{five} 7. 4122_{five} 9. 1443_{five}
11. a. base 9   b. base 3   c. base 5
    Combination is 935
★13. An example:

23_{five}
34_{five}
$\overline{202_{\text{five}}}$
124_{five}
$\overline{1442_{\text{five}}}$

or $23_{\text{five}} \times 34_{\text{five}} = (20_{\text{five}} + 3) \times (30_{\text{five}} + 4)$
$= (20_{\text{five}} \times 30_{\text{five}}) + (20_{\text{five}} \times 4)$
$+ (3 \times 30_{\text{five}}) + (3 \times 4)$
$= 1100_{\text{five}} + 130_{\text{five}} + 140_{\text{five}} + 22_{\text{five}}$
$= 1442_{\text{five}}$

Exercises 1.6

1. $\begin{array}{r} 3\,17 \\ \cancel{4}7 \\ -29 \\ \hline 18 \end{array}$
3. $\begin{array}{r} 9 \\ 2\,\cancel{10}\,17 \\ \cancel{3}\cancel{0}7 \\ -128 \\ \hline 179 \end{array}$
5. 24_{five}
7. 24_{five}

Answers

9. 334_{five}

★11. $12_{five}\overline{)3241_{five}}$ quotient 223_{five} Remainder 10_{five}

$$\begin{array}{r} 24_{five} \\ \hline 34_{five} \\ 24_{five} \\ \hline 101_{five} \\ 41_{five} \\ \hline 10_{five} \end{array}$$

Exercises 1.7

1. 0, 1
3. 16
5. 64
7. 1
9. .125
11. 5
13. 25
15. 43
17. 1101_{two}
19. 100001_{two}
21. 1111111_{two}
23. 10000010_{two}
25. 100000000_{two}
27. a. 5   b. 6   c. 9
   Combination is 569

Exercises 1.8

1. 111_{two}
3. 1010_{two}
5. 1000000_{two}
7. 10110_{two}
9. 100101010_{two}
11. 111_{two}
13. 100100_{two}
15. 11000100_{two}

CHAPTER 2

Exercises 2.1

1. Answers will vary. Some examples:
   A set of pots and pans . . .
   A set of twins . . .
3. The set of even natural numbers less than ten
   The set of the first four even natural numbers
7. {1}, not empty   9. ∅
11. It is not clear *which* 36 even numbers are being considered.
13. T   15. F   17. F   19. T
21. a. 22 is a member of the original set.
    b. 101 ∉ of the set
       200 ∈ of the set
       3601 ∉ of the set
23. T   25. F   27. 49   29. 54
31. 26   33. 22   35. 7, 9, 11: odd natural numbers
37. 25, 36, 49: squares of natural numbers (or integers)
39. 125, 216, 343: cubes of natural numbers
41. 81, 243, 729: increasing powers of 3—i.e., the nth number is 3^n
43. 720, 5040, 40320: multiply preceding number by one more than multiplied by before—i.e., × 2, × 3, × 4, etc.

Answers

Exercises 2.2

1. No 3. Yes 5. No
7. $A = \{Z, W, Y\}$
   $B = \{A, G, D\}$
   $n(A) = 3; n(B) = 3$. The robots are equivalent. Thus, a repair can be made.
9. k in A matched with $k^2 - 1$ in B.
11. $\frac{1}{k}$ in A matched with $\frac{1}{k+1}$ in B.
13. k in A matched with k^4 in B.
15. k in A matched with $2k + 1$ in B.
17. k in A matched with $2k^2$ in B.
19. k in A matched with $k^2 - 1$ in B.
★21. Let $N = \{1, 2, 3, \ldots\}$
   $S = \{1, 2, \ldots n\}$
   and $T = \{n + 1, n + 2, n + 3, \ldots\}$
   k in N is matched with $n + k$ in T.

Exercises 2.3

1. Answers will vary. Some examples:
   The set of all freshmen
   The set of all students over nineteen years of age
3. Yes, every member of S is a member of W and $S \neq W$.
5. w in W is matched with w^2 in S.   7. a. T   b. T   c. F   d. F
9. 11, 13, 15, 17, 19, and 21:
   Group I : 9 and 12
   Group II : 10, 19, and 21
   Group III: 11, 20, and 22
   Group IV: 13, 15, and 18
   Group V : 14, 16, and 17
23. The universal set is the total number of pages in the book:
   $A = \{1, 2, 3, 4, \ldots, 398, 399, 400\}$
   But subset $J = \{98, 99, 100, \ldots, 121, 122\}$ pages to be rewritten. Thus, $J \subseteq A$, and pages 98 to 122 inclusive, or 25 pages must be rewritten.
25. $\emptyset, \{4\}, \{7\}, \{4, 7\}$, four subsets
27. F   29. T   31. T   ★33. T
35. Yes; every element of the finite set can be a member of an infinite set—e.g., $\{2, 4, 6\} \subset \{2, 4, 6, 8, \ldots\}$.
37. No, they do not have the same elements or the same number of elements.
★39. One, the number 0.
★41. $\{0\}$ is not empty. It contains the number 0.

Exercises 2.4

1. Not disjoint   3. Not disjoint   5. Not disjoint   7. Disjoint
9. Not disjoint   11. Disjoint   13. Not disjoint

Answers

15. a. Disjoint (broken)   b. Not disjoint (intact)
    c. Not disjoint (intact); you have apparently broken only one bone in your foot, and the doctor decides it will mend by itself.
17. M and P are not disjoint.
    $M \cap P = \{3 \text{ common elements}\}$
    M and P are nonempty sets.
19. a. A b. B 21. a. A b. B
23. a. A b. B 25. a. A b. B
27. A has all of its elements in common with itself.
29. The empty set and A have no elements in common.
Note: Problems 25–28 are also true for $A = \emptyset$.
31. Yes, no elements combined so still no elements.
33. $\overline{B} = \{2, 4, 6, 8\}$
35. $\overline{D} = \emptyset$
37. Yes, no elements in $\emptyset$ so *all* elements in U.
★39. Yes, example: Let $U = \{0, 1, 2, \ldots\}$
                    and $N = \{1, 2, 3, \ldots\}$
                    Then $N = \{0\}$, a finite set

Exercises 2.5

1. Consider $\angle ABC$. $\overrightarrow{AB}$ and $\overrightarrow{AC}$ intersect at a single point.
   The *intersection of* $\overrightarrow{AB}$ and $\overrightarrow{AC}$ is $\{A\}$.
3. $\overline{AB} \cup \overline{BC} \cup \overline{CD} \cup \overline{DA}$ = rectangle $ABCD$ if any intersecting pair of segments are perpendicular.
7. A set containing one point.
★ 9. No, they must be in different planes (skew lines).

Exercises 2.6

1.

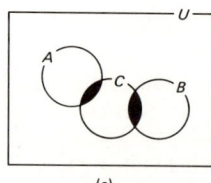

(a)

3.

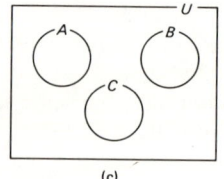

(c)

5. a. $M \not\in X$ $X \subseteq U$ b. $M \in X$ $M \subseteq U$ c. $M \in X$ $X \subseteq U$
      $M \not\in Y$ $Y \subseteq U$ $M \in Y$ $X \subseteq U$ $M \in Y$ $Y \subseteq U$
                      $M \in U$ $Y \subseteq U$ $X \cup Y$

Answers

CHAPTER 3

Exercises 3.1

1. $2 \cdot 1342$
3. $1763 = 43 \times 41$. Fred is wrong.
5. 3093 is odd and the square of an odd number is odd.
7. Yes, $41 \cdot 7 = 287$.
9. No, there is no natural number k such that $11 \cdot k = 259$.
11. 3
13. $\{3, 6, 9, 12, 15, \ldots\}$ and $\{1, 2, 4, 5, 7, \ldots\}$
15. Yes, 4, 8, 12, 16   17. $6 = 2 \cdot 3$ 19. $102 = 17 \cdot 3 \cdot 2$

Exercises 3.2

1. a. $2 \cdot 3 + 1 = 7$, prime
   b. $2 \cdot 3 \cdot 5 + 1 = 31$, prime
   c. $2 \cdot 3 \cdot 5 \cdot 7 + 1 = 211$, prime
   d. $2 \cdot 3 \cdot 5 \cdot 7 \cdot 11 + 1 = 2311$, prime
   e. $2 \cdot 3 \cdot 5 \cdot 7 \cdot 11 \cdot 13 + 1 = 30{,}031 = 59 \cdot 509$, composite
   f. $2 \cdot 3 \cdot 5 \cdot 7 \cdot 11 \cdot 13 \cdot 17 + 1 = 19 \cdot 26{,}869$, composite
3. Let $n = 82$. $82^2 + 82 + 41 = 41(82 \cdot 2 + 2 + 1) = 41 \cdot 167 = 6847$.
5. 911   7. 151

Exercises 3.3

1. $41 = 6 \cdot 7 - 1$ 3. $71 = 6 \cdot 12 - 1$
5. $181 = 6 \cdot 30 + 1$ 7. $709 = 6 \cdot 118 + 1$
11. 29, 31; 41, 43; 59, 61; 71, 73; 89, 91
13. $50 = 13 + 37$ 15. $224 = 193 + 31$
17. $2 < 3 < 4$
    $3 < 5 < 9$
    $4 < 5 < 16$
    $5 < 7 < 25$
    $6 < 7 < 36$
    $7 < 11 < 49$
    $8 < 11 < 64$
    $9 < 11 < 81$
    $10 < 11 < 100$
19. a. Reversible (11)   b. Reversible (17)   c. Not reversible (34)
    d. Not reversible (14) e. Reversible (79)

Exercises 3.4

1. 1, 3, 9   3. 1, 3, 17   5. Deficient
7. Deficient   9. Abundant
11. $1 + 2 + 4 + 8 + 16 + 31 + 62 + 124 + 248 = 496$

Answers

13. a. Important person (perfect) b. Average citizen (abundant)
    c. Criminal (deficient)   d. Politician (prime)
    e. Average citizen (abundant)

Exercises 3.5

1. Let p and m be prime numbers. The divisors of p are 1 and p, and the divisors of m are 1 and m. So the greatest common divisor of p and m is 1.
3. 1
5. Let k be the common divisor of 48 and 30. $48 = kq$ and $30 = kq'$; for some natural numbers q and q', $q > q'$.
   $18 = 48 - 30 = kq - kq' = k(q - q')$
   Since $q - q'$ is a natural number, k is a divisor of 18.
7. 13
9. a. GCD = 6   b. GCD = 9   c. GCD = 8   (b is correct)

Exercises 3.6

1. 30, 60, 90
3. 99, 198, 297
5. $36 = 1 \cdot 72 - 1 \cdot 36$
7. $30 = 1 \cdot 210 - 2 \cdot 90$

Exercises 3.7

1. 2, 3; 2 divides 8 and 6, 3 divides 6.
3. 2, 4, 8, 16 and 32 are all relatively prime to 5.
5. 6 divides $48 = 16 \cdot 3$, but 6 does not divide 16 or 3.

Exercises 3.8

3. 36 ft   5. 210 $2 \cdot 3 \cdot 5 \cdot 7$ 7. 172 $2 \cdot 2 \cdot 43$
9. $12 = 2 \cdot 2 \cdot 3$ and $52 = 2 \cdot 2 = 4$, so GCD(12, 52) = $2 \cdot 2 = 4$.
11. $24 = 2 \cdot 2 \cdot 2 \cdot 3$ and $60 = 2 \cdot 2 \cdot 3 \cdot 5$, so GCD(24, 60) = $2 \cdot 2 \cdot 3 = 12$.
13. $8 = 2 \times 2 \times 2$
    $9 = 3 \times 3$
    The GCD is therefore 1 since the two numbers are relatively prime. If they sell one hat, they must buy one sweater; if they sell three hats, they must buy three sweaters.

Exercises 3.9

1–23. Divisors shown:
1. No given divisors   3. 2, 3   5. 2, 3, 4, 9
7. 3, 9   9. 2, 4, 5   11. 2, 3, 4, 9
13. 2, 3, 4, 5, 9   15. 2, 3   17. 2, 3, 11
19. 2
25. a. If 12 divides k, then $k = 12a = 3 \cdot 4a$, for some natural number a. $k \div 3 = 4a$ and $k \div 4 = 3a$, so k is divisible by both 3 and 4.
    Conversely, if 4 divides k, then $k = 4m$, and if 3 divides k, then 3 divides $4m$.

Answers

3 does not divide 4, so by Theorem 3.4, 3 divides m. So $m = 3m'$ and $k = 4m = 4 \cdot 3m' = 12m'$. Therefore, k is divisible by 12.
b. Exercises 4, 5, 6, 11, 13

27. If x and y are each divisible by z, then $x = mz$ and $y = m'z$ for natural numbers m and m'. $x + y = mz + m'z = (m + m')z$. $m + m'$ is a natural number, so z divides $x + y$.

29. Apply the divisibility rule for 11 and notice with the even number of 9s you will get an alternating-digit sum of 0, which is divisible by 11.

31. $425{,}823 = 425(1000) + 823$
    $= 425(999 + 1) + 823$
    $= \underbrace{425 \cdot 999}_{\text{divisible by 37}} + \underbrace{(425 + 823)}_{\text{test}}$

  In general, if x, y, and z are three-digit numbers $x(1{,}000{,}000) + y(1000) + z$
  $= x(999{,}999 + 1) + y(999 + 1) + z$
  $= \underbrace{x \cdot 999{,}999 + y \cdot 999}_{\text{divisible by 37}} + \underbrace{x + y + z}_{\text{test}}$

33. Yes, $37 \cdot 12$ 35. Yes, $37 \cdot 14$

37. 207 can be divided by 9 exactly 23 times. All other divisors leave a remainder. Therefore, you will invite 9 people.

★39.   365
     +230
      595
    −660
    −65, not divisible by 7

★41. $90 - 6 = 84 = 7 \cdot 12$, so 903 divisible by 7.

★43. 86,10̸7   ★45. 306,50̸1
       14   2
    859̸6   3064̸8
      12   16
     84̸7   304̸8
      14   16
      70, divisible by 7   28̸
                                                      16
                                                12, not divisible by 7

★47. Use both divisibility by 9 and divisibility by 11 rules.

Exercises 3.10

                    Nines
                    Excesses
1.   5682   3
   +4371   +6
   10053 $9 \to 0$

Answers   356

$$
\begin{array}{r}
\text{Elevens} \\
\text{Excesses}
\end{array}
$$

5. (1) $2 - 8 + 6 - 5 = -5 \to 6$
   $\underline{1 - 7 + 3 - 4 = -7 \to 4}$
   $3 - 5 + 0 + 1 = -1 \to 10$

$$
\begin{array}{r}
\text{Elevens} \\
\text{Excesses}
\end{array}
$$

(2) $6 - 9 + 0 - 1 = -4 \to 7$
    $\underline{1 - 6 + 9 =   4}$
    $5 - 3 + 1 =   3$

(3) $5 - 6 + 3 = 2$
    $\underline{3 - 9 = -6 = 5}$
    $5 - 4 + 9 - 3 + 3 = 10$

(4) $ 4 - 2 = 2$
    $3 - 5 + 1 = -1$
    $ 3 - 2 = 1$
    $5 - 9 + 6 - 3 = -1$

$$
\begin{array}{c}
2 \\
10 \times 10 \\
5
\end{array}
$$

Relation: $24 \times 153 + 23 = 3695$

Excesses: $2 \times -1 + 1 = -1$

CHAPTER 4

Exercises 4.1

1. $\{21\}$ 3. $\{12\}$ 5. $\{13\}$
7. $t = V/10 + \frac{1}{8} = \frac{70}{10} + \frac{1}{8} = 7\frac{1}{8}$ sec
9. $10 - 12 = -2$. -2 is not a whole number. Therefore, the set is not closed under subtraction.
11. $7 \div 3 = 2\frac{1}{3}$; $2\frac{1}{3}$ is not a whole number.
13. $3 + 7 = 10$; 10 is not prime.
15. $8 + 15 = 23$; 23 is not composite.
17. Commutative property of addition
19. Closure property of multiplication
21. Associative property of addition
23. Commutative property of multiplication
25. Distributive property
27. $(125 + 137) + 63 = 262 + 63 = 325$
    $125 + (137 + 63) = 125 + 200 = 325$
29. There are an infinite number of pairs to verify.

Exercises 4.2

1. No, no   3. $\{-4\}$ 5. $\{-11\}$ 7. $\{10\}$
9. $0 = -123 + t$. Therefore, $t = 123$ minutes before detonation.
11. 12   13. 25   17. -3 19. 14
21. -4 23. 6   25. 30   27. 0
29. -40 31. 15   33. 151

Answers

35. Property of additive inverse   **37.** Associative property
39. There are no numbers which when added give 0. There isn't even a 0!
41. $300 overdrawn = -300. Absolute value of $|-300| = 300$ = positive value. Therefore, you could be absolutely $300 in credit (theoretically).

Exercises 4.3

1. When it is addition: $10 - (-8) = 10 + 8 = 18$. So $10 - (-8) = 18$.
3. 62 **5.** -50 **7.** 30 **9.** -12 **11.** 10 **13.** 67
15. T **17.** F **19.** T **21.** T **23.** T **25.** T
27. F **29.** F **31.** T **33.** T **35.** T **37.** Yes

Exercises 4.4

1. -5 **3.** 0 **5.** -14 **7.** 0 **9.** 30 **11.** 60
13. $-\$10 \times -\$15 = \$150$. Two wrongs sometimes make a right.
15. T **17.** T **19.** T
21. $-5 \cdot (-8) = [5 \cdot (-1)] \cdot [8 \cdot (-1)]$
$= 5 \cdot [(-1) \cdot 8 \cdot (-1)]$
$= 5 \cdot [8 \cdot (-1) \cdot (-1)]$
$= (5 \cdot 8) \cdot [(-1) \cdot (-1)]$
$= 40 \cdot 1$
$= 40$

23. $ab = ab \cdot 1$ Identity property of multiplication
$= ab \cdot [(-1)(-1)]$ $(-1) \cdot (-1) = 1$
$= a(-1) \cdot b(-1)$ Commutative and associative properties
$= (-a)(-b) > 0$ $a < 0$ implies $-a > 0$
            $b < 0$ implies $-b > 0$
and the product of two positive integers is a positive integer.

25. This was shown in Problem 21; i.e., it is a direct result of Problem 20.

Exercises 4.5

1. $\frac{9}{5}$ **3.** $\frac{13}{2}$ **5.** $\frac{3}{11}$
7. Old equation: New equation:
$3Q = 21H$ $3Q = 20 \times 25$
$H = 25$ ft $Q = \frac{500}{3} = 166.6\overline{6}$
$3Q = 525$
$Q = 175$

9. Multiplicative identity: for any rational number $\frac{a}{b}$, $\frac{a}{b} \times 1 = 1 \times \frac{a}{b} = \frac{a}{b}$.

11. F   **13.** T   **15.** F   **17.** $\frac{7}{6}$ **19.** $\frac{22}{15}$ **21.** $\frac{29}{6}$
23. $\frac{8}{15}$ **25.** $\frac{4}{15}$ **27.** $\frac{2}{15}$ **29. a.** Correct   **b.** Alias   **c.** Correct
31. Distributive property of multiplication over addition
33. Closure of rationals under division
35. Closure of rationals under multiplication
37. Additive identity

Answers

★39. For any rational numbers $\frac{a}{b}$ and $\frac{c}{d}$, $\frac{a}{b} \times \frac{c}{d}$ is a rational number.

Proof: 1) $\frac{a}{b} \times \frac{c}{d} = \frac{ac}{bd}$, by the definition of multiplication of rationals.

2) ac and bd are integers, by the closure of integers under multiplication.

3) $\frac{ac}{bd}$ is a rational number, by the definition of rational numbers.

★41. For any rational numbers $\frac{a}{b}, \frac{c}{d}$, and $\frac{e}{f}$, $\left(\frac{a}{b} + \frac{c}{d}\right) + \frac{e}{f} = \frac{a}{b} + \left(\frac{c}{d} + \frac{e}{f}\right)$.

Proof: $\left(\frac{a}{b} + \frac{c}{d}\right) + \frac{e}{f} = \frac{ad + bc}{bd} + \frac{e}{f} = \frac{f(ad + bc) + (bd)e}{(bd)f}$

$= \frac{(fad + fbc) + bde}{bdf} = \frac{fad + (fbc + bde)}{bdf} = \frac{fad}{bdf} + \frac{fbc + bde}{bdf}$

$= \frac{a}{b} + \frac{b(fc + de)}{bdf} = \frac{a}{b} + \frac{fc + de}{df} = \frac{a}{b} + \left(\frac{c}{d} + \frac{e}{f}\right)$

$= \frac{a(cf + de)}{bdf} = \frac{acf + ade}{bdf} = \frac{acf}{bdf} + \frac{ade}{bdf} = \frac{ac}{bd} + \frac{ae}{bf}$

$= \frac{a}{b} \cdot \frac{c}{d} + \frac{a}{b} \cdot \frac{e}{f}$

43. Set of rational numbers. See Theorem 4.1.

Exercises 4.6

1. $\frac{2}{5}$ **3.** $\frac{2}{3}$ **5.** $\frac{3}{7}$ **7.** $\frac{-4}{5}$ **9.** $\frac{-2}{5}$ **11.** F
13. T **15.** T **17.** F **19.** F **21.** F
23. a. Does not match b. Corresponds c. Corresponds
25. $\frac{3}{5}$ **27.** $-\frac{4}{9}$

Exercises 4.7

1. a. $0.\overline{8}$ b. 0.4 c. 0.25 **3.** $.1\overline{6}$ **5.** $.\overline{13}$
7. $.\overline{384612}$ **9.** $.375$ **11.** $\frac{3}{25}$ **13.** $\frac{8}{9}$ **15.** $\frac{13}{99}$
17. a. $\frac{22}{7}$ b. $\frac{11}{19}$ c. $\frac{13}{30}$

Exercises 4.8

1. $-9, -7$ **3.** $-50, -48$ **5.** a. 1,000,001 b. -19 c. 1
7. $\frac{31}{45}$ **9.** a. $\frac{161}{12960}$ b. $\frac{8}{30}$ c. $\frac{249}{336}$ **11.** 3
13. 2 **15.** $0 < 3 < 9$; $5 < 5\frac{1}{3} < 6$; $-2 < 2 < 10$; $8 < 8\frac{2}{3} < 10$
17. $\frac{5}{12}$ **19.** 1 **21.** $\frac{94}{126}$

CHAPTER 5

Exercises 5.1

1. 125   3. −125   5. 324   7. 12   9. .1   11. 4
13. a. 4 sq in.   b. 16 sq in.   c. 64 sq in.
15.

Side	Area	Volume
2	4	8
4	16	64
8	64	512

If the side is doubled, then the area is 4 times the original area and the volume is 8 times the original. Note that 4 and 8 are equal to 2^2 and 2^3.

17. $\frac{1}{9}$ 19. 1   21. $\frac{1}{4}$ 23. −.001   25. −32
27. 1   29. 0   31. x^{12} 33. m^3n 35. x^5yz^8
37. $v = (w)^2 - 300{,}000$
$v = (6 \times 10^2)(6 \times 10^2) - 300{,}000$
$= 36 \times 10^4$ or $360{,}000 - 300{,}000$
$= 60{,}000$
$= 6 \times 10^4$
The spacecraft will achieve orbit.
39. Positive   41. Positive   43. Positive   45. Positive   47. Negative
49. Negative   51. a. No defect   b. Defect   c. No defect
★53. Greater than
★55. Greater than; $(-\frac{1}{2})^2 = \frac{1}{4}$, $(-\frac{1}{3})^2 = \frac{1}{9}$
★57. Greater than; $(-2)^2 = 4$, $(-5)^2 = 25$
★59. $(-x)^2(-x)^6 = (-x)(-x)(-x)(-x)(-x)(-x)(-x)(-x)$
$= (-1)(-1)(-1)(-1)(-1)(-1)(-1)(-1)xxxxxxxx$
$= 1 \cdot x^8 = x^8$
61. $-(x)^2[-(x)^6] = (-1)(x^2)(-1)(x^6)$
$= (-1)(-1)(x^2)(x^6)$
$= 1 \cdot x^8 = x^8$

Exercises 5.2

1. y^{12} 3. m^{18} 5. 2^8 7. $(-5)^{15}$ 9. 4^4 11. $.7^{36}$
13. a. $(x^3)^4$ b. $(y^2)^4$ c. $(z^3)^3$ d. $(q^2)^5$
15. Positive   17. Negative   19. Negative
21. The additive inverse of a positive number is negative, but the square of a negative number is positive; hence, $-(5)^2 \neq (-5)^2$. Also, the cube of a negative number is negative; hence $-(5)^3 = (-5)^3$.
23. (1) $6^2 = 36$ 25. (1) $(-3)^3 = -27$
    (2) $3^2 \times 2^2 = 9 \times 4 = 36$ (2) $(-1)^3 \times 3^3 = (-1) \times 27 = -27$
27. (1) $2^3 = 8$
    (2) $(-1)^3 \times (-2)^3 = (-1) \times (-8) = 8$
29. a. x^{21} b. $(x^3)^{27}$ c. y^{20}
30–41. The following are true: 31, 33, 35, 37.

43. Use the x, y, and m from Problem 42.
$(xy)^m = 144$, $(-x)^m(-y)^m = (-3)^2(-4)^2$
$= 9 \cdot 16 = 144$

45. F **47.** T **49.** F **51.** T
53. $27x^3y^3$ **55.** $-8x^3$ **57.** x^3 **59.** $x^2y^2a^2$

Exercises 5.3

1. $\dfrac{x^3}{y^3}$ **3.** $\dfrac{-x^5}{y^5}$ **5.** $\dfrac{-64m^3}{n^3}$ **7.** $\dfrac{16x^4y^4}{81}$ **9.** $\dfrac{e^2d^2}{x^2y^2}$

11. a. x^8/x^6 b. x^{25}/x^{20} c. z^{16}/z^{12}
13. If n is an even number, it is true, or if $x = 0$ it is true.
15. x^2 **17.** x^2y^5 **19.** -5 **21.** $(a+b)^3$ **23.** $x - y$
25. T   **27.** F   **29.** F   **31.** F   **33.** F
35. F   **37.** F   **39.** a. x^{-5} b. x^2 c. x^6y^6

Exercises 5.4

1. x^4 **3.** $(a+b)^5$ **5.** $\dfrac{y^4}{x^7}$ **7.** n **9.** $\dfrac{1}{x^8}$ **11.** $\dfrac{1}{y^6}$

13. a. x^{-7} b. x^{-13} c. x^{-1} d. x^{-40} or x^{-4} e. $(x+y)^{-2}$

15. $z^{-(-4)} = \dfrac{1}{z^{-4}} = \dfrac{1}{z^{-4}} = \dfrac{1}{\frac{1}{16}} = 16$

$z^{-(-4)} = z^4 = 16$

b. $-4^{-2} = \dfrac{1}{-4^2} = \dfrac{1}{16}$; $-4^{-2} = \dfrac{1}{-4^2} = \dfrac{1}{16}$

17. $x^{-m} = (-4)^{-2} = \dfrac{1}{(-4)^2} = \dfrac{1}{x^m}$ **19.** $a = -b$ **21.** $a = -2b$ **23.** $x = -y$

Exercises 5.5

1. $6\sqrt{2}$ **3.** $4\sqrt{3}$ **5.** $a^3\sqrt{a}$ **7.** $\dfrac{\sqrt{30}}{6}$ **9.** $4\sqrt{7}$

11. $\dfrac{\sqrt{3}}{3}$ **13.** $30\sqrt{2}$ **15.** -1 **17.** $\dfrac{-3}{x^4}$ **19.** $\dfrac{\sqrt{10}}{5}$

21. $\dfrac{-\sqrt{2}}{2}$ **23.** $x^2 - y$ **25.** $x^2 - \dfrac{1}{y}$ **27.** $2\sqrt{2} - 2$ **29.** $\dfrac{1}{3}$

31. $\dfrac{\sqrt{15} + 3}{2}$ **33.** $\dfrac{x + \sqrt{xy}}{x - y}$

35. $\dfrac{-8 - 2\sqrt{3}}{13}$ **37.** $\dfrac{-\sqrt{2} - \sqrt{3} + \sqrt{10} + \sqrt{15}}{4}$

39. $\dfrac{\sqrt{35} + \sqrt{10} - \sqrt{21} - \sqrt{6}}{2}$ **41.** a. $\dfrac{\sqrt{21}}{7}$ b. $\dfrac{8}{3}$ c. $\dfrac{\sqrt{240}}{24}$

Answers

CHAPTER 6

Exercises 6.1

The following exercises are irrational numbers: 2, 3, 6, 8.
9. a. Well behaved.   b. Trouble maker   c. Well behaved
   d. Trouble maker   e. Trouble maker
11. $\sqrt{5}, -\sqrt{5}$ 13. $\sqrt{15}, -\sqrt{15}$ 15. $\sqrt{7}, -\sqrt{7}$ 17. $\sqrt{34}$
19. $2\sqrt{2}$ 21. 3
25.

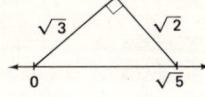

27. 5 ft

Exercises 6.2

1. F   3. T   5. F   7. T   9. F
13. $\sqrt{2} + (-\sqrt{2}) = 0$, a rational number.
15. $\sqrt{2} \cdot \sqrt{2} = \sqrt{4} = 2$, a rational number.
17. $\frac{1}{7}$ 19. $\frac{7}{31}$ 21. $\frac{3}{40}$ 23. $\frac{4}{21}$

Exercises 6.3

1. $\sqrt{3}$, .10120012000120 ...
3. Square the guesses:
   a. 10.24   b. 10.89   c. 11.0224   d. 11.0091   e. 11.0058
   f. 11.0025   g. 11.0012   h. 10.9998
   The correct answer is h, only 0.0002 difference from 11.
5. .1012001200012...   7. a. .1012001200012...   b. Yes
   +.8987998799987...   +.2312551255512...
   ─────────────────   ─────────────────
    .9999999999999...   .2324552455524...
9. $1 + \sqrt{5}, \frac{1}{2} + \sqrt{5}, \frac{1}{3} + \sqrt{5}, \frac{1}{4} + \sqrt{5}, \ldots$

Exercises 6.4

1. Not reflexive or transitive; it is symmetric; not an equivalence relation.
3. Not reflexive, symmetric or transitive; it is not an equivalence relation.
5. Not reflexive or transitive; it is symmetric; not an equivalence relation.
7. Not reflexive or transitive; it is symmetric; not an equivalence relation.
9. It is reflexive, symmetric and transitive; it is an equivalence relation.

Exercises 6.5

1. $\sqrt{x}$ is not a real number.   3. $x \cdot 1 = 1 \cdot x = x$
5. -3 7. 0   9. $\sqrt{10}$

Answers

11. There exists a unique real number $\dfrac{1}{x}$ such that $x \cdot \dfrac{1}{x} = 1$.

13. $-\dfrac{7}{4}$ **15.** $\sqrt{5}$ **17.** $-\dfrac{1}{2.5} = -\dfrac{2}{5} = -.4$

19. a. $\dfrac{1}{2}$ b. $\dfrac{9}{8}$ c. 1000   d. $\dfrac{3}{10}$ e. $\sqrt{\dfrac{2}{15}}$

21. No; the set of integers does not have multiplicative inverses.

Exercises 6.6

1. $x + 9 < -2$ **3.** $1 + 3x > 3$ **5.** $2(x + 3) > x - 4$

7. $2(2x + 5) < x - 6$

9. a. $\{x \mid x < -11\}$ b. $\{x \mid x < \tfrac{4}{3}\}$ c. $\{x \mid x > \tfrac{2}{3}\}$

     d. $\{x \mid x > -\tfrac{11}{15}\}$ e. $\{x \mid x > -10\}$ f. $\{x \mid x > -\tfrac{4}{13}\}$

     g. $\{x \mid x < -\tfrac{16}{3}\}$ h. $\{x \mid x < -\tfrac{16}{9}\}$

11. $9 < 12 < 16$ **13.** $0 \leq 1 \leq 5$ **15.** $r < s < t$

17. $r \leq s \leq t$

19. $xz < yz, z > 0$ Given

     $xz + s = yz, s > 0$ Definition of $<$

     $s = yz - sz$ Additive property for equations

     $s = z(y - x)$ Distributive property

     $y - x = w > 0$ $s > 0, z > 0$

     $xz + wz = yz$ Substituting $s = wz$ in step 2

     $(x + w)z = yz$ Distributive property

     $x + w = y$ Cancellation property of multiplication

     $x < y$ Definition of $<$

Exercises 6.7

1. $6x - 1 > 3x + 7$
$6x - 1 + (-3x) > 3x + 7 + (-3x)$
$3x - 1 > 7$
$x - \tfrac{1}{3} > \tfrac{7}{3}$
$x > \tfrac{8}{3}$

The minimum deposit must always be greater than $\$2\tfrac{2}{3}$, so you will be able to please your aunt.

3. $\{9\}$ **5.** $\{-2\}$ **7.** $\{\tfrac{1}{4}\}$ **9.** $\{\tfrac{10}{13}\}$

11. $\{x \mid x > 6\}$ **13.** $\{x \mid x > 12\}$ **15.** $\{x \mid x > -2\}$ **17.** $\{x \mid x < -\tfrac{3}{4}\}$

19. $\{x \mid x < -8\}$ **21.** $\{x \mid x > \tfrac{21}{4}\}$ **23.** $\{1\}$ **25.** $\{1\}$

27. $\{5\}$ **29.** $\{5\}$ **31.** $\{-1\}$ **33.** $\{0\}$

35. $\sqrt{x - 9} = 3$ **37.** $\sqrt{2x + 3} = 5; \{11\}$
     $\sqrt{x} = 12$
     $x = 144$ miles

CHAPTER 7

Exercises 7.1

1. Topology, nonmetric geometry, metric geometry, deductive geometry, motion geometry.

Answers

Exercises 7.2

1. Point
3. Points A, B, and C; ray $\vec{AB}$ and ray $\vec{AC}$; segments $\overline{AB}$ and $\overline{AC}$; angle A
5. a. $\overline{BD}$ b. $\overleftrightarrow{FB}$ c. $\emptyset$ d. $\{D\}$ e. $\emptyset$
   f. $\vec{AG}$ g. $\overleftrightarrow{FB}$ h. $\triangle ABC$ i. $R_{\triangle ADC}$ j. $\emptyset$

Exercises 7.3

1. The set of all points
3. Tetrahedron (4 faces), cube (6 faces), octahedron (8 faces), dodecohedron (12 faces), icosahedron (20 faces)
5. A circular region and a subset of a circular region called a sector
★ 7. A line and two half-planes; a line corresponds to the vertex; the two half-planes

Exercises 7.4

1. 100   3. 1000   5. 100,000   7. 0.1   9. 0.001
11. 0.000001   13. 10^{-1} 15. 10^3 17. Answers will vary.
19. $P = 121$ mm   21. $P = 148$ mm   23. $P = 159$ mm

Exercises 7.5

1. 1 sq in.   3. 14 sq cm   5. 16 sq cm   7. 600 sq mm
9. 254.3 sq mm (approx.)   11. a. no   b. yes

Exercises 7.6

1. a. No   b. No   c. No   d. Yes   3. An infinite number

Exercises 7.7

1. a. Reflection   b. Rotation   c. Translation
   d. Reflection   e. Reflection   f. Rotation
3. They preserve congruence—that is, the shape and size of figures are not changed.

CHAPTER 8

Exercises 8.1

1. For any two whole numbers, a and b, $a + b$ is a whole number.
3. 8 hr   5. a. 1   b. 2   c. 2   7. 1   9. 2
11. 1   13. Yes   15. One, the number 0

Exercises 8.2

1. Subtraction is a counterclockwise move on the clock.
3. For a and $b \in S_3$, $a \times b$ means $\underbrace{b + b + \ldots + b}_{a \text{ times}}$
   $2 \times 2 = 2 + 2 = 1$ in S_3

Answers

5. The multiplicative inverse of 1 is 1.
   The multiplicative inverse of 2 is 2.
7. a. 2 sq units   b. 0 sq units   c. 2 sq units

Exercises 8.3

Addition Table for S_5

+	0	1	2	3	4
0	0	1	2	3	4
1	1	2	3	4	0
2	2	3	4	0	1
3	3	4	0	1	2
4	4	0	1	2	3

Addition Is Associative

2. $(2 + 3) + 3 = 0 + 3 = 3$
   $2 + (3 + 3) = 2 + 1 = 3$
3. $(4 + 1) + 2 = 0 + 2 = 2$
   $4 + (1 + 2) = 4 + 3 = 2$
4. $(1 + 1) + 4 = 2 + 4 = 1$
   $1 + (1 + 4) = 1 + 0 = 1$
5. $(4 + 3) + 3 = 2 + 3 = 0$
   $4 + (3 + 3) = 4 + 1 = 0$
6. $(2 + 3) + 4 = 0 + 4 = 4$
   $2 + (3 + 4) = 2 + 2 = 4$

Multiplication Table for S_5

×	0	1	2	3	4
0	0	0	0	0	0
1	0	1	2	3	4
2	0	2	4	1	3
3	0	3	1	4	2
4	0	4	3	2	1

Multiplication Is Associative

2. $(0 \times 4) \times 2 = 0 \times 2 = 0$
   $0 \times (4 \times 2) = 0 \times 3 = 0$
3. $(1 \times 3) \times 3 = 3 \times 3 = 4$
   $1 \times (3 \times 3) = 1 \times 4 = 4$
4. $(3 \times 1) \times 4 = 3 \times 4 = 2$
   $3 \times (1 \times 4) = 3 \times 4 = 2$
5. $(2 \times 3) \times 4 = 1 \times 4 = 4$
   $2 \times (3 \times 4) = 2 \times 2 = 4$
6. $(3 \times 4) \times 4 = 2 \times 4 = 3$
   $3 \times (4 \times 4) = 3 \times 1 = 3$

Multiplication Is Distributive Over Addition in S_5

2. $0 \times (3 + 4) = 0 \times 2 = 0; (0 \times 3) + (0 \times 4) = 0 + 0 = 0$
3. $2 \times (4 + 1) = 2 \times 0 = 0; (2 \times 4) + (2 \times 1) = 3 + 2 = 0$
4. $3 \times (0 + 1) = 3 \times 1 = 3; (3 \times 0) + (3 \times 1) = 0 + 3 = 3$
5. $1 \times (4 + 2) = 1 \times 1 = 1; (1 \times 4) + (1 \times 2) = 4 + 2 = 1$
6. $4 \times (4 + 0) = 4 \times 4 = 1; (4 \times 4) + (4 \times 0) = 1 + 0 = 1$

1.   i. Closure property of addition: For any a and b in S_5, $(a + b) \in S_5$.
   Example: $4 + 4 = 3$ in S_5.
   ii. Closure property of multiplication: For any a and b in S_5, $(ab) \in S_5$.
   Example: $4 \times 3 = 2$ in S_5.
   iii. Commutative property of addition: For any a and b in S_5, $a + b = b + a$.
   Example: $4 + 3 = 2 = 3 + 4$ in S_5.
   iv. Commutative property of multiplication: For any a and b in S_5, $ab = ba$.
   Example: $2 \times 3 = 1 = 3 \times 2$ in S_5.

Answers

v. Associative property of addition: For any numbers a, b, and c in S_5, $(a + b) + c = a + (b + c)$. Example: $(2 + 3) + 4 = 0 + 4 = 4 = 2 + 2 = 2 + (3 + 4)$.
vi. Associative property of multiplication: For any numbers a, b, and c in S_5, $(ab)c = a(bc)$. Example: $(2 \cdot 4) \cdot 3 = 3 \cdot 3 = 4 = 2 \cdot 2 = 2 \cdot (4 \cdot 3)$.
vii. Distributive property of multiplication over addition: For any numbers a, b, and c in S_5, $a(b + c) = (ab) + (ac)$. Example: $2(3 + 4) = 2 \cdot 2 = 4 = 1 + 3 = (2 \cdot 3) + (2 \cdot 4)$.
viii. Property of additive identity: 0 is the additive identity in S_5 since for every $a \in S_5$, $a + 0 = a$. Example: $4 + 0 = 4$.
ix. Property of multiplicative identity: 1 is the multiplicative identity in S_5 since for every $a \in S_5$, $a \times 1 = a$. Example: $3 \times 1 = 3$.
x. Property of additive inverse: For every a in S_5 there is an additive inverse of a, say b, such that $a + b = 0$. Example: $3 + 2 = 0$, 3 and 2 are additive inverses of each other in S_5.
xi. Property of multiplicative inverse: For every number $a \neq 0$ in S_5 there is a multiplicative inverse, say c, such that $(ac) = 1$. Example: $4 \times 4 = 1$, 4 is its own multiplicative inverse.

3. $4; 6 - 2 = 4$ 5. $2; 1 - 4 = -3 = 2$ 7. $2; 4 - 2 = 2$
9. $1; 4 - 3 = 1$
11. Distributive property of multiplication over subtraction
13. Yes; the element b since $b \, \beta \, a = a$, $b \, \beta \, b = b$, etc.
15. Yes; $a \, \beta \, (b \, \alpha \, c) = a \, \beta \, d = a$ and $(a \, \beta \, b) \, \alpha \, (a \, \beta \, c) = a \, \alpha \, a = a$
$c \, \beta \, (e \, \alpha \, d) = c \, \beta \, c = e$ and $(c \, \beta \, e) \, \alpha \, (c \, \beta \, d) = d \, \alpha \, b = e$

Exercises 8.4

Addition Table for S_4

+	0	1	2	3
0	0	1	2	3
1	1	2	3	0
2	2	3	0	1
3	3	0	1	2

Addition Is Associative in S_4

2. $(3 + 3) + 2 = 2 + 2 = 0$
   $3 + (3 + 2) = 3 + 1 = 0$
3. $(0 + 3) + 1 = 3 + 1 = 0$
   $0 + (3 + 1) = 0 + 0 = 0$

Multiplication Is Associative in S_4

2. $(2 \times 0) \times 3 = 0 \times 3 = 0; 2 \times (0 \times 3) = 2 \times 0 = 0$
3. $(2 \times 3) \times 1 = 2 \times 1 = 2; 2 \times (3 \times 1) = 2 \times 3 = 2$
4. $(3 \times 3) \times 2 = 1 \times 2 = 2; 3 \times (3 \times 2) = 3 \times 2 = 2$

Multiplication Is Distributive Over Addition in S_4

2. $3 \times (0 + 2) = 3 \times 2 = 2; (3 \times 0) + (3 \times 2) = 0 + 2 = 2$
3. $2 \times (2 + 2) = 2 \times 0 = 0; (2 \times 2) + (2 \times 2) = 0 + 0 = 0$
4. $3 \times (2 + 2) = 3 \times 0 = 0; (3 \times 2) + (3 \times 2) = 2 + 2 = 0$
5. $2 \times (3 + 2) = 2 \times 1 = 2; (2 \times 3) + (2 \times 2) = 2 + 0 = 2$
6. $1 \times (2 + 3) = 1 \times 1 = 1; (1 \times 2) + (1 \times 3) = 2 + 3 = 1$
7. $0 \times (1 + 2) = 0 \times 3 = 0; (0 \times 1) + (0 \times 2) = 0 + 0 = 0$

Answers

1. 0   **3.** 3   **5.** 1   **7.** 2   **9.** 1

11. In S_4, 2 can never be reduced to 1 since it does not have a multiplicative inverse. If the 2 can never become 1, Tripos and Pentos cannot join forces to conquer Earth.

13. 1   **15.** 1   **17.** 3

19.

+	0	1	2	3	4	5
0	0	1	2	3	4	5
1	1	2	3	4	5	0
2	2	3	4	5	0	1
3	3	4	5	0	1	2
4	4	5	0	1	2	3
5	5	0	1	2	3	4

21. $5 + 1 = 0; 0 + 0 = 0$
23. $5 + 4 = 3; 1 + 2 = 3$
25. Yes
27. 0
29. 4
31. 2

33.

×	0	1	2	3	4	5
0	0	0	0	0	0	0
1	0	1	2	3	4	5
2	0	2	4	0	2	4
3	0	3	0	3	0	3
4	0	4	2	0	4	2
5	0	5	4	3	2	1

35.

×	0	1	2	3	4	5
0	0	0	0	0	0	0
1	0	1	2	3	4	5
2	0	2	4	0	2	4
3	0	3	0	3	0	3
4	0	4	2	0	4	2
5	0	5	4	3	2	1

37. 1   **39.** None   **41.** None

43. 2, 3, and 4 have no multiplicative inverses.
45. $1 \times 4, 2 \times 2, 2 \times 5, 4 \times 1, 4 \times 4, 5 \times 2$
47. $3 \cdot 4 = 0; 3 + 3 = 0$ **49.** $0 \cdot 1 = 0; 0 + 0 = 0$
51. $5 \cdot 3 = 3; 3 + 0 = 3$ **53.** $5 \cdot 4 = 2; 1 + 1 = 2$
55. S_6 has no multiplicative inverses for all of its elements.
57. 7   **59.** 15   **61.** Two, 1 and 5   **63.** Yes, 0
65. 2, 5   **67.** b   **69.** $(a * b) * c = a * c = b$

Exercises 8.5

1. {2}   **3.** {1}   **5.** {3}   **7.** {1}   **9.** {4}
11. {2, 3}   **13.** {2, 3}   **15.** {0}   **17.** $2x + 1 = 3$
Therefore, $x = 1$ (100 yards)

19.

+	0	1	2	3	4	5	6
0	0	1	2	3	4	5	6
1	1	2	3	4	5	6	0
2	2	3	4	5	6	0	1
3	3	4	5	6	0	1	2
4	4	5	6	0	1	2	3
5	5	6	0	1	2	3	4
6	6	0	1	2	3	4	5

×	0	1	2	3	4	5	6
0	0	0	0	0	0	0	0
1	0	1	2	3	4	5	6
2	0	2	4	6	1	3	5
3	0	3	6	2	5	1	4
4	0	4	1	5	2	6	3
5	0	5	3	1	6	4	2
6	0	6	5	4	3	2	1

21. 4   **23.** 2   **25.** 4   **27.** 5   **29.** 2   **31.** 4
33. 5   **35.** 6   **37.** 2   **39.** 4   **41.** {3}   **43.** {2}

45. $\{3\}$ **47.** $\{6\}$ **49.** $\{4\}$ **51.** $\{6\}$
53. S_7 = numeration system $6x - 1 = 3; x = 3$

Exercises 8.6
1. -6 **3.** 5   **5.** $1, \frac{2}{3}, 3, \sqrt{2}$ **7.** F
9. T   **11.** T   **13.** MATH IS FUN
15. $x = 2$ **17.** $x = 4$ **19.** $x = -5$
$y = 6$ $y = -8$ $y = -2$
$w = 12$ $w = 3$ $w = \frac{3}{2}$
$z = 20$ $z = 26$ $z = 0$

Exercises 8.7
1. $\begin{pmatrix} -4 & -3 \\ -4 & -\frac{1}{4} \end{pmatrix}$ **3.** $\begin{pmatrix} -\frac{1}{2} & -\frac{2}{9} \\ \frac{1}{6} & \frac{1}{8} \end{pmatrix}$ **5.** $\begin{pmatrix} -.11 & .36 \\ .41 & -.12 \end{pmatrix}$

7. $\begin{pmatrix} 110_{\text{seven}} & 110_{\text{seven}} \\ 100_{\text{seven}} & 142_{\text{seven}} \end{pmatrix}$ **9.** $\begin{pmatrix} 5x+4 & 5x-10 \\ 4x-6 & 3x-10 \end{pmatrix}$ **11.** $\begin{pmatrix} \frac{7x-11}{12} & \frac{5x-7}{6} \\ \frac{5x-11}{6} & \frac{7x}{12} \end{pmatrix}$

13. $\begin{pmatrix} 2 & 2 \\ 11 & -3 \end{pmatrix}$ **15.** $\begin{pmatrix} -1 & 1 \\ -1 & 0 \end{pmatrix}$ **17.** $\begin{pmatrix} -\frac{3}{14} & -\frac{7}{12} \\ -\frac{1}{10} & -\frac{13}{6} \end{pmatrix}$

19. $\begin{pmatrix} x-2 & -x+1 \\ 2x-2 & x-4 \end{pmatrix}$ **21.** $x = -4$ **23.** $x = \frac{1}{3}$
$y = 4$ $y = \frac{1}{2}$
$w = -\frac{1}{3}$ $w = 2$
$z = -\frac{3}{2}$ $z = -1$

25. $a_{12} = -2$
$a_{21} = 3(z+1) = -11$
$z = \frac{8}{3}$
$a_{21} = 11$ but $z = \frac{8}{3}$
$a_{11} = 2(y-1) = -10$
$y - 1 = -5$
$y = -4$
$a_{11} = -10$ but $y = -4$
$b_{22} = 4(q+3) = -18$
$q + 3 = -\frac{9}{2}$
$q = -\frac{15}{2}$
$b_{22} = -18$ but $q = -\frac{15}{2}$

27. $\begin{pmatrix} -\frac{1}{2} & \frac{1}{3} \\ -\frac{2}{5} & \frac{4}{7} \end{pmatrix}$

29. $\begin{pmatrix} -\sqrt{2} & -3-\sqrt{3} \\ -1+\sqrt{2} & -1-\sqrt{5} \end{pmatrix}$

31. $\begin{pmatrix} -x+1 & -2y-3 \\ -z+1 & -3+w \end{pmatrix}$

33. $\begin{pmatrix} -x-y & x+y \\ -2x+y & -x+2y \end{pmatrix}$

Exercises 8.8
1. $\begin{pmatrix} 0 & 1 \\ -4 & 8 \end{pmatrix}$ **3.** $\begin{pmatrix} -10 & 5 \\ 0 & 15 \end{pmatrix}$ **5.** $\begin{pmatrix} -1 & 0 \\ \frac{1}{2} & -\frac{2}{3} \end{pmatrix}$

7. $\begin{pmatrix} -6 & -2 \\ 1 & 5 \end{pmatrix}$ **9.** $\begin{pmatrix} 0 & 0 \\ 0 & 0 \end{pmatrix}$ **11.** $\begin{pmatrix} 0 & 0 \\ 0 & 0 \end{pmatrix}$ **13.** $\begin{pmatrix} -15 & 20 \\ 10 & 0 \end{pmatrix}$

Answers

15. $(bc) \times A = (bc) \times \begin{pmatrix} a_{11} & a_{12} \\ a_{21} & a_{22} \end{pmatrix}$

$= \begin{pmatrix} (bc)a_{11} & (bc)a_{12} \\ (bc)a_{21} & (bc)a_{22} \end{pmatrix}$

$= \begin{pmatrix} b(ca_{11}) & b(ca_{12}) \\ b(ca_{21}) & b(ca_{22}) \end{pmatrix}$

$= b \times \begin{pmatrix} ca_{11} & ca_{12} \\ ca_{21} & ca_{22} \end{pmatrix}$

$= b \times \left[c \times \begin{pmatrix} a_{11} & a_{12} \\ a_{21} & a_{22} \end{pmatrix} \right] = b(c \times A)$

17. The answer is always $\begin{pmatrix} -3 & 4 \\ 2 & 0 \end{pmatrix}$.

19. $0 \times A = 0 \times \begin{pmatrix} a_{11} & a_{12} \\ a_{21} & a_{22} \end{pmatrix} = \begin{pmatrix} 0 \cdot a_{11} & 0 \cdot a_{12} \\ 0 \cdot a_{21} & 0 \cdot a_{22} \end{pmatrix}$

$= \begin{pmatrix} a_{11} \cdot 0 & a_{12} \cdot 0 \\ a_{21} \cdot 0 & a_{22} \cdot 0 \end{pmatrix}$ So $0 \times A = A \times 0$

$= \begin{pmatrix} 0 & 0 \\ 0 & 0 \end{pmatrix}$

21. $-1 \times A = -1 \begin{pmatrix} a_{11} & a_{12} \\ a_{21} & a_{22} \end{pmatrix} = \begin{pmatrix} -1 \cdot a_{11} & -1 \cdot a_{12} \\ -1 \cdot a_{21} & -1 \cdot a_{22} \end{pmatrix}$

$= \begin{pmatrix} a_{11}(-1) & a_{12}(-1) \\ a_{21}(-1) & a_{22}(-1) \end{pmatrix} = \begin{pmatrix} a_{11} & a_{12} \\ a_{21} & a_{22} \end{pmatrix} \times (-1) = A \times (-1)$

Also $\begin{pmatrix} -1 \cdot a_{11} & -1 \cdot a_{21} \\ -1 \cdot a_{21} & -1 \cdot a_{22} \end{pmatrix} = \begin{pmatrix} -a_{11} & -a_{21} \\ -a_{21} & -a_{21} \end{pmatrix} = -A$

23. $\begin{pmatrix} 0 & -1 \\ 8 & -2 \end{pmatrix}$ **25.** $\begin{pmatrix} 2 & 2 \\ 1 & 1 \end{pmatrix}$ **27.** $\begin{pmatrix} 5 & 0 \\ 6 & 0 \end{pmatrix}$ **29.** $\begin{pmatrix} 2 & -2 \\ 1 & 3 \end{pmatrix}$

31. $\begin{pmatrix} 4 & 0 \\ -1 & 0 \end{pmatrix}$ **33.** $\begin{pmatrix} -3 & 0 \\ 1 & 0 \end{pmatrix}$ **35.** $\begin{pmatrix} 0 & 0 \\ 0 & 0 \end{pmatrix}$ **37.** $\begin{pmatrix} -4 & -4 \\ -3 & -3 \end{pmatrix}$

39. $\begin{pmatrix} -3 & 0 \\ -3 & 0 \end{pmatrix}$ **41.** $\begin{pmatrix} -1 & -2 \\ -3 & -4 \end{pmatrix}$ **43.** $\begin{pmatrix} -4 & 0 \\ -1 & 0 \end{pmatrix}$ **45.** $\begin{pmatrix} -2 & 0 \\ -5 & 0 \end{pmatrix}$

47. a. $\begin{pmatrix} -3 & 0 \\ -3 & 0 \end{pmatrix}$ b. $\begin{pmatrix} 0 & -5 \\ -1 & -2 \end{pmatrix}$ c. $\begin{pmatrix} -1 & -2 \\ -3 & -4 \end{pmatrix}$ d. $\begin{pmatrix} -1 & 0 \\ -1 & -2 \end{pmatrix}$

e. $\begin{pmatrix} -4 & 0 \\ -1 & 0 \end{pmatrix}$ f. $\begin{pmatrix} 0 & -2 \\ 0 & -1 \end{pmatrix}$ g. $\begin{pmatrix} -2 & 0 \\ -5 & 0 \end{pmatrix}$ h. $\begin{pmatrix} 0 & -6 \\ 0 & -1 \end{pmatrix}$

49. Prove: $(A + B) \times C = (A \times C) + (B \times C)$

$(A + B) \times C = \left[\begin{pmatrix} a_{11} & a_{12} \\ a_{21} & a_{22} \end{pmatrix} + \begin{pmatrix} b_{11} & b_{12} \\ b_{21} & b_{22} \end{pmatrix} \right] \begin{pmatrix} c_{11} & c_{12} \\ c_{21} & c_{22} \end{pmatrix}$

$= \begin{pmatrix} a_{11} + b_{11} & a_{12} + b_{12} \\ a_{21} + b_{21} & a_{22} + b_{22} \end{pmatrix} \begin{pmatrix} c_{11} & c_{12} \\ c_{21} & c_{22} \end{pmatrix}$

$= \begin{pmatrix} (a_{11} + b_{11})c_{11} + (a_{12} + b_{12})c_{21} & (a_{11} + b_{11})c_{12} + (a_{12} + b_{12})c_{22} \\ (a_{21} + b_{21})c_{11} + (a_{22} + b_{22})c_{21} & (a_{21} + b_{21})c_{12} + (a_{22} + b_{22})c_{22} \end{pmatrix}$

Answers

$$= \begin{pmatrix} a_{11}c_{11}+b_{11}c_{11}+a_{12}c_{21}+b_{12}c_{21} & a_{11}c_{12}+b_{11}c_{12}+a_{12}c_{22}+b_{12}c_{22} \\ a_{21}c_{11}+b_{21}c_{11}+a_{22}c_{21}+b_{22}c_{21} & a_{21}c_{12}+b_{21}c_{12}+a_{22}c_{22}+b_{22}c_{22} \end{pmatrix}$$

$$= \begin{pmatrix} a_{11}c_{11} + a_{12}c_{21} & a_{11}c_{12} + a_{12}c_{22} \\ a_{21}c_{11} + a_{22}c_{21} & a_{21}c_{12} + a_{22}c_{22} \end{pmatrix}$$

$$+ \begin{pmatrix} b_{11}c_{11} + b_{12}c_{21} & b_{11}c_{12} + b_{12}c_{22} \\ b_{21}c_{11} + b_{22}c_{21} & b_{21}c_{12} + b_{22}c_{22} \end{pmatrix}$$

$$= \begin{pmatrix} a_{11} & a_{12} \\ a_{21} & a_{22} \end{pmatrix} \begin{pmatrix} c_{11} & c_{12} \\ c_{21} & c_{22} \end{pmatrix} + \begin{pmatrix} b_{11} & b_{12} \\ b_{21} & b_{22} \end{pmatrix} \begin{pmatrix} c_{11} & c_{12} \\ c_{21} & c_{22} \end{pmatrix}$$

$$= (A \times C) + (B \times C)$$

Exercises 8.9

1. Answers will vary.

CHAPTER 9

Exercises 9.1

1. 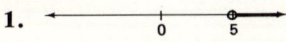 **3.**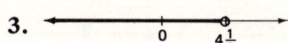

5. 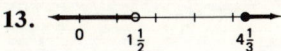 **7.**

11. **13.**

15. **17.**

19. **21.**

23. **27.** $x < -2$

29. $x < 11$ **33.** $x < -\frac{17}{2}$

35. $\{x \mid x > -2 \text{ or } x < -5\}$

37. $\{x \mid x < 0\}$

Answers

39. $\{x \mid x < -3\}$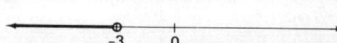

41. $x > -1$

43. $\{x \mid x > \frac{1}{2} \text{ or } x < \frac{1}{2}\}$

45. ϕ

Exercises 9.2

2–7.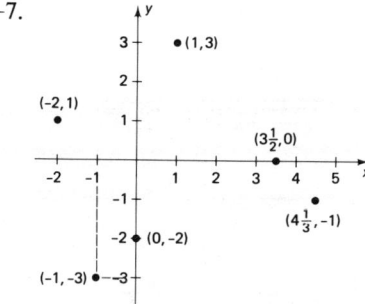

9. 5

11. 4

13.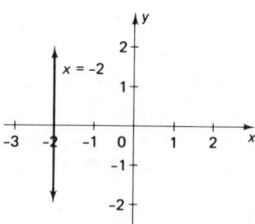

15.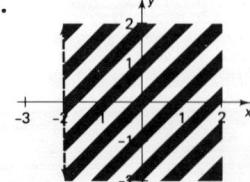

Answers

17.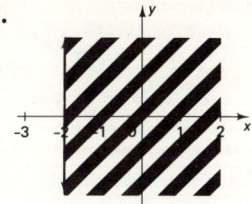

19.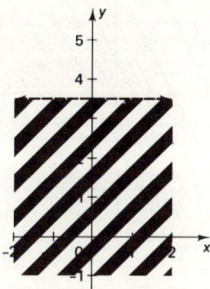

23. ∅   **25.** The real plane, $R \times R$.

Exercises 9.3

9. $x = 0$ **11.** $x = c$, where c is some real number

Exercises 9.4

	Equation	Slope	y-intercept
1.	$y = -x + 6$	-1	6
3.	$y = 2x + 5$	2	5
5.	$y = \frac{3}{4}x - 2$	$\frac{3}{4}$	-2
7.	$y = 2x - \frac{4}{3}$	2	$-\frac{4}{3}$

9. a. $4y = 3x - 8$ b. $y + 3x = -4$ c. $5x = 2y + 7$
$y = \frac{3}{4}x - 2 = +\frac{3}{4}$ $y = 3x + 4 = 3$ $5x - 7 = 2y = \frac{5}{2}$

10–15.

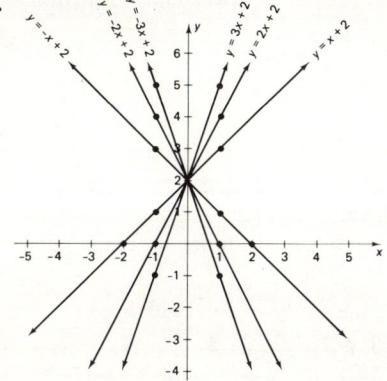

Answers

17. The y-intercept for each line is 2.
19.–24.

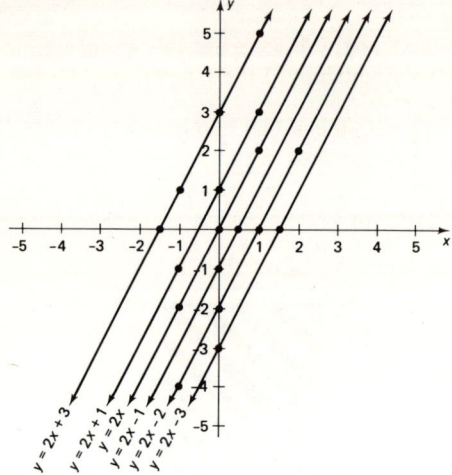

25. The lines are parallel.
27. The y-intercept of each line in problems 19–23 above:
   19. 3   **21.** 0   **23.** -2

Exercises 9.5

11. $.10x + .08y \leq 8.00$ **13.** $2x + 4y \geq 120$

Exercises 9.6

1. Read domain first:
MEET AT THE OLD TREE
Then range:
16 OR MORE WILL COME
3. Function
5. Not a function
7. Function
9. Not a function
11. Function   **13.** Not a function   **15.** Not a function
17. Function   **19.** Function   **21.** Function
23. Function   **25.** Function   **27.** Not a function

CHAPTER 10

Exercises 10.1

1. F   **3.** Open   **5.** T   **7.** Open   **9.** T
11. T   **13.** F   **15.** Open   **17.** F
19. $x^2/z^3 = 33K^8$ is an open statement. Therefore, the doors are really open.

Exercises 10.2

1. $\forall x, x + 5x = 6x$ **3.** $\exists n, n + 1 = 4$ **5.** $\exists y, y(y - 1) = 0$

Answers

7. $\forall x, y \in R \; 3(x + y) = 3x + 3y$ 9. $\exists x, x^2 = x$
11. $x = 3$ 15. $n = 10$
17. a. A round hole fits into a square peg; 0 is a permissible replacement.
    b. A square peg will not fit into a round hole; 1 is not a permissible replacement.
19. If $x = 0$, $\dfrac{x}{x-1} = \dfrac{0}{0-1} = \dfrac{0}{-1} = 0$, 0 is a real number.

Exercises 10.3

1. War is not bitter.
3. It is not true that each triangle has three sides.
5. $21 \neq 13$ 7. a. War is bitter.   b. Yes   c. Yes   9. $\neq$

Statement	Truth-Value	Negation	Truth-Value
11. $8 < 3$	F	$8 \not< 3$	T
13. $0 \not< 2$	F	$0 < 2$	T
15. $1 > 5$	F	$1 \not> 5$	T
17. $3 \not> 7$	T	$3 > 7$	F

19. There is a student in this course that is not an A student.

Exercises 10.4

1. My students and I do not argue or fight; it's just that they cannot understand me.
3. F   5. F   7. F   9. The moon is not shining and I am sad.
11. The moon is not shining and I am not sad.
13. It is not true that the moon is not shining and I am sad.
15. F   17. T   19. T   21. F   23. T   25. T
27. F   29. F   31. F   33. T   35. T   37. F
39. T

Exercises 10.5

1.

p	q	$\sim p$	$p \to q$	$\sim p \vee q$
T	T	F	T	T
T	F	F	F	F
F	T	T	T	T
F	F	T	T	T

$p \to q$ and $\sim p \vee q$ are logically equivalent statements.

3. If it is a triangle, then it has three sides.
5. If it is a parallelogram, then it has opposite sides parallel.
7. a. If John can walk, then he can go.
   b. If $\dfrac{3x}{2} = \dfrac{9}{2}$, then $x = 3$.
   c. If you can decipher this, then you understand the principle.

9.

p	q	$\sim p$	$\sim q$	$p \vee q$	$\sim(p \vee q)$	$(\sim p) \wedge (\sim q)$
T	T	F	F	T	F	F
T	F	F	T	T	F	F
F	T	T	F	T	F	F
F	F	T	T	F	T	T

11. $5 \neq 4$ and $4 \not> 1$

Exercises 10.6

1. If each set has the same number of elements, then two finite sets are matching sets.
3. If at least two sides of a triangle have the same measure, then it is an isosceles triangle.
5. If a triangle is nonequilateral, then it is a scalene triangle.
7. Let p be $1 > 2$ and q be $1 + 1 = 2$.
   $p \to q$: "If $1 > 2$ then $1 + 1 = 2$" is true.
   $\sim p \to \sim q$: "If $1 \not> 2$ then $1 + 1 \neq 2$" is false.
9. Let p be $2 + 1 = 3$ and q be $2 < 3$.
   $p \to q$: "If $2 + 1 = 3$ then $2 < 3$" is true.
   $\sim p \to \sim q$: "If $2 + 1 \neq 3$ then $2 \not< 3$" is true.
11. If you do a problem, it is hard.
    If you do not do a problem, it is not hard. (Problem 10 is the converse of Problem 11.)
13. Let p be "a triangle has four sides" and q be "a quadrilateral has four angles."
    $p \to q$: "If a triangle has four sides then a quadrilateral has four angles" is true.
    $\sim p \to \sim q$: "If a triangle does not have four sides then a quadrilateral does not have four angles" is false.
15. Let p be "all angles of a rectangle are right angles" and q be "a rectangle is a parallelogram."
    $p \to q$: "If all angles of a rectangle are right angles then a rectangle is a parallelogram" is true.
    $\sim p \to \sim q$: "If all angles of a rectangle are not right angles then a rectangle is not a parallelogram" is true.

17.

p	q	$p \to q$	$\sim p$	$\sim q$	$\sim p \to \sim q$	$\sim q \to \sim p$
T	T	T	F	F	T	T
T	F	F	F	T	T	F
F	T	T	T	F	F	F
F	F	T	T	T	T	T

$p \to q$ and $\sim q \to \sim p$ are logically equivalent.

Exercises 10.7

1–2.

p	$\sim p$	$p \wedge \sim p$	$p \vee \sim p$
T	F	F	T
F	T	F	T

3–6.

p	q	$p \wedge q$	$p \vee q$
T	T	T	T
T	F	F	T
F	T	F	T
F	F	F	F

7. $(5 \neq 6) \rightarrow (3 \neq 4)$
9. $(9 \neq 2 \times 4\tfrac{1}{2}) \rightarrow (3 \neq 4)$
11. $(4 \neq 3 + 1) \rightarrow (2 \neq 1 + 1)$
13. $(2 \not> 6) \rightarrow (1 \not> 5)$
15. (7) T, T (8) T, T (9) T, T (10) F, F (11) T, T (12) F, F
(13) T, T (14) T, T

17.

p	$\sim p$	$\sim\sim p$	$p \leftrightarrow \sim\sim p$
T	F	T	T
F	T	F	T

Since p and $\sim\sim p$ have the same truth values they are logically equivalent or $p \leftrightarrow \sim\sim p$ is a tautology.

19.

p	q	$p \wedge q$	$q \wedge p$	$(p \wedge q) \leftrightarrow (q \wedge p)$
T	T	T	T	T
T	F	F	F	T
F	T	F	F	T
F	F	F	F	T

21.

p	q	r	$(p \vee q)$	$(p \vee q) \vee r$	$(q \vee r)$	$p \vee (q \vee r)$	$(p \vee q) \vee r \leftrightarrow p \vee (q \vee r)$
T	T	T	T	T	T	T	T
T	T	F	T	T	T	T	T
T	F	T	T	T	T	T	T
T	T	T	T	T	F	T	T
F	T	T	T	T	T	T	T
F	T	F	T	T	T	T	T
F	F	T	F	T	T	T	T
F	F	F	F	F	F	F	T

Answers 376

23.

p	q	r	$q\vee r$	$p\wedge(q\vee r)$	$p\wedge q$	$p\wedge r$	$(p\wedge q)\vee(p\wedge r)$	$p\wedge(q\vee r)\leftrightarrow(p\wedge q)\vee(p\wedge r)$
T	T	T	T	T	T	T	T	T
T	T	F	T	T	T	F	T	T
T	F	T	T	T	F	T	T	T
T	F	F	F	F	F	F	F	T
F	T	T	T	F	F	F	F	T
F	T	F	T	F	F	F	F	T
F	F	T	T	F	F	F	F	T
F	F	F	F	F	F	F	F	T

25.

p	q	$p\rightarrow q$	$\sim p$	$\sim p \vee q$	$(p\rightarrow q)\leftrightarrow(\sim p \vee q)$
T	T	T	F	T	T
T	F	F	F	F	T
F	T	T	T	T	T
F	F	T	T	T	T

27.

p	q	$p\leftrightarrow q$	$p\rightarrow q$	$q\rightarrow p$	$(p\rightarrow q)\wedge(q\rightarrow p)$	$(p\leftrightarrow q)\leftrightarrow(p\rightarrow q)\wedge(q\rightarrow p)$
T	T	T	T	T	T	T
T	F	F	F	T	F	T
F	T	F	T	F	F	T
F	F	T	T	T	T	T

29.

p	q	$p\vee q$	$\sim(p\vee q)$	$\sim p$	$\sim q$	$\sim p \vee \sim q$	$\sim(p\vee q)\leftrightarrow\sim p\wedge\sim q$
T	T	T	F	F	F	F	T
T	F	T	F	F	T	F	T
F	T	T	F	T	F	F	T
F	F	F	T	T	T	T	T

Exercises 10.8

1. Valid, by Exercise 18 of Section 10.7 (*modus ponens*).
3. Invalid, this is using a converse argument.
5. Valid, chain of implications as in Example 5 in this section.
7. Valid, similar to Exercises 4 and 5; that is,
   $(p\rightarrow q) \wedge (\sim r \rightarrow \sim q)$ is the same as
   $(p\rightarrow q) \wedge (q\rightarrow r)$ so
   $(p\rightarrow r)$
9. Invalid, this is not a correct use of a chain of implications.

Answers

11. p: You play tennis
    q: You are fit.
    r: You win matches.
13. It is not true, but it is valid.

CHAPTER 11

Exercises 11.1

1. See page 307.
3. If a girl is married at the age of 19 what are the chances of her celebrating her twenty-fifth anniversary?
5. Sample space: the set of all possible outcomes of an experiment.
7. {T, H} (T indicates tails up, H indicates heads up)
9. Completely alters the probability to one of certainty. Stated mathematically, $P_{(\text{Weighted side})} = 1$
11. $P(H)$ = the probability that heads will turn up = $\frac{1}{2}$
    $P(H)$ = the probability that tails will turn up = $\frac{1}{2}$

Exercises 11.2

1. a. Once: H, T = $\frac{1}{2}$
   b. Twice: HH, HT, TH, TT = $\frac{1}{4}$
   c. Three times: HHH, HHT, HTT, HTH, THT, TTH, TTT = $\frac{1}{8}$
3. {HH, HT, TT, TH}
5. {HHHH, HHHT, HHTH, HTHH, HHTT, HTHT, HTTH, HTTT, TTTT, TTTH, TTHT, THTT, TTHH, THTH, THHT, THHH}
7. $2^5 = 32$ 9. 2^n 11. a. $\frac{1}{4}$ b. Yes c. 1
13. $\frac{1}{16}$ 15. $\frac{1}{64}$ 17. $\frac{1}{32}$

Exercises 11.3

1. $\frac{1}{4}$ 3. 1 5. 0 7. 0 9. $\frac{1}{4}$ 11. $\frac{3}{11}$
13. $\frac{1}{5}$ 15. $\frac{1}{2}$ 17. $\frac{1}{6}$ 19. 1 21. $\frac{5}{36}$ 23. $\frac{11}{12}$
25. $\frac{1}{6}$ 27. $\frac{1}{5}$ 29. $\frac{1}{2}$ 31. $\frac{2}{5}$ 33. $\frac{3}{10}$ 35. $\frac{1}{5}$
37. $\frac{1}{216}$ 39. $\frac{56}{216} = \frac{7}{27}$ 41. 6

Exercises 11.4

1. 12
3. $\bar{E}$ is the event that neither 3 nor 4 will land up on the second roll of the die.
5. $\frac{2}{3}$ 7. $\frac{1}{52}$ 9. F 11. F 13. F 15. T
17. T 19. T 21. $\frac{10}{17}$ 23. $\frac{11}{17}$ 25. 1 27. $\frac{26}{52} = \frac{1}{2}$
29. $\frac{16}{52}, \frac{4}{13}$ 31. $\frac{14}{25}$ 33. $\frac{12}{25}$ 35. $\frac{12}{25}$ 37. $\frac{31}{36}$ 39. $\frac{1}{36}$
41. $\frac{35}{36}$ 43. $\frac{33}{36}, \frac{11}{12}$ 45. 0 ★47. $\frac{1}{4}$

Answers

Exercises 11.5

1. 45 3. 30 different ways 5. 1500 7. 750 ways
9. $26 \cdot 25 \cdot 24 \cdot 10 \cdot 9 = 1,404,000$

★11. $\dfrac{52!}{5! \cdot 47!} = 2,597,960$ 13. 240 different arrangements

Exercises 11.6

1. $1; the player breaks even.
3. $\$\frac{5}{6}$ (approximately $.83); the player loses about 17 cents for each dollar bet.
5. $\$\frac{3}{4} = \$.75$; you would lose 25 cents for each dollar bet.
7. You lose $2.52 for a $1 bet.
9. The more you bet, the more you lose.
11. $\frac{81}{216} \cdot 10 = \frac{30}{8}$ (or $3.75); in the long run, you would lose $1.25 for each $5 bet.

CHAPTER 12

Exercises 12.1

1. To describe some quantitative characteristics of a situation
3. To make inferences about an entire population using a sample of the population
5. Answers will vary.

Exercises 12.2

1. 13 3. 7 5. Median 7. These are the raw data; mean = 40.
9. 36 ★11. No 13. 7

Exercises 12.3

1. 77 3. 73 5. 13 7. 20 9. 74.2
11. Mode = 58; variance = 278.56

Exercises 12.4

1. Sam is taller than approximately 92% of the 19-year-olds.
3. 50 5. 18 7. 73.75 9. 95th

Index

Absolute value, 93–94
Addition, 92
   in base five, 12
   in base two, 19–20
   of rational numbers, 104
   in S_3, 209
   of 2-by-2 matrices, 227–28
Additive identity, 88, 161, 206, 228
Additive inverse, 93, 106, 162, 206, 210, 212, 228–29
Angle, 42, 180
   dihedral, 186
Antecedent (hypothesis), 290
Area, 191
   of a circle, 193
   of a parallelogram, 192
   of a rectangle, 191
   of a trapezoid, 193
   of a triangle, 192, 193
Arithmetic mean, 117
Associative property, 88
   of addition, 161, 209
   of multiplication, 161, 205, 209
Axis, 185, 247

Bar graph, 342
Base, 6, 122
Binary numeration system, 16–18
Binary operation, 38, 88

Cancellation property, 158
Cartesian coordinate system, 247–51
Casting out, 82–85
Cayley, Arthur, 223
Closure property, 88
   of addition, 161, 205, 209
   of multiplication, 161, 209

Combination, 326
Commutative property, 88
   of addition, 161, 205
   of multiplication, 161, 209, 233
Complement, 39, 45
   of a set, 317
Cone, 185
Conjunction, 285
Consequent (conclusion), 290
Contrapositive, 296, 301
Converse of implication, 294
Coordinate, 248
Counting, 324–27
Cubic unit, 193
Cylinder, 184–85

Decimal numeral, 114
   nonrepeating, 154–55
   nonterminating and nonrepeating, 154
   repeating nonterminating, 113
   terminating, 112
Decimal numeration system, 7
Deductive system, 179, 197
de Fermat, Pierre, 307
de Méré, Chevalier, 307
Disjoint set, 38, 45, 249
Disjunction, 286
Distributive property, 88
   of multiplication over addition, 161, 209
Divisibility rules, 77–79
   by 11, 78–79
   by 4, 77
   by 3 and 9, 77–78
   by 2, 5, and 10, 77
Division, 49, 64–65
   of square roots, 141

Index

Divisor, 51–52
　greatest common, 65–66, 68–69, 75
　proper, 62

Element (of set), 27
Endpoint, 41
Equality, 35, 157
Equation, linear, 169
　quadratic, 169
　radical, 172
　solving, 158, 169
　Solving in S_5, 221
Equivalence, logical, 291–92
Eratosthenes, 57
Euclid, 63, 65, 179
Euler circle, 44
Euler, Leonhard, 44
Event, 308
Exponent, 6, 121–24

Factor tree, 73
Factorial (!), 326
Field, 162, 216–17
Function, 269–71
Fundamental Theorem of Arithmetic, 73, 74

Geometry, 41, 177
　deductive, 178, 197
　metric, 178, 187
　motion, 178, 199
　nonmetric, 178
Gillies, Donald B., 56
Graph, of equation, 256, 262
　of inequality, 241–45
Greatest common divisor (GCD), 65–66, 68–69, 75

Heisenberger, Werner, 223
Hero's formula, 193
Histogram, 342
Homeomorphism, 178

Implication (conditional), 289
　truth-table for, 290
　converse of, 293–94
Inequality, 158–59
　graphing, 241–45
　solving, 164–67, 170–71
Integer, 91, 168
　discrete, 116
　multiplication of, 98–101
　negative, 91

Interger (*continued*)
　positive, 91
　subtraction of, 96–97
Intercept, 254, 255, 259
Intersection, 39
Inverse, 294–95
Irrational number, 147–48, 154–55

Law of detachment (*modus ponens*), 299
Law of the excluded middle, 297
Least common multiple (LCM), 68–70
Line, 42, 180, 253
　slope of, 256–59
Linear programming, 263
Logic, 277

Mathematical expectation, 329
Matrix, addition of, 227–28
　multiplication of, 232–33
　subtraction of, 228–29
　2-by-2, 224–25, 227–29
Matrix algebra, 223
Mean, 335
Median, 335
Metric system, 187–88
Mode, 335
Modulo, 204
Multiplication, 98–101, 158
　of rational numbers, 104
　in S_3, 209
　of square roots, 140
　of two matrices, 232–33
Multiplicative identity, 88, 161, 208, 210
Multiplicative inverse, 107, 162, 208, 210, 213, 217

Negation, 282
Niebaum, Jerome, 60
Nines excess, 82
Number, absolute value of, 93–94
　abundant, 63
　amicable, 63
　complex, 161
　composite, 52, 54, 73
　deficient, 62
　even, 49, 50
　irrational, 147–48, 154–55
　natural, 26, 168
　odd, 49, 50–51
　perfect, 62
　prime, 52, 54–56

Index

Number, absolute value of (*continued*)
   rational, 104–06, 109–10, 114, 117, 146–47, 168
   real, 168
   whole, 168
Number line, 91, 147–48, 151, 240–41
Numeral, 7
Numeration system, 7

Odd number, 49, 50–51
Odds, 329

Parallel lines, 197
Pascal, Blaise, 307
Percentile rank, 339–41
Pi (π), 189
Point, 179
   noncollinear, 197
   noncoplanar, 197
Polygon, 184, 261
Polyhedron, 184
Postulate, 197
Power, 122–24
   of a power, 127, 137
   of a product, 128
   product of powers, 136–37
   of a quotient, 131
   quotient of powers, 132–33, 135, 136
Powers of ten, 7, 9
Prime number, 52, 54–56
   relatively, 65, 151
   reversible, 60
   sieve method of determining, 57–58
   symmetrical, 60
   twin, 59
Probability, *a priori*, 307
   in dice games, 311–14, 317–21
   principles of, 308, 310, 312, 317, 319, 320
   statistical, 307
Proof, 49
Pythagorean Theorem, 147

Quantifier, 279–80
Quinary system, 9–10

Radius, 185
Ray, 42, 180
Reciprocal, 162
Rectangular coordinate system, 247–51
Reflection (flip), 199
Reflexive property, 157, 159
Relation, 269

Rotation (turn), 200
S_3, 207–10
S_4, 216–18
S_5, 211–14, 221
Sample space, 307, 311–14
Segment, 41, 179
Set, 26, 27, 179
   complement of, 39, 45
   disjoint, 38, 45
   element of, 27
   empty (null), 28
   equivalent, 31
   finite, 28
   infinite, 28, 155
   intersection of, 39
   replacement, 277, 282
   subset, 34
   union of, 39
   universal, 34
   well-defined, 27
Sieve method, 57–58
Singulary operations, 38
Slope, 256–57
Solution set, 221
Space, 183
Sphere, 185
Square root, 139–41
   on number line, 147–48
Standard deviation, 337
Statistics, 333–34
Statement, biconditional, 292
   compound, 285
   conditional, 290
   false, 279
   open, 277
   true, 278, 302
Subset, 34–35
Subtraction, 96–97
   in base five, 14
   in base two, 21
   of 2-by-2 matrices, 228–29
Symmetric property, 157, 159
System, 204

Tautology, 296, 299–301
Theorem, 50, 197
Topology, 177
Transitive property, 157, 159, 166
Transformation (rigid motion), 199
Translation (slice), 199
Trapezoid, 193
Triangle, 180

Trichotomy Property, 160
Truth-table, for conjunction, 286
  for converse of implication, 294
  for disjunction, 287
  for implication, 290
  for negation, 283

Union, 39

Validity, 302
Venn diagram, 44–45
Venn, John, 44
Variable, 277
Variance, 337
von Leibniz, Gottfried Wilhelm, 22

Zero power of ten, 7